T0262367

Carbon Nanotubes and their Composites: Processing and Applications

Carbon Nanotubes and their Composites: Processing and Applications

Edited by **Lindy Bowman**

NY RESEARCH
P R E S S

New York

Published by NY Research Press,
23 West, 55th Street, Suite 816,
New York, NY 10019, USA
www.nyresearchpress.com

Carbon Nanotubes and their Composites:
Processing and Applications
Edited by Lindy Bowman

International Standard Book Number: 978-1-63238-069-2 (Hardback)

Contents

Preface

The demand for innovative industrial applications of carbon nanotubes are growing significantly. Carbon nanotubes are large molecules of pure carbon that are long and thin and are shaped like tubes of nanometer scale diameter with quasi-one-dimensional structure. In the past 20 years, carbon nanotubes have attracted a lot of attention from chemists, electronic device engineers, physicists, and material scientists, due to their fine optical, mechanical, structural, chemical and electronic properties. This book encompasses latest research topics about the synthesis technologies of carbon nanotubes and nanotube-based composites. It will serve as a helpful source of information for engineers, researchers and students engaged in the field of carbon nanotubes.

The researches compiled throughout the book are authentic and of high quality, combining several disciplines and from very diverse regions from around the world. Drawing on the contributions of many researchers from diverse countries, the book's objective is to provide the readers with the latest achievements in the area of research. This book will surely be a source of knowledge to all interested and researching the field.

In the end, I would like to express my deep sense of gratitude to all the authors for meeting the set deadlines in completing and submitting their research chapters. I would also like to thank the publisher for the support offered to us throughout the course of the book. Finally, I extend my sincere thanks to my family for being a constant source of inspiration and encouragement.

Editor

Syntheses of Carbon Nanotubes and Their Composites

Fabrication, Purification and Characterization of Carbon Nanotubes: Arc-Discharge in Liquid Media (ADLM)

Mohsen Jahanshahi and Asieh Dehghani Kiadehi

Additional information is available at the end of the chapter

1. Introduction

Carbon nanotubes (CNTs) were first discovered by Iijima in 1991 [1]. CNTs have sparked great interest in many scientific fields such as physics, chemistry, and electrical engineering [2, 3]. CNTs are composed of graphene sheets rolled into closed concentric cylinders with diameter of the order of nanometers and length of micrometers. CNTs are in two kinds, based on number of walls, the single-walled and multi-walled.

The diameter of single walled carbon nanotubes (SWNTs) ranges from 0.4 nm to 3nm and the length can be more than 10 mm that makes SWNTs good experimental templates to study one-dimensional mesoscopic physics system [3]. These unique properties have been the engines of the rapid development in scientific studies in carbon based mesoscopic physics and numerous applications such as high performance field effect transistors [4-9], single-electron transistors [10, 11], atomic force microscope tips [12], field emitters [13, 14], chemical/biochemical sensors [15-18], hydrogen storage [19].

There are three important methods to produce high quality CNT namely laser [20], arc discharge [21, 22], and Chemical Vapor Deposition (CVD) [23, 24]. Recently, arc discharge in liquid media has been developed to synthesize several types of nano-carbon structures such as: carbon onions, carbon nanohorns and carbon nanotubes. This is a low cost technique as it does not require expensive apparatus [25,26].

However, several techniques such as oxidation, nitric acid reflux, HCl reflux, organic functionalization, filtration, mechanical purification and chromatographic purification have been developed that separate the amorphous carbons and catalyst nanoparticles from the CNTs while a significant amount of CNTs are also destroyed during these purification processes [27].

In this review paper, synthesis, purification and structural characterization of CNTs based on arc discharge in liquid media are reviewed and discussed. In addition, several parameters such as: voltage difference between electrodes, current, type and ratio of catalysts, electrical conductivity, concentration, type and temperature of plasma solution, as well as thermal conductivity on carbon nanotubes production are investigated.

2. Synthesis of CNTs

2.1. Laser vaporization

The laser vaporization method was developed for CNT production by Smalley's group [28, 29]. The laser is suitable for materials with high boiling temperature elements such as carbon because of its high energy density. The quantities of CNTs, in this method are large. Smalley's group further developed the laser method also known as the laser-furnace method [28]. Fullerenes with a soccer ball structure are produced only at high furnace temperatures, underlining the importance of annealing for nanostructures [28]. These discoveries were applied to produce CNTs [29] in 1996, especially SWNTs. A beam of high power laser impinges on a graphite target sitting in a furnace at high temperature as Figure1 shows.

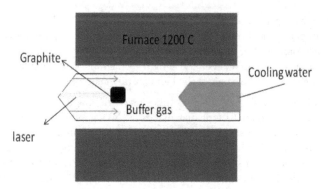

Figure 1. Schematic drawing of a laser obtain.

The target is vaporized in high-temperature buffer gas like Ar and formed CNTs. The produced CNTs are conveyed by the buffer gas to the trap, where they are collected. Then CNTs can be found in the soot at cold end.

This method has several advantages, such as high-quality CNTs production, diameter control, investigation of growth dynamics, and the production of new materials. High-quality CNTs with minimal defects and contaminants, such as amorphous carbon and catalytic metals, have been produced using the laser-furnace method together with purification processes [30-32] but the laser has sufficiently high energy density to vaporize target at the molecular level. The graphite vapor is converted into amorphous carbon as the starting material of

CNTs [33-35]. However, the laser technique is not economically advantageous, since the process involves high purity graphite rods, high power lasers and low yield of CNTs.

2.2. Chemical vapor deposition (CVD)

The chemical vapor deposition (CVD) is another method for producing CNTs in which a hydrocarbon vapor is thermally decomposed in the presence of a metal catalyst. In this method, carbon source is placed in gas phase in reaction chamber as shown in Figure 2. The synthesis is achieved by breaking the gaseous carbon molecules, such as methane, carbon monoxide and acetylene, into reactive atomic carbon in a high temperature furnace and sometime helped by plasma to enhance the generation of atomic carbon [36].

This carbon will get diffused towards substrate, which is coated with catalyst and nanotubes grow over this metal catalyst. Temperature used for synthesis of nanotube is 650 – 900 ⁰C range and the typical yield is 30% [36].

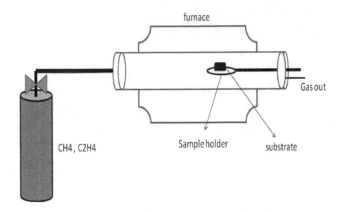

Figure 2. Schematic diagram of a CVD setup.

In fact, CVD has been used for producing [37-39] carbon filaments and fibers since 1959. Figure 2 shows a schematic diagram of the setup used for CNT growth by CVD in its simplest form. CNTs grow over the catalyst and are collected upon cooling the system to room temperature. The catalyst material may be different, solid, liquid, or gas and can be placed inside the furnace or fed in from outside [40, 41].

2.3. Arc Discharge

A schematic diagram of the arc discharge apparatus for producing CNTs is shown in Figure 3. In this method, two graphite electrodes are installed vertically, and the distance between the two rod tips is usually in the range of 1–2 mm. The anode and cathode are made of pure graphite (those are, with a purity of 99.999%).

The anode is drilled, and the hole is filled with catalyst metal powder then the chamber is connected to a vacuum line with a diffusion pump and to a gas supply [43]. Like the anode in a DC electric arc discharge reactor, CNT is synthesized of graphite rod. After the evacuation of the chamber by a diffusion pump, rarefied ambient gas is introduced [43].

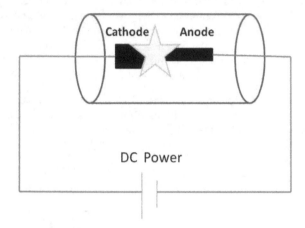

Figure 3. Schematic diagram of apparatus for preparing CNTs.

When a dc arc discharge is applied between the two graphite rods, the anode is consumed, and fullerene is formed in the chamber soot [43]. The mass production of multi wall carbon nanotubes (MWCNTs) by this dc arc discharge evaporation was first achieved by Ebbesen and Ajayan [44].

3. Arc-Discharge in Liquid Media (ADLM)

The traditional arc-plasma growth method for CNTs necessitates complex gas-handling equipment, a sealed reaction chamber, a liquid-cooled system and time consuming purge cycles. The act of extraction of nanotube's product is so complicated [45]. To be compared, the growth of arc plasma in water requires simple operation and equipment which had made the process of CNTs production noticeable [47].

Ishigami et al. [48] developed a simple arc method in liquid nitrogen for the first time that allow for the continuous synthesis of high-quality CNTs. The materials obtained are mainly MWCNTs, amorphous carbon, graphitic particles and carbonaceous material [48, 49]. Subsequently, an aqueous arc-discharge (arc-water) method was developed. Lange et al. [50] generated onions, nanotubes and encapsulates by arc discharge in water. Figure 4 is shown produced CNTs in LiCl 0.25 N media [51].

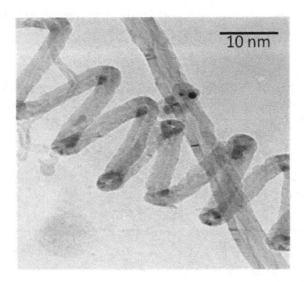

Figure 4. Carbon nanotubes produced in LiCl 0.25 N [51].

The arc discharge in liquid is initiated between two high purity graphite electrodes. Figure 5 is shown schematic device of arc discharge in liquid. Both electrodes are submerged in the liquid in a beaker. At first, the electrodes touch each other and are connected with a direct current (DC) power supply.

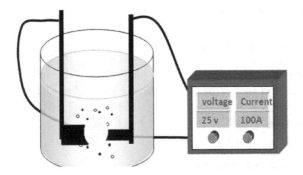

Figure 5. Schematic device of arc discharge in liquid.

The cathode is usually 20 mm in diameter, while the anode is 6 mm in diameter. Then the arc discharge is initiated by slowly detaching the moveable anode from the cathode. The arc gap is kept at the proper value (about 1 mm) that the continuous arc discharge could be obtained [52].

Recently carbon nanotubes (CNTs) were fabricated successfully with arc discharge in solution by a novel full automatic set up [51].

The arc discharge and consequently consumption of anode result to increase the distance between the two electrodes and degrees the voltage between them as well. In order to remain constant voltage and gap between the two electrodes, the program automatically will compare the initial voltage with the voltage of two electrodes with an accuracy of 0.1 V. Based on the calculated difference the program calculates the proportional coefficient for the proportional controller. Figure 6 is schematic of the apparatus used for automatic arc discharge in solution [51].

During the arc discharge, the gap between the two electrodes is maintained at approximately 1 mm, while the synthesis time is 60 s [51].

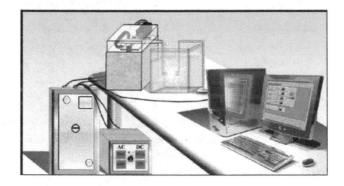

Figure 6. Schematic of the apparatus used for automatic arc discharge in solution [51].

3.1. Catalyst Materials and their ratio

A metal catalyst is necessary for the growth of the CNTs in all the methods used for synthesis of CNTs. Catalysts use to prepare CNTs usually include transition metals as a single or mixture of two catalysts such as a single, Fe, Co, Ni or Mo [53] or mixture of two catalysts such as FeNi [54], PtRh [55] and NiY [22]. Hsin et al. [57] firstly reported the production of metal-containing CNTs by arc discharge in solution.

The catalyst activation is determined in relation to the melting temperature and the boiling temperature thus the melting and boiling temperature of a catalyst can be one of the vital factors in the synthesis of SWCNTs [51].

CNTs are synthesized while the anode is filled by divergent single or bimetallic catalysts [51]. Scanning electron micros copies (SEM) show that the fabricated CNTs without any catalyst, figure 7, is in a very short long, disordered and is faulty grown.

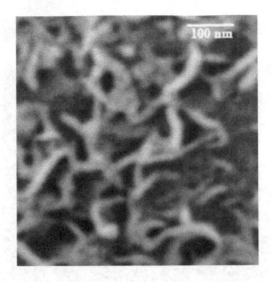

Figure 7. SEM image of the product which was fabricated without catalyst [51]

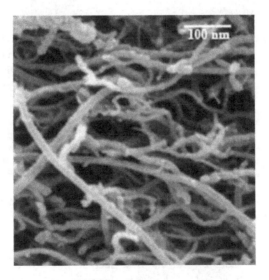

Figure 8. SEM image of the product which is fabricated with 5% Ni as catalyst [51].

Ni catalyst, figure 8, motivate a production of elongated CNTs and springy CNTs with a relatively good yield while Fe catalysts, figure 9, promote a production of CNTs with short length and defect structure and the yield is moderate.

Figure 9. SEM image of the product which was fabricated with 5% Fe as catalyst [51].

Jahanshahi et al. showed that Mo catalysts motivate a production of CNTs with long length and high crystalline structure but with a wide diameter while it has a relatively good yield. In contrast, Mo-Ni bimetallic catalyst cause the production of CNTs with long length, narrow distribution diameters and crystalline structure without any defect and follow with a good yield [51].

3.2. Plasma Solution Temperature

The effect of solution temperature on the synthesis of CNTs and the structure of fabricated CNTs was investigated by Dehghani et al [57]. Scanning electron microscopes and transient electron microscopes (TEMs), figure 10, shows that the fabricated CNTs below zero as thermal condition is not suitable for synthesizing CNTs by the arc discharge method in liquid and CNTs cannot grow under low temperatures, especially below zero. High temperature is also not suitable for synthesizing CNTs by the arc discharge method in liquid media.

In contrast, observations show that in the environment with (25 °C) temperature, long CNTs are formed with narrow distribution diameter, complete clean, flat surface and arranged structure. Constant temperature around 25°C is the best thermal condition for synthesizing CNTs by the arc discharge method in liquid [57].

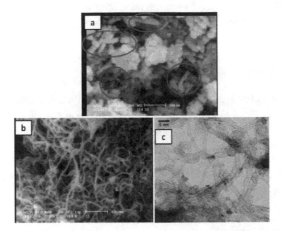

Figure 10. SEM image of the product which was fabricated a) below zero temperature b) at a high temperature (80°C), c) at the environment temperature (25 °C)[57].

3.3. Voltage difference between electrodes

The voltage effects on the production of the nanostructures by applying a variety of voltage values in different experiments were investigated by Jahanshahi et al [58]. The SEMs of the synthesized materials, figure 11 shows the formation of fullerene at a voltage of 10 V, while both CNTs and fullerenes are fabricated at a voltage of 20 V.

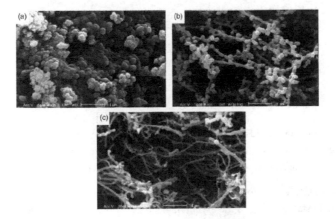

Figure 11. a) SEM images of the produced sample by arc discharge at a voltage of 10 V. (b) SEM images of the produced sample by arc discharge at a voltage of 20 V. (c) SEM images of the produced sample by arc discharge at a voltage of 30 V [58].

On the contrary, the elongated CNTs were synthesized with high quality at a voltage of 30 V. The results show that the rate of production efficiency and anode consumption is increased by increasing the voltage amount [58].

3.4. Plasma Solution Concentration

Liquid nitrogen provides a good environment for the MWCN synthesis, but the strong evaporation cause by the operation of the arc discharge does not allow a good thermal exchange between the synthesized material and its surroundings.

Arc discharge in deionised water and liquid nitrogen are erratic due to their electrical insulation [50]. The electrical conductivity of LiCl solution is also better than deionised water and liquid nitrogen [59].

Figure 12 shows TEM image of the as-grown MWCN synthesized in LiCl. Investigators have demonstrated the possibility of producing carbon nanostructures in the liquid phase (water, hydrocarbons, dichloromethane, CCl_4, in liquid gases) [61].

Figure 12. TEM image of the as-grown MWCN synthesized in LiCl [58].

In contrast, liquid water besides providing a suitable environment also provides the thermal conditions necessary to retain good quality CNTs in the raw material, while the reactivity of the water with hot carbon does not appear to have any major effects on the reaction [59].

Figure 13 shows CNTs are produced in NaCl [62]. Nevertheless, arc discharge in NaCl solution is extremely stable owing to the excellent electrical conductivity induced by Cl⁻ and Na⁺ ions. Too many Na⁺ ions would hinder carbon ions flying from anode towards the center of cathode. Researchers found that perhaps this is another reason that the length of SWCNT is short and only a single SWCNT [61].

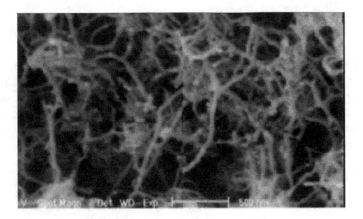

Figure 13. Production of CNTs in NaCl solution [62].

The optimized conditions to synthesize large quantities of SWCNT by applying arc discharge in NaCl solution deserve further investigation. Arc discharge in NaCl solution provides a very simple and cheap method to synthesize CNTs [60].

3.5. Discharge current

Discharge current is another important parameter influencing the products of arc discharge. If the catalyst percentage is 1 mol% Fe, and the discharge current is intentionally reduced to 20 A, the arc became very unstable, and disappear when the voltage is increased to 28 V [63].

3.6. The solution electrical conductivity effect

The effect of electrical conductivity of liquid on CNTs production might be important. A series of experiments carried out and the products were fabricated using arc discharge between two graphite electrodes submerged in different aqueous solutions of NaCl, KCl as well as LiCl. In comparative studies, CNTs were synthesized under different electrical conductivity conditions, and the results were analyzed, compared and discussed. The scanning electron microscopy (SEM), transmission electron microscopy (TEM) and Raman spectroscopy were employed to study the morphology of these carbon nanostructures and reported. LiCl 0.25 N (with 22.7mS as electrical conductivity) media when applied as solution, high-crystalline and a longed multi walled carbon nanotubes, single walled carbon nanotubes and springy carbon nanotubes (SCNTs) were synthesized. This study is one of the first one have demonstrated application of an arc discharge in liquid media with electrical conductivity effects upon CNT preparation and deserves further study (Dehghani and Jahanshahi, (2012); unpublished data).

4. Purification of fabricated CNTs

CNTs usually contain a large amount of impurities such as metal particles, amorphous carbon, and multi shell. There are different steps in purification of CNTs. Purification of CNTs is a process that separates nanotubes from non-nanotube impurities included in the raw products, or from nanotubes with undesired numbers of walls. Purification has been an important synthetic effort since the discovery of carbon nanotubes, and there are many publications discussing different aspects of the purification process. Good review articles on the purification of CNTs are available in the recent literature [64, 65].

The current industrial methods applied oxidation and acid-refluxing techniques that affect the structure of tubes. Purification difficulties are great because of insolubility of CNT and the limitation of liquid chromatography.

CNT purification step (depending on the type of the purification) removes amorphous carbon from CNTs, improves surface area, decomposes functional groups blocking the entrance of the pores or induces additional functional groups.

Most of these techniques are combined with each other to improve the purification and to remove different impurities at the same time. These techniques are as follow:

4.1. Oxidation

Oxidation is a way to remove CNTs impurities. In this way CNTs and impurities are oxidized. The damage to CNTs is less than the damage to the impurities. This technique is more preferable with regard to the impurities that are commonly metal catalysts which act as oxidizing catalysts [66, 67].

Altogether, the efficiency and yield of the procedure are highly depending on a lot of factors, such as metal content, oxidation time, environment, oxidizing agent and temperature [67].

4.2. Acid treatment

Refluxing the sample in acid is effective in reducing the amount of metal particles and amorphous carbon. Different used acids are hydrochloric acid (HCl), nitric acid (HNO_3) and sulphuric acid (H_2SO_4), while HCl is identified to be the ideal refluxing acid. When a treatment in HNO3 had been used the acid had an effect on the metal catalyst only, and no effects was observed on the CNTs and the other carbon particles. [66-69]. Figure 14 shows the SEM images of CNTs after and before purification stage with HCl [51].

If a treatment in HCl is used, the acid has also a little effect on the CNTs and other carbon particles [69, 70]. A review of literature demonstrates the effects that key variables like acid types and concentration & temperature have on the acid treatment [69, 70].

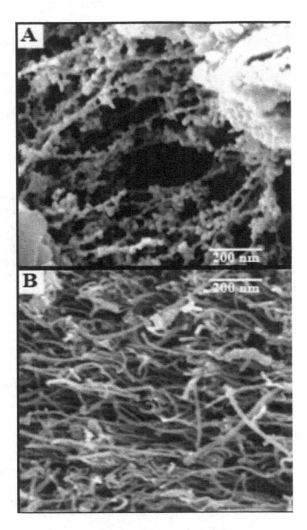

Figure 14. The SEM images of CNTs (A) after (B) before purification stages with HCl [51].

4.3. Annealing and thermal treatment

High temperature has effect on the productions and paralizes the graphitic carbon and the short fullerenes. When high temperature is used, the metal will be melted and can also be removed [69].

4.4. Ultrasonication

This technique is based on the separation of particles due to ultrasonic vibrations and also agglomerates of different nanoparticles will be more dispersed by this method. The separation of the particles is highly dependable on the surfactant, solvent and reagents which are used [67-70].

When an acid is used, the purity of the CNTs depends on the sonication time. During the tubes vibration to the acid for a short time, only the metal is solvated, but in a more extended period, the CNTs are also chemically cut. [69].

4.5. Micro-filtration

Micro-filtration is based on particle size. Usually CNTs and a small amount of carbon nanoparticles are trapped in a filter. The other nanoparticles (catalyst metal, fullerenes and carbon nanoparticles) are passing through the filter [65, 69, 70, 72].

A special form of filtration is cross flow filtration. Through a bore of fiber, the filtrate is pumped down at head pressure from a reservoir and the major fraction of the fast flowing solution is reverted to the same reservoir in order to be cycled through the fiber again. A fast hydrodynamic flow down the fiber bore sweeps the membrane surface and prevents building up of a filter cake [67].

5. Morphological and structural characterizations

To investigate the morphological and structural characterizations of the CNTs, a reduced number of techniques can be used. It is very important to characterize and determine the quality and properties of the CNTs, since its applications will require certification of properties and functions [74].

However, only few techniques are able to characterize CNTs at the individual level such as scanning tunneling microscopy (STM) and transmission electronic microscopy (TEM). X-ray photoelectron spectroscopy is required to determine the chemical structure of CNTs in spite of the fact that Raman spectroscopy is mostly introduced as global characterization technique.

5.1. Electron microscopy (SEM & TEM)

The morphology, dimensions and orientation of CNTs can be easily revealed by using scanning (SEM) and Transmission Electron Microscopes (TEM) which have high resolution. [70-75] (Figs. 15).

Therefore, the TEM technique is applied as a method for measurement of the outer and inner radius and linear electron absorption coefficient of CNTs [76].This method is used to study CNTs before and after annealing and notice a significant increase of the electron absorption coefficient. The inter shell spacing of MWNTs was studied by Kiang et al. [77] using high resolution TEM images.

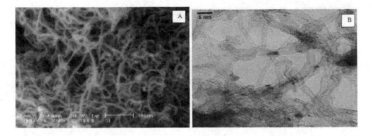

Figure 15. Electron micrographs of CNT (A) SEM of the CNT. (B) TEM of the CNT [57].

5.2. X-ray diffraction (XRD)

This technique is used to obtain some information on the interlayer spacing, the structural strain and the impurities. However, in comparing CNTs with x-ray incident beam, CNTs have multiple orientations. This leads to a statistical characterization of CNTs [78].

5.3. Raman spectroscopy

Raman spectroscopy is one of the most powerful tools for characterization of CNTs. Without sample preparation, a fast and nondestructive analysis is possible. All allotropic forms of carbon are active in Raman spectroscopy [79]. The position, width, and relative intensity of bands are modified according to the carbon forms [80].

A Raman spectrum of a purified sample (after applying the purification procedure) is shown in figure 16. The peaks at 1380 cm^{-1} and 1572 cm^{-1} correspond to disorder (D-band) and graphite (G-band) bands, respectively. The former is an indication of the presence of defective material and the latter one refers to the well-ordered graphite [62].

The most characteristic features are summarized as following:

1. Low-frequency peak <200 cm-1 characteristic of the SWNT, whose frequency is depended on the diameter of the tube mainly (RBM: radical breathing mode).

2. D line mode (disorder line), which is a large structure assign of residual ill-organized graphite.

3. High-frequency bunch that is called G band and is a characteristic of CNTs. This bunch has the ability to be superimposed with the G-line of residual graphite [81].

Raman spectroscopy is considered an extremely powerful tool for characterizing CNT, which gives qualitative and quantitative information on its diameter, electronic structure, purity and crystalline, and distinguishes metallic and semiconducting material as well as chirality.

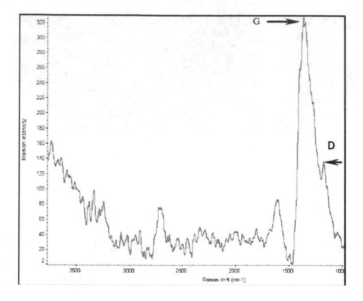

Figure 16. Raman spectrum showing the most characteristic features of CNTs produced by arc discharge method in liquid followed by acid treatment [62].

6. Conclusion

Carbon nanotubes (CNTs), a new structure of carbon element, are composed of graphen sheets rolled into closed concentric cylinders with diameter of the order of nanometers and length of micrometers. CNTs are attracted significant attention because of their unique physical and mechanical properties. These properties have been the engines of the rapid development in scientific studies in numerous applications such as in fuel cell and electrocatalyst, nanobiosensors, gas adsorptions and membrane separation [82-88].

Three methods, laser, arc discharge and chemical vapor deposition are used to synthesize CNTs. The laser method is also known as the laser-furnace method. The quantities CNTs in this method are large but this technique is not economically advantageous, since the process demanded considerable power. The chemical vapor deposition is another method for producing CNTs. It could produce CNTs at temperatures above 700 ^{0}C in large quantities, but the walls of the CNTs frequently contain many defects. Traditional arc discharge requires a complicated vacuum and heat exchange system. The yields of the laser and traditional arc discharge methods are very low (mg/h). From the application perspective, researchers are continuously trying to devise improved methods for CNTs fabrication.

Arc discharge in liquid media is a new method of synthesizing CNTs developed recently. All that is required is a dc power supply and an open vessel full of liquid nitrogen, deionized water or aqueous solution. This method is not requiring vacuum equipment, reacted gases, a high temperature furnace and a heat exchange system. Consequently, this method is extremely simple and cheap.

As it has been deeply investigated above, synthesis, purification and characterization of CNTs based on arc discharge in liquid media were described and discussed in this review paper. The observations of CNT growth under electron microscopy and other analytical techniques by different groups suggested that the mechanism are extremely sensitive to each fabrication parameter such as voltage difference between electrodes, current, type and ratio of catalysts, electrical conductivity, concentration, type and temperature of plasma solution and thermal conductivity. All these parameters were reviewed and studied herein. To the best of our knowledge the current review is the first one has discussed all aspects of arc discharge method in liquid media for CNT preparation and this technique deserves further attention.

Acknowledgements

The authors are grateful to Prof. M. Shariaty Niasar (Tehran Uiversity, Iran), Prof. J. Raoof (The University of Mazandaran, Iran), Dr. H. Molavi and PhD student Mrs R. Jabari Sheresht for their previous collaborations and productive discussion during preparation of this paper.

Author details

Mohsen Jahanshahi* and Asieh Dehghani Kiadehi

*Address all correspondence to: mjahan@nit.ac.ir

Nanotechnology Research Institute and Faculty of Chemical Engineering Babol University of Technology, Babol,, Iran

References

[1] Iijima, S. (1991). Helical Microtubules of Graphitic Carbon. *Nature*, 354, 56-58.

[2] Dresselhaus, M. S., Dresselhaus, G., & Eklund, P. C. (1996). Science of fullerenes and carbon nanotubes., Academic Press, New York.

[3] Dekker, C. (1999). Carbon nanotubes as molecular quantum wires. *Physics Today*, 52, 22-30.

[4] Tans, S. J., Verschueren, A. R. M., & Dekker, C. (1998). Room-temperature transistor based on a single carbon nanotube. *Nature*, 393, 49-52.

[5]   Javey, A., Kim, H., Brink, M., Wang, Q., Ural, A., Guo, J., Mc Intyre, P., Mc Euen, P., Lundstrom, M., & Dai, H. (2002). *Nature Materials*, 1, 241-246.

[6]   Javey, A., Guo, J., Wang, Q., Lundstrom, M., & Dai, H. J. (2003). Ballistic Carbon Nanotube Transistors. *Nature*, 424, 654-657.

[7]   Rosenblatt, S., Yaish, Y., Park, J., Gore, J., Sazonova, V., & Mc Euen, P. L. (2002). High Performance Electrolyte Gated Carbon Nanotube Transistors. *Nano Letters*, 2, 869-872.

[8]   Lu, C., Fu, Q., Huang, S., & Liu, J. (2004). *Nano Letters*, 4-623.

[9]   Huang, S. M., Woodson, M., Smalley, R., & Liu, J. (2004). Growth Mechanism of Oriented Long Single Walled Carbon Nanotubes Using "Fast-Heating" Chemical Vapor Deposition Process. *Nano Letters*, 4, 1025-1028.

[10] Tan, S. J., Devoret, M. H., Dai, H. J., Thess, A., Smalley, R. E., Geerligs, L. J., & Dekker, C. (1997). *Nature*, 386-474.

[11] Bockrath, M., Cobden, D. H., & Mc Euen, P. L. (2000). Individual Single-Wall Carbon Nanotubes As Quantum Wires. *Science*, 290, 1552-1555.

[12] Dai, H. J., Hafner, J. H., Rinzler, A. G., Colbert, D. T., & Smalley, R. E. (1996). Nanotubes As Nanoprobes in Scanning Probe Microscopy. *Nature*, 384, 147-150.

[13] Wong, S. S., Harper, J. D., Lansbury, P. T., & Lieber, C. M. (1998). *Am J. Chem. Soc.*, 120, 603.

[14] Rinzler, A., & Hafner, J. (1995). Unraveling Nanotubes- Field-Emission from an Atomic Wire. *Science*, 269, 1550-1553.

[15] Collins, P. G., Bradley, K., Ishigami, M., & Zettl, A. (2000). Extreme oxygen sensitivity of electronic properties of carbon nanotubes. *Science*, 287, 1801-1804.

[16] Kong, J., Franklin, N. R., Zhou, C., Peng, S., Cho, J. J., & Dai, H. (2000). Nanotube molecular wires as chemical sensors. *Science*, 287, 622-625.

[17] Chen, R. J., Bangsaruntip, S., Drouvalakis, K. A., Kam, N. W. S., Shim, M., Li, Y. M., Kim, W., Utz, P. J. H., Dai, J., & Natl, P. (2003). Nanotube molecular wires as chemical sensors. *Acad. Sci. USA*, 100, 4984-4989.

[18] Star, A., Gabriel, J. C. P., Bradley, K., & Grüner, G. (2003). Electronic detection of specific protein binding using nanotube FET devices. *Nano Letters*, 3, 459-463.

[19] Kuchta, B., Firlej, L., Pfeifer, P., & Wexler, C. (2010). Numerical Estimation of Hydrogen Storage Limits in Carbon Based Nanospaces. *Carbon*, 48, 223-231.

[20] Scott, C., Arepalli, S., Nikolaev, P., & Smalley, R. E. (2001). *Applied Physics A: Materials Science & Processing*, 72, 573-580.

[21] Bethune, D., Kiang, C., De Vries, M., Gorman, G., Savoy, R., & Beyer, R. (1993). *Nature*, 363, 605-607.

[22] Journet, C., Maser, W. K., Bernier, P., Loiseau, A., Chapelle, M., Lefrant, S., Deniard, P., Lee, R., & Fischer, J. E. (1997). Large scale production of single wall carbon nanotubes by the electric arc technique. *Nature*, 388, 756-758.

[23] Cassell, A. M., Raymakers, J. A., Kong, J., & Dai, H. (1999). Solvation of fluorinated single-wall carbon nanotubes in alcohol solvents. *Journal of Physical Chemistry B*, 103, 6484-6492.

[24] Liu, J., Fan, S., & Dai, H. (2004). *MRS Bulletin*, 4, 224-250.

[25] Jabari, Seresht. R., & Jahanshahi, M. (2010). *Fullerenes Nanotubes Carbon Nanostruct*, 2, 1-12.

[26] Biró, L., Horváth, Z., Szalmás, L., Kertész, K., Wéber, F., Juhász, G., Radnóczi, G., & Gyulai, J. (2003). Continuous carbon nanotube production in underwater ac electric arc. *Chemical physics letters*, 372, 399-402.

[27] Feng, Y., & Zhou, G. (2003). Removal of some impurities from carbon nanotubes. *Chemical physics letters*, 375, 645-648.

[28] Guo, T., Diener, M., & Chai, Y. (1992). Uranium Stabilization of C28- a Tetravalent Fullerene. *Science*, 257, 1661-1664.

[29] Thess, A., & Lee, R. (1996). Crystalline Ropes of Metallic Carbon Nanotubes. *Science*, 273, 483-487.

[30] Bandow, S., & Rao, A. (1997). Purification of single-wall carbon nanotubes by microfiltration. *Journal of Physical Chemistry B*, 101, 8839-8842.

[31] Chiang, I., & Brinson, B. (2001). *Journal of Physical Chemistry B*, 105, 8297-8301.

[32] Ishii, H., & Kataura, H. (2003). Direct observation of Tomonaga-Luttinger-liquid state in carbon nanotubes at low temperatures. *Nature*, 426, 540-544.

[33] Puretzky, A., & Geohegan, D. (2000). *Applied Physics A: Materials Science & Processing*, 70, 153-160.

[34] Sen, R., & Ohtsuka, Y. (2000). Time period for the growth of single-wall carbon nanotubes in the. *Chemical physics letters*, 332, 467-473.

[35] Kokai, F., & Takahashi, K. (2000). *Journal of Physical Chemistry B*, 104, 6777-6784.

[36] Li, Y., & Mann, D. (2004). *Nano Letters*, 4, 317-321.

[37] Walker, P., & Rakszawski, J. (1959). *Journal of Phys. Chem*, 63, 133-140.

[38] Dresselhaus, M., & Dresselhaus, G. (1988). Physical Properties of Carbon Nanotubes : Synthesis, Structure, Properties and Applications. Springer-Verlag, Berlin.

[39] Endo, M. (1988). Grow Carbon Fibers in the Vapor Phase. *Chemtech*, 18, 568-576.

[40] Baker, R. T. K., & Harris, P. S. (1978). Marcel Dekker, New York.

[41] Tibbetts, G. G. (1984). Why are carbon filaments tubular? *Journal of crystal growth*, 66, 632-638.

[42] Seresht, R. J., Jahanshahi, M., & Yazdani, M. (2009). Parametric study on the synthesis of single wall carbon nanotube by gas arc-discharge method with multiple linear regressions and artificial neural network. *International Journal of Nanoscience*, 8, 243-249.

[43] Saito, Y., & Inagaki, M. (1992). Yield of fullerenes generated by contact arc method under He and Ar-dependence on carbon nanotube. *Chemical physics letters*, 200, 643-648.

[44] Ebbesen, T., & Ajayan, P. (1992). *Nature*, 358, 220-222.

[45] Belin, T., & Epron, F. (2005). Characterization methods of carbon nanotubes: a review. *Materials Science and Engineering B*, 119, 105-118.

[46] Jahanshahi, M., Raoof, J., Hajizadeh, S., & Jabari, Seresht. R. (2007). *Nanotechnol. Appl.*, 929, 71-77.

[47] Bera, D., & Johnston, G. (2006). A parametric study on the nanotube structure through arc-discharge in water. *Nanotechnology*, 17, 1722-1730.

[48] Ishigami, M., & Cumings, J. (2000). A simple method for the continuous production of carbon nanotubes. *Chemical physics letters*, 319, 457-459.

[49] Jung, S. H., & Kim, M. R. (2003). Applied Physics A: Materials Science & Processing . 76, 285-286.

[50] Lange, H., Sioda, M., Huczko, A., Zhu, Y. Q., Kroto, H. W., & Walton, D. R. M. (2003). Nanocarbon production by are discharge in water. *Carbon*, 41, 1617-1623.

[51] Jahanshahi, M., & Seresht, R. J. (2009). Catalysts effects on the production of carbon nanotubes by an automatic arc discharge set up in solution. *physica status solidi (c)*, 6, 2174-2178.

[52] Xing, G., & Jia, S. (2007). Influence of transverse magnetic field on the formation of carbon nanotube. *Carbon*, 45, 2584-2588.

[53] Xu, B., Guo, Ju., Wang, X., Liu, X., & Ichinose, H. (2006). Synthesis of carbon nanocapsules containing Fe, Ni or Co by arc discharge in aqueous solution. *Carbon*, 44-2631.

[54] Seraphin, S., & Zhou, D. (1994). *Applied Physics Letters*, 64, 2087-2089.

[55] Saito, Y., & Tani, Y. (1998). High yield of single-wall carbon nanotubes by arc discharge using Rh-Pt mixed catalysts. *Chemical physics letters*, 294, 593-598.

[56] Hsin, Y. L., Hwang, K. C., Chen, R. R., & Kai, J. J. (2001). Production and in-situ Metal Filling of Carbon Nanotubes in Water. *J. Advanced Materials*, 13, 830-833.

[57] Dehghani, Kiadehi. A., Jahanshahi, M., Mozdianfard, M. R., Vakili-Nezhaad, G. H. R., & Jabari, Seresht. R. (2011). Influence of the solution temperature on carbon nano-

tube formation by arc discharge method. *Journal of Experimental Nanoscience*, 4, 432-440.

[58] Jahanshahi, M., Raoof, J., & Seresht, R. J. (2009). Voltage effects on production of nanocarbons by a unique arc-discharge set-up in solution. *Journal of Experimental Nanoscience*, 4, 331-339.

[59] Lide, D. R. (2003). *CRC handbook of chemistry and physics; CRC Pr I Llc.*

[60] Wang, Sh., Chang, M., Lan, K. M., Wu, Ch., Cheng, J., & Chang, H. (2005). *Carbon*, 43, 1778-1814.

[61] Schur, D. V., Dubovoy, A. G., Zaginaichenko, S., Yu, Adejev. V. M., Kotko, A. V., Bo-golepov, V. A., Savenko, A. F., & Zolotarenko, A. D. (2007). *Carbon*, 45, 1322-1329.

[62] Jahanshahi, M., Raoof, J., Hajizadeh, S., & Seresht, R. J. (2009). Synthesis and subse-quent purification of carbon nanotubes by arc discharge in NaCl solution. *Phys. Sta-tus Solidi A*, 1, 101-105.

[63] Wang, S. D., Chang, M. H., Cheng, J. J., Chang, H. K., & Lan, K. M. D. (2005). *Carbon*, 43, 1317-1339.

[64] Park, T. J., Banerjee, S., Hemraj-Benny, T., & Wong, S. S. (2006). Mater J. Chem. , 16, 141-154.

[65] Haddon, R. C., Sippel, J., Rinzler, A. G., & Papadimitrakopoulos, F. (2004). Purifica-tion and separation of carbon nanotubes. *MRS Bull.*, 29, 252-259.

[66] Hajime, G., Terumi, F., Yoshiya, F., & Toshiyuki, O. (2002). Method of purifying sin-gle wall carbon nanotubes from metal catalyst impurities. Honda Giken Kogyo Ka-bushiki Kaisha, Japan.

[67] Borowiak-Palen, E., & Pichler, T. (2002). *Chemical physics letters*, 363, 567-572.

[68] Farkas, E., Anderson, M. E., Chen, Z. H., & Rinzler, A. G. (2002). Length sorting cut single wall carbon nanotubes by high performance liquid chromatography. *Chem. Phys. Lett.*, 363, 111-116.

[69] Kajiura, H., Tsutsui, S., Huang, H. J., & Murakami, Y. (2002). High-quality single-walled purification from arc-produced soot. *Chem Phys Lett.*, 364, 586-92.

[70] Chiang, I., & Brinson, B. (2001). *Journal of Physical Chemistry B*, 105, 8297-8301.

[71] Bandow, S., & Rao, A. (1997). *Journal of Physical Chemistry B*, 101, 8839-8842.

[72] Moon, J. M., & An, K. H. (2001). High-yield purification process of singlewalled car-bon nanotubes. *Journal of Physical Chemistry B*, 105, 5677-5681.

[73] Jahanshahi, M., Tobi, F., & Kiani, F. (2005). Carbon-Nanotube Based Nanobiosen-sors:. Paper presented at Electrochemical Pretreatment. The 10th Iranian Chemical Engineering Conference,, Zahedan, Iran.

[74] Mawhinney, D. B., Naumenko, V., Kuznetsova, A., Yates, J. T., Liu, J., & Smalley, R. E. (2000). Surface defect site density on single walled carbon nanotubes by titration. *Chem. Phys. Lett.*, 324, 213-216.

[75] Li, W., Wen, J., & Tu, Y. (2001). *Ren Z. Appl. Phys. A*, 73, 259-264.

[76] Gommes, C., & Blacher, S. (2003). *Carbon*, 41, 2561-2572.

[77] Kiang, C. H., & Endo, M. (1998). Size effect in carbon nanotubes. *Physical review letters*, 81, 1869-1872.

[78] Zhu, W., Miser, D., Chan, W., & Hajaligol, M. (2003). *Mater. Chem. Phys*, 82, 638-647.

[79] Arepalli, S., & Nikolaev, P. (2004). Protocol for the characterization of SWCNT material quality. *Carbon*, 42, 1783-1791.

[80] Ferrari, A., & Robertson, J. (2000). *Physical Review B*, 61, 14095.

[81] Hiura, H., Ebbesen, T. W., Tanigaki, K., & Takahashi, H. (1993). Single-walled carbon nanotubes produced by electric arc. *Chem. Phys. Lett.*, 202, 509.

[82] Raoof, J. B., Jahanshahi, M., & Momeni, Ahangar. S. (2010). Nickel Particles Dispersed into Poly (o-anisidine) and Poly (oanisidine)/Multi-walled Carbon Nanotube Modified Glassy Carbon Electrodes for Electrocatalytic Oxidation of Methanol. *Int. J. Electrochem. Sci.*, 5, 517-530.

[83] Toubi, F., Jahanshahi, M., Rostami, A. A., & Hajizadeh, S. (2009). Voltametric tests on different carbon nanotubes as nanobiosensor devices. *D. Biochem Process Biotech. Mol. Biol.*, 2, 71-74.

[84] Khalili, S., Ghoreyshi, A. A., & Jahanshahi, M. (2012). Equilibrium, kinetic and thermodynamic studies of hydrogen adsorption on multi-walled carbon nanotubes. *Iranica Journal of Energy & Environment*, 1, 69-75.

[85] Delavar, M., Ghoreyshi, A. A., Jahanshahi, M., Khalili, S., & Nabian, N. (2012). The effect of chemical treatment on adsorption of natural gas by multi-walled carbon nanotubes: sorption equilibria and thermodynamic studies. Chemical Industry & Chemical Engineering Quarterly Inpress.

[86] Delavar, M., Ghoreyshi, A. A., Jahanshahi, M., & Nabian, N. (2012). Journal of Experimental Nanoscience Inpress.

[87] Delavar, M., Ghoreyshi, A. A., Jahanshahi, M., Khalili, S., & Nabian, N. (2012). Equilibria and kinetics of natural gas adsorption on multi-walled carbon nanotube material. *Green Chemistry*, 2, 4490-4497.

[88] Rahimpour, A, Jahanshahi, M, Khalili, S, Mollahosseini, A, Zirepour, A, & Rajaeian, B. (2012). Novel functionalized carbon nanotubes for improving the surface properties and performance of polyethersulfone (PES) membrane. *J. Desalination.*, 286, 99-107.

Chapter 2

Polymer Nanocomposites Containing Functionalised Multiwalled Carbon NanoTubes: a Particular Attention to Polyolefin Based Materials

Emmanuel Beyou, Sohaib Akbar,
Philippe Chaumont and Philippe Cassagnau

Additional information is available at the end of the chapter

Introduction

Incorporation of carbon nanotubes (CNTs) into a polymer matrix is a very attractive way to combine the mechanical and electrical properties of individual nanotubes with the advantages of plastics. Carbon nanotubes are the third allotropic form of carbon and were synthesized for the first time by Iijima in 1991 [1]. Their exceptional properties depend on the structural perfection and high aspect ratio (typically ca 100-300). Two types of CNTs are distinguished : single-walled CNTs (SWCNTs) consist of a single graphene sheet wrapped into cylindrical tubes with diameters ranging from 0.7 to 2nm and have lengths of micrometers while multi-walled CNTs (MWCNTs) consist of sets of concentric SWCNTs having larger diameters [2-5]. The unique properties of individual CNTs make them the ideal reinforcing agents in a number of applications [6-9] but the low compatibility of CNTs set a strong limitation to disperse them in a polymer matrix. Indeed, carbon nanotubes form clusters as very long bundles due to the high surface energy and the stabilization by numerous of π–π electron interactions among the tubes. Non covalent methods for preparing polymer/CNTs nanocomposites have been explored to achieve good dispersion and load transfer [10-12]. The non-covalent approaches to prepare polymer/CNTs composites via processes such as solution mixing [13,14], melt mixing [14,15], surfactant modification [16], polymer wrapping [17], polymer absorption [18] and in situ polymerization [19, 20] are simple and convenient but interaction between the two components remains weak. Relatively uniform dispersion of CNTs can be achieved in polar polymers such as nylon, polycarbonate and polyimide because of the strong interaction between the polar moiety of the polymer chains and the surface of the CNTs [21-24]. Moreover, it was found that MWNTs disperse well in PS and form a network-like structure due to π-stacking interactions with aromatic groups of the PS chains [25]. However, it is difficult to disperse CNTs within a non polar polymer matrix such as polyolefins. To gain the advantages of CNTs at its best, one needs: (i) high interfacial area between nanotubes and polymer; and, (ii) strong interfacial interaction. Unfortunately the solvent technique does not help much in achieving these targets and, as a result, a nano-composite having properties much inferior to theoretical expectations are obtained. For example, the mechanical properties of polyethylene (PE) reinforced by carbon nanotubes do not improve significantly because the weak polymer-CNT interfacial adhesion prevents efficient stress transfer from the polymer matrix to CNT [26-28]. A strategy for enhancing the compatibility between nanotubes and polyolefins consists in functionalising the sidewalls of CNT to introduce reactive moieties and to disrupt the rope structure. Functional moieties are attached to open ends and sidewalls to improve the solubility of nanotubes [29-32] while the covalent polymer grafting approaches, including 'grafting to' [33-36] and 'grafting from' [37-39] that create chemical linkages between polymer and CNTs, can significantly improve dispersion and change their rheological behaviour. First, methods used for processing CNTs-based nanocomposites and for the functionalisation of carbon nanotubes (CNTs) with polymers will be described. This is followed by a review of the surface chemistry of carbon nanotubes in order to perform their dispersion in polyolefin matrix. Finally, general trends of the viscoelastic properties of CNTs/ polyolefin composites are discussed.

1. Methods to process polymer/carbon nanotubes composites

Similar to the case of carbon nanotube/solvent suspensions, pristine carbon nanotubes have not yet been shown to be soluble in polymers illustrating the extreme difficulty of overcoming the inherent thermodynamic drive of nanotubes to bundle [40]. Several processing methods available for fabricating CNT/polymer composites based on either thermoplastic or thermosetting matrices mainly include solution mixing, melt blending, and in situ polymerisation (figure 1) [41, 42].

1.1. Solution blending

The most common method for preparing polymer nanotube composites has been to mix the nanotubes and polymer in a suitable solvent before evaporating the solvent to form a composite film (Figure 1a). One of the benefits of this method is that agitation of the nanotubes powder in a solvent facilitates nanotubes' de-aggregation and dispersion. Almost all solution processing methods are based on a general theme which can be summarised as:

1. Dispersion of nanotubes in either a solvent or polymer solution by energetic agitation.

2. Mixing of nanotubes and polymer in solution by energetic agitation.

3. Controlled evaporation of solvent leaving a composite film.

In general, agitation is provided by magnetic stirring, shear mixing, reflux or ultrasonication. Sonication can be provided in two forms, mild sonication in a bath or high-power sonication using a tip or horn. An early example of solution based composite formation is described by Jin *et al* [43]. By this method, high loading levels of up to 50wt% and reasonably good dispersions were achieved. A number of papers have discussed dispersion of nanotubes in polymer solutions [44-46]. This can result in good dispersion even when the nanotubes cannot be dispersed in the neat solvent. Coleman *et al* [44] used sonication to disperse catalytic MWCNT in polyvinylalcohol/H_2O solutions, resulting in a MWCNT dispersion that was stable indefinitely. Films could be easily formed by drop-casting with microscopy studies showing very good dispersion. Cadek *et al* [46] showed that this procedure could also be applied to arc discharge MWCNTs, double walled nanotubes (DWNTs) and High-Pressure CO Conversion (HiPCO) SWCNTs. They also showed that this procedure could be used to purify arc-MWCNTs by selective sedimentation during composite production.

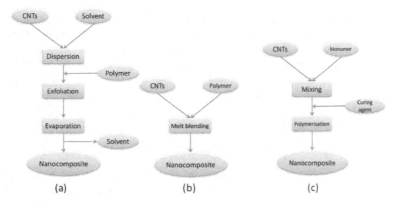

Figure 1. Schematic representation of different steps of polymer/CNTs composite processing: solution mixing (a); melt mixing (b); *in situ* polymerisation (c).

1.2. Melt mixing

While solution processing is a valuable technique for both nanotube dispersion and composite formation, it is completely unsuitable for the many polymer types that are insoluble. Melt processing is a common alternative method, which is particularly useful for dealing with thermoplastic polymers (Figure 1b). This range of techniques makes use of the fact that thermoplastic polymers soften when heated. Amorphous polymers can be processed above their glass transition temperature while semi-crystalline polymers need to be heated above their melt temperature to induce sufficient softening. Advantages of this technique are its speed and simplicity, not to mention its compatibility with standard industrial techniques [47, 48]. Any additives, such as carbon nanotubes can be mixed into the melt by shear mixing. However, Bulk samples can then be fabricated by techniques such as compression

moulding, injection moulding or extrusion. However it is important that processing conditions are optimised for the whole range of polymer–nanotube combinations. High temperature and shear forces in the polymer fluid are able to break the carbon nanotubes bundles
and CNTs can additionally affect melt properties such as viscosity, resulting in unexpected
polymer degradation [49]. Andrews and co-workers [50] showed that commercial polymers
such as high impact polystyrene, polypropylene and acrylonitrile–butadiene–styrene (ABS)
could be melt processed with CVD-MWCNT to form composites. The polymers were blended with nanotubes at high loading level in a high shear mixer to form master batches. An
example of using combined techniques was demonstrated by Tang *et al* [51]. High density
polyethylene pellets and nanotubes were melted in a beaker, then mixed and compressed.
The resulting solid was broken up and added to a twin screw extruder at 170°C and extrud-
ed through a slit die. The resulting film was then compression moulded to form a thin film.

1.3. In Situ Polymerisation

This fabrication strategy starts by dispersing carbon nanotubes in vinyl monomers followed
by polymerising the monomers (Figure 1c). This method produces polymer-grafted CNTs
mixed with free polymer chains resulting in a homogeneous dispersion of CNTs. In situ radical polymerisation was applied for the synthesis of PMMA-based composites by Jia *et al*
[52] using a radical initiator and the authors suggested that π-bonds of the CNT graphitic
network were opened by the radical fragments of initiator and therefore the carbon nanostructures could participate in PMMA polymerisation by acting as efficient radical scaveng-
ers. Dubois et al [53] applied the in situ polymerization to olefin monomers by anchoring
methylaluminoxane, a commonly used co-catalyst in metallocene-based olefin polymerization onto carbon nanotubes surface. Then, the metallocene catalyst was added to the surface-
activated CNTs and the course of ethylene polymerization was found to be similar to the
one without the presence of pristine MWCNTs. Epoxy nanocomposites comprise the majori-
ty of reports using in situ polymerisation methods [54, 55], where the nanotubes are first dispersed in the resin followed by curing the resin with the hardener. Zhu *et al* [56] prepared
epoxy nanocomposites by this technique using end-cap carboxylated SWCNTs and an esterification reaction to produce a composite with improved tensile modulus (E is 30% higher
with 1 wt % SWCNT).

1.4. Novel methods

Rather than avoid the high viscosities of nanotube/polymer composites, some researchers have
decreased the temperature to increase viscosity to the point of processing in the solid state. Solid-state mechanochemical pulverisation processes (using pan milling [57] or twin-screw pulverisation [58]) have mixed MWCNTs with polymer matrices. Pulverisation methods can be
used alone or followed by melt mixing. Nanocomposites prepared in this manner have the advantage of possibly grafting the polymer on the nanotubes, which account in part for the observed good dispersion, improved interfacial adhesion, and improved tensile modulus.

Polymer Nanocomposites Containing Functionalised Multiwalled Carbon NanoTubes:
a Particular Attention to Polyolefin Based Materials

29

An innovative latex fabrication method for making nanotube/polymer composites has been used by first dispersing nanotubes in water (SWCNT require a surfactant, MWCNT do not) and then adding a suspension of latex nanoparticles [59,60]. For example, PEG-based amphiphilic molecule containing aromatic thiophene rings, namely, oligothiophene-terminated poly(ethylene glycol) (TN-PEG) was synthesized, and its ability to disperse and stabilize pristine carbon nanotubes in water was shown.This promising method can be applied to polymers that can be synthesised by emulsion polymerisation or formed into artificial latexes, e.g., by applying high-shear conditions [61].

Finally, to obtain nanotube/polymer composites with very high nanotube loadings, Vigolo et al [62] developed a "coagulation spinning" method to produce composite fibers comprising predominately nanotubes. This method disperses SWCNT using a surfactant solution, coagulates the nanotubes into a mesh by wet spinning it into an aqueous poly(vinyl alcohol) solution, and converts the mesh into a solid fiber by a slow draw process. In addition, Mamedov et al [63] developed a fabrication method based on sequential layering of chemically modified nanotubes and polyelectrolytes to reduce phase separation and prepared composites with SWCNT loading as high as 50 wt %.

2. Surface modifications of carbon nanotubes with polymers

CNTs are considered ideal materials for reinforcing fibres due to their exceptional mechanical properties. Therefore, nanotube–polymer composites have potential applications in aerospace science, where lightweight robust materials are needed [64]. It is widely recognised that the fabrication of high performance nanotube–polymer composites depends on the efficient load transfer from the host matrix to the tubes. The load transfer requires homogeneous dispersion of the filler and strong interfacial bonding between the two components [65]. A dispersion of CNT bundles is called "macrodispersion" whereas a dispersion of individual nonbundled CNT is called a nanodispersion [66, 67]. To address these issues, several strategies for the synthesis of such composites have been developed. Currently, these strategies involve physical mixing in solution, *in situ* polymerisation of monomers in the presence of nanotubes, surfactant-assisted processing of composites, and chemical functionalisation of the incorporated tubes. As mentioned earlier, in many applications it is necessary to tailor the chemical nature of the nanotube's walls in order to take advantage of their unique properties. For this purpose, two main approaches for the surface modification of CNTs are adopted i.e. covalent and noncovalent, depending on whether or not covalent bonding between the CNTs and the functional groups and/or modifier molecules is involved in the modification surface process. Figure 2 depicts a typical representation of such surface modifications.

2.1. Noncovalent attachment of polymers

The noncovalent attachment, controlled by thermodynamic criteria [68], which for some polymer chains is called wrapping, can alter the nature of the nanotube's surface and make it more compatible with the polymer matrix. Non-covalent surface modifications are based mainly on weak interactions, such as van der Waals, $\pi-\pi$ and hydrophobic interactions, between CNTs and modifier molecules. Non-covalent surface modifications are advantageous in that they conserve sp^2-conjugated structures and preserve the electronic performance of CNTs. The main potential disadvantage of noncovalent attachment is that the forces between the wrapping molecule and the nanotube might be weak, thus as a filler in a composite the efficiency of the load transfer might be low.

Non-covalent modification approaches typically use organic mediating molecules that range from low molecular weight molecules to supra- molecules to polymers.

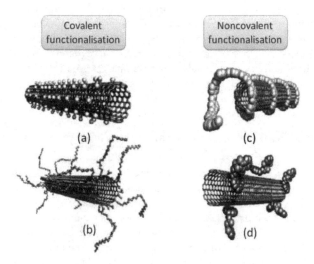

Figure 2. Different routes for nanotubes' functionalisation: sidewall covalent functionalisation (a); defect-group covalent functionalisation (b); noncovalent polymer wrapping (c); noncovalent pi-stacking (d).

2.1.1. Polymer wrapping

O'Connell et al. [68] reported that nanotubes could be reversibly solubilised in water by noncovalently associating them with a variety of linear polymers such as polyvinyl pyrrolidone (PVP) and polystyrene sulfonate (PSS). They demonstrated that the association between the polymer and the nanotubes is robust, not dependent upon the presence of excess polymer in solution, and is uniform along the sides of the tubes (Figure 1c). A general thermodynamic driving force for such wrapping in an aqueous environment has been identified [68].

Conjugated luminescent polymer poly-{(m-phenylenevinylene)-co-[(1,5-dioctyloxy-p-phenylene)-vinylene]} (PmPV) and its derivatives [69-71] have been successfully used for the wrapping around nanotubes on account of stabilising noncovalent bonding interactions, presumably as a result of $\pi-\pi$ stacking (Figure 1d) and van der Waals interactions between PmPV and the surfaces of the nanotubes. Star et al [72] also synthesised the Stilbenoid dendrimers, a hyperbranched variant of the PmPV polymer, which exhibits an appropriate degree of branching, and it was found to be more efficient at breaking up nanotube bundles, provided it is employed at higher polymer-to-nanotube ratios than was the "parent" PmPV polymer.

In addition, the behavior of single walled and multi walled carbon nanotubes in aqueous solutions of Gum Arabic, a natural polysaccharide, has been described by Nativ-Roth et al [73]. They observed that while the as-prepared nanotube powders contain highly entangled ropes and bundles, the dispersions are mainly composed of individual tubes suggesting that the ability of Gum Arabic to exfoliate the bundles, and stabilize individual tubes in aqueous dispersions, can be utilized in the preparation of carbon nanotube-polymer composites. In the latter case, the dispersing polymer acts as a compatabilizer and as an adhesion promoter leading to strengthening of the matrix-nanotube interface.

It is clear from these accounts that noncovalent functionalisation of carbon nanotubes can be achieved without disrupting the primary structure of the nanotubes themselves.

2.1.2. Polymer absorption

Xia et al [74] has described a method to prepare polymer-encapsulated MWCNTs : it has been successfully prepared through ultrasonically initiated in situ emulsion polymerisations of n-butyl acrylate (BA) and methyl methacrylate (MMA) in presence of MWCNT. By employing the multiple effects of ultrasound, i.e., dispersion, pulverizing, activation, and initiation, the aggregation and entanglement of carbon nanotubes in aqueous solution can be broken down, while in situ polymerization of monomer BA or MMA on the surface of MWCNTs proceeds and the MWCNTs are coated by the formed polymer.

The hydrophilic regions of surfactants interact with polar solvent molecules, and the hydrophobic regions can adsorb onto nanotube surfaces [75]. Thus, the process of dispersing CNTs from aggregates, bundles, or ropes into separated individual CNTs depends strongly on the length of the hydrophobic regions and the types of hydrophilic groups of the surfactant. A topological, noncovalent solution to improving the dispersion of SWNTs by encasing them in cross-linkable surfactant micelles was demonstrated by Kang and Taton [16]. SWCNTs were dispersed in the dimethylformamide (DMF) solutions of amphiphilic poly(styrene)-block-poly(acrylic acid) copolymer. Water was added to the solutions and the poly(styrene)-block-poly(acrylic acid) copolymer wrapped the SWCNTs and formed micelle. Then the PAA blocks of the micellar shells were permanently crosslinked by addition of a water-soluble diamine linker and a carbodiimide activator. This encapsulation significantly enhances the dispersion of SWCNTs in a wide variety of polar and nonpolar solvents and polymer matrices [76]. Encapsulated SWNTs can be used as an alternative starting material to pure SWNTs for the production and investigation of nanotube composite materials.

2.2. Covalent attachment of polymers

Functionalisation of carbon nanotubes with polymers is a key issue to improve the interfacial interaction between CNTs and the polymer matrix when processing polymer/CNT nanocomposites. The covalent reaction of CNT with polymers is important because the long polymer chains help to dissolve the tubes into a wide range of solvents even at a low degree of functionalisation. There are two main methodologies for the covalent attachment of polymeric substances to the surface of nanotubes, which are defined as "grafting to" and 'grafting from' methods [76, 77]. The former relies on the synthesis of a polymer with a specific molecular weight followed by end group transformation. Subsequently, this polymer chain is attached to the graphitic surface of CNT. A disadvantage of this method is that the grafted polymer contents are limited because of high steric hindrance of macromolecules. The 'grafting from' method involved the immobilisation of initiators onto the substrate followed in situ surface polymerization to generate grafted polymer chains. Because the covalent attachment of the surface modifiers involves the partial disruption of the sidewall sp2 hybridization system, covalently modified CNTs inevitably lose some degree of their electrical and/or electronic performance properties [78].

2.2.1. 'Grafting to' method

Since the curvature of the carbon nanostructures imparts a significant strain upon the sp^2 hybridised carbon atoms that make up their framework, the energy barrier required to convert these atoms to sp^3 hybridisation is lower than that of the flat graphene sheets, making them susceptible to various addition reactions. Therefore, to exploit this chemistry, it is only necessary to produce a polymer-centred transient in the presence of CNT material. Alternatively, defect sites on the surface of oxidized CNTs, as open-ended nanostructures with terminal carboxylic acid groups, allow covalent linkages of oligomer or polymer chains. So, the 'grafting to' method involves the chemical reaction between as-prepared or commercially available polymers with reactive end groups and nanotubes' surface functional groups or the termination of growing polymer radical, cation and anion formed during the polymerization of various monomers in the presence of CNTs or the deactivation of living polymer chain ends with the CNT surface.

For example, oxidized SWCNTs were grafted with amino-terminated poly (N-isopropylacrylamide) (PNIPAAm) by carbodiimide-activated reaction, which yielded a 8wt% polymer content[77]. In a different approach, oxidized MWCNTs were attached ontopolyacrylonitrile(PAN) nanoparticles through the reaction of the reduced cyano-groups of the polymer and the carboxylic moieties of CNT surface [79]. In addition, the amidation reaction was used for grafting of oligo-hydroxyamides to MWCNTs as described in figure 3 [80].

Ester-based linkages have been used by Baskaran et al. [81] by performing the reaction of hydroxy-terminated PS with thionyl chloridetreated MWCNTs, resulting in a hybrid containing 86wt% of CNTs. The esterification reaction was also used for grafting polyethylene glycol(PEG) chains to acylchloride-activated SWCNTs [82]. Silicone-functionalised CNT derivatives were prepared by opening terminal epoxy groups of functionalised polydimethylsiloxanes (PDMS) by the carboxylic groups of acid-treated MWCNTs [83]. Another example of the "grafting to" approach has been reported by Qin et al. [84] through the grafting of polystyrene with azide end group onto SWCNTs (Figure 4).

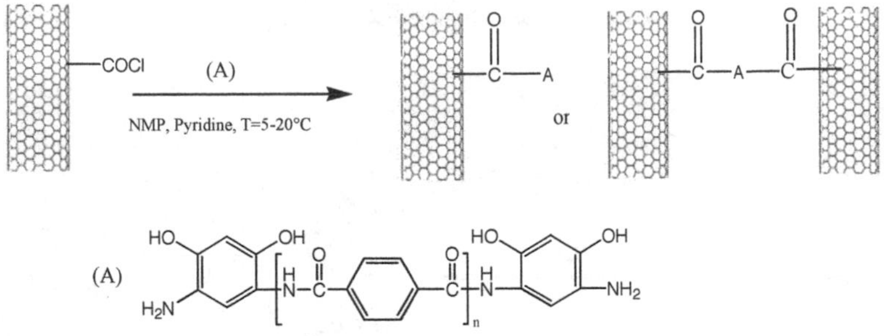

Figure 3. Synthesis of oligo-hydroxyamide-grafted MWNT. Reproduced from [80] with permission of Elsevier.

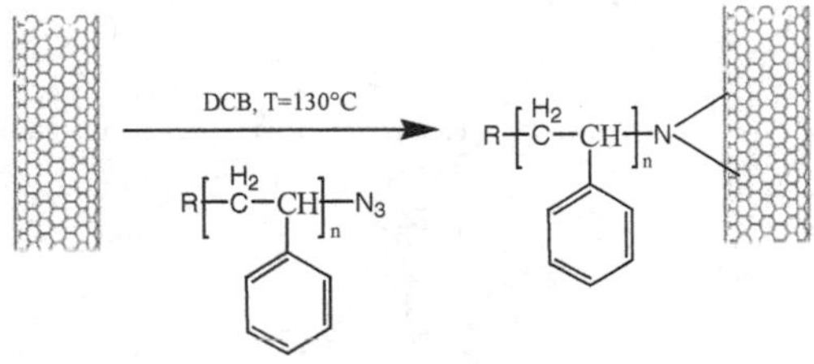

Figure 4. reaction of azide-terminatedpolystyreneonto CNTs surface. Reproduced from [80] with permission of ACS publications.

In an analogous approach, alkyne-decorated SWCNTs and PS-N₃were coupled via [3+1] Huisgen cycloaddition between the alkyne and azide end groups [85]. A new method was developed by Hung et al. [86] for preparing polystyrene-functionalized multiple-walled carbon nanotubes (MWNTs) through the termination of anionically synthesized living polystyryllithium with the acyl chloride functionalities on the MWNTs. The acyl chloride functionalities on the MWNTs were in turn obtained by the formation of carboxyls via chemical oxidation and their conversion into acyl chlorides (Figure 5).

Lou et al. [87] reported the radical grafting of polyvinylpyridine chains onto the surface of nanotubes through the thermolysis of poly (2-vinylpyridine) terminated with a radical-stabilizing nitroxide (Figure 6), resulting in grafting densities up to 12 wt.-%.

2.2.2. 'Grafting from' technique

Mostly, it involves the polymerisation of monomers from surface-derived initiators on either MWCNTs or SWCNTs. These initiators are covalently attached using the various functionalisation reactions developed for small molecules [77]. Then, the polymer is bound via in situ radical, cationic, anionic, ring opening and condensation polymerizations. The advantage of 'grafting from' approach is that the polymer growth is not limited by steric hindrance, allowing high molecular weight polymers to be efficiently grafted as well as quite high grafting density [9].

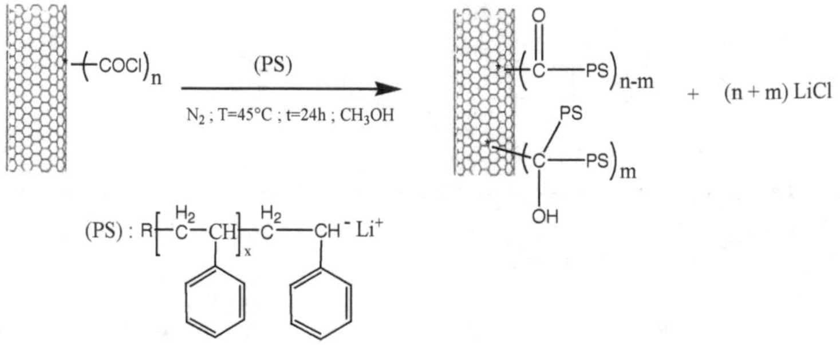

Figure 5. substitution reaction of living polystyryllithium anions with acyl choride-modified CNTs. Reproduced from [86] with permission of John Wiley and Sons.

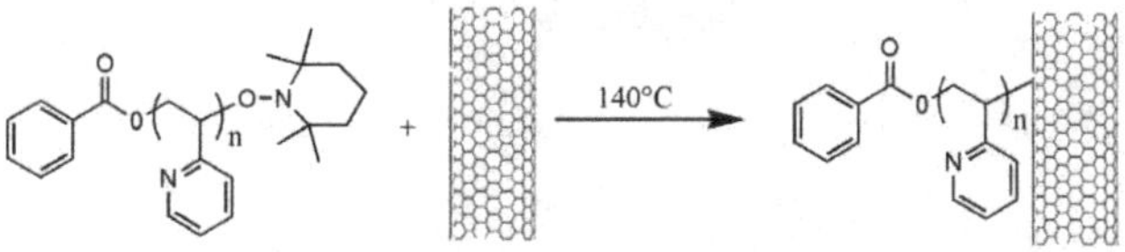

Figure 6. Radical grafting of TEMPO-end capped PVP to MWCNTs. Reproduced from [87] with permission of Elsevier.

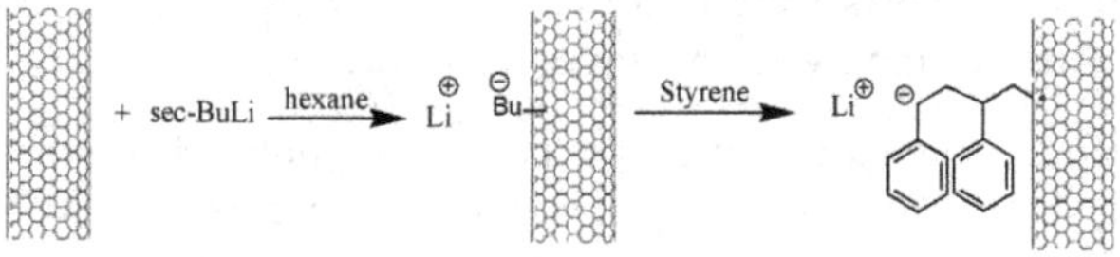

Figure 7. Anionic polymerisation of styrene onto carbon nanotubes

For example, Viswanathan et al [88]. have developed a procedure based on the SWCNT surface treatment with butyllithium providing initiating sites for the anionic polymerization of styrene (Figure 6).

The latter procedure eliminates the need for nanotube pretreatment prior to functionalization and allows attachment of polymer molecules to pristine tubes without altering their
original structure.

In addition, polyethylenimine has been grafted onto the surface of MWNTs by performing a
cationic polymerization of aziridine in the presence of amine-functionalized MWNTs (NH_2–
MWNTs [89]. The grafting of PEI was realized through two mechanisms, the activated monomer mechanism (AMM) or the activated chain mechanism (ACM), by which protonated
aziridine monomers or the terminal iminium ion groups of propagation chains, respectively,
are transferred to amines on the surface of MWNTs [89].

Bonduel et al. [90] reported the homogeneous surface coating of long carbon nanotubes by
in situ polymerization of ethylene as catalyzed directly from the nanotube surface-treated by
a highly active metallocene-based complex. It allowed for the break-up of the native nanotube bundles leading, upon further melt blending with HDPE, to high-performance polyole-
finic nanocomposites [90]. In another attempt, an easy method for preparing polystyrene-
grafted multi-walled carbon nanotubes (MWCNTs) with high graft yields was developed by
using free radical graft polymerisation from photoinduced surface initiating groups on
MWCNTs [91]. Conventional microscopy, including optical, atomic force, sanning electronic
microscopy (SEM), and transmission electronic microscopy (TEM), reveal the dispersion
state or quality of CNTs within a very limited area of a given nanofiller composite [67]. High
resolution-TEM image of the MWCNTs-PS (Figure 8) shows that the surface of the
MWCNTs-PS is covered with 4–5nm thick amorphous PS layers while the wall surface of
purified MWCNTs was smooth, without any detectable polymer Layer [91].

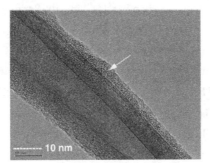

Figure 8. HR-TEM image of PS-g-MWCNT. Reprinted from [91] with permission of Elsevier.

Controlled radical polymerisation techniques such as nitroxide mediated polymerisation
(NMP), atom transfer radical polymerisation (ATRP) and radical adition fragmentation
transfer (RAFT) have been also used to graft polymer chains from the CNT surface [92-103]
(see figure 9 as example).

3. Carbon nanotubes nanocomposites based on Polyolefins

Polyethylene (PE) is one of the most widely used commercial polymer due to the excellent combination of low coefficient of friction, chemical stability and excellent moisture barrier properties [104]. The combination of a soft polymer matrix such as PE with nanosized rigid filler particles may provide new nanocomposite materials with largely improved modulus and strength. To improve the stiffness and rigidity of PE, CNTs can be used to make CNT/PE composites [104-107]. The mechanical properties of polyethylene (PE) reinforced by carbon nanotubes do not improve significantly because the weak polymer-CNT interfacial adhesion prevents efficient stress transfer from the polymer matrix to CNT [108-110]. The lack of functional groups and polarity of PE backbone results in incompatibility between PE and other materials such as glass fibres, clays, metals, pigments, fillers, and most polymers. A strategy for enhancing the compatibility between nanotubes and polyolefins consists in functionalising the sidewalls of CNT with polymers either by a 'grafting to' or a 'grafting from' approach. As discussed before, the "grafting from" approach involves the growth of polymers from CNT surfaces via in situ polymerisation of olefins initiated from chemical species immobilised on the CNT. As an example, Ziegler-Natta or metallocene catalysts for ethylene polymerisation can be immobilised on nanotubes to grow PE chains from their surface. However, covalent linkages or strong interactions between PE chains and nanotubes cannot be created during polymerisation [90, 111-113]. The "grafting to" technique involves the use of addition reactions between the polymer with reactive groups and the CNT surface. However, the synthesis of end-functionalized polyethylene (PE), which is necessary in the "grafting to" approach, is difficult [114]. Another promising route for a chemical modification of MWCNTs by PE is to use free radical initiators such as peroxides. The general mechanism of free radical grafting of vinyl compound from hydrocarbon chains detailed by Russell [115], Chung [116] and Moad [117] seems to express a widespread view. The grafting reaction starts with hydrogen abstraction by alkoxyl radicals generated from thermal decomposition of the peroxide. Then, the active species generated onto the hydrocarbon backbone react with unsaturated bonds located on the MWCNTs surface. This chemical modification is thus conceivable during reactive extrusion because the radicals' lifetimes (in the range of few milliseconds) are compatible with typical residence time in an extruder (around one minute).

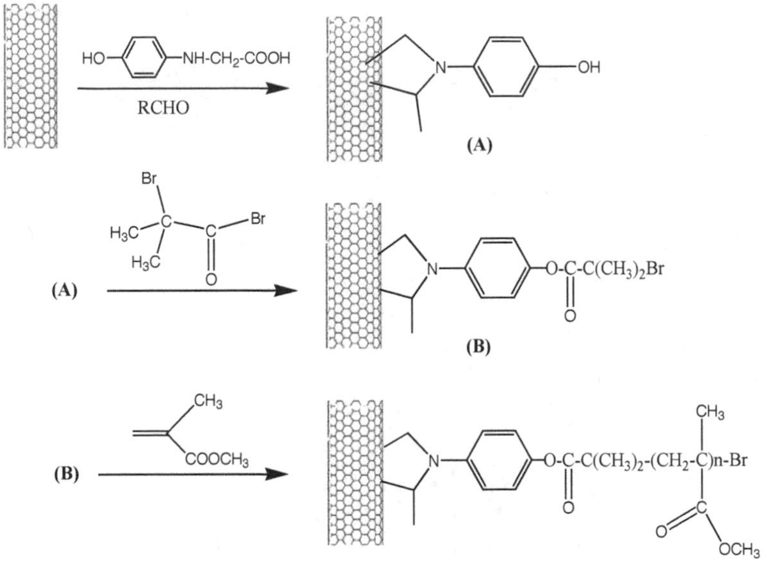

Figure 9. ATRP 'grafting from' modification approach. Reproduced from [92] with permission of ACS publications.

3.1. Radical grafting of polyethylene onto MWCNTs

The main drawback of the free radical grafting is the low selectivity of the radical center, specially at high temperatures (in the range of 150-200°C, required for extrusion of polyethylene), leading to side reactions such as coupling and chain scission [115, 118]. Moreover, performing this chemical modification by reactive processing brings in many constraints inherent to the processing (e.g. short reaction time, viscous dissipation and high temperature). For instance, the difference of viscosity between the monomer and the molten polymer could enhance these side reactions. So, to separate these physical influences from the chemical modification, the grafting reaction can be predicted with a model compound approach based on a radical grafting reaction between peroxide-derived alkoxyl radicals, and a low molar mass alkane representing characteristics moieties of PE.

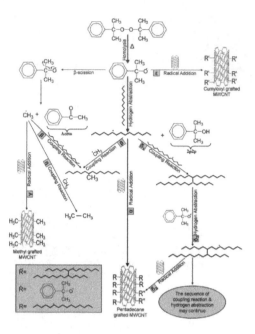

Figure 10. General reactive pathways of free radical grafting of pentadecane onto MWCNTs ; Reproduced from [119] with permission of Elsevier.

3.1.1. A model compound approach through the use of pentadecane

Covalent functionalization of pentadecane-decorated multiwalled carbon nanotubes (MWCNTs) has been studied as a model compound approach for the grafting of poly (ethylene-co-1-octene) onto MWCNTs by reactive extrusion [119]. It was accomplished through radical addition onto unsaturated bonds located on the MWCNTs' surface using dicumyl peroxide as hydrogen abstractor. Pentadecane ($C_{15}H_{32}$) has been resorted as model for polyethylene because high boiling points of long chain alkanes permit study under high temperature conditions, typically over 150°C. It also gives clues about low viscosity at 150°C, on top of that the formed products in the grafting experiment can hence be analysed more easily than in the polymer melt. Figure 10 sums up main reactive pathways of free radical grafting of pentadecane onto MWCNTs with dicumyl peroxide as initiator. The hydrogen abstraction reactions from alkyl hydrocarbon bonds was studied starting from the reaction of DCP-derived radicals with pentadecane.

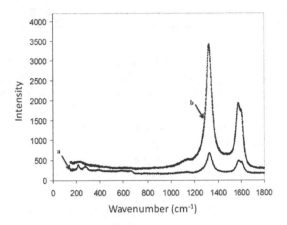

Figure 11. Raman spectra of: p-MWCNTs (a) and penta-g-MWCNTs (b).

However, the alkoxy radicals can undergo additional reactions including β-scission leading to the formation of methyl radicals [117]. These latter preferentially induce coupling reaction (Fig. 10, route b and h) or attack onto the sp2 carbon of the MWCNTs (Fig. 10, route g) whereas cumyloxyl radicals are more prone to hydrogen abstraction from pentadecane [120]. The formed pentadecyl radicals through hydrogen abstraction are able to react with MWCNTs by radical addition onto sp2 carbon of theMWCNTs (Fig.10, main route a) and with other radical species via the common radical coupling reactions (Fig. 10, routes d1 and b). According to the results of Johnston [121,122], based on a study of the crosslinking reaction of poly (ethylene-co-1-octene) in the presence of DCP at 160 °C, coupling reactions are four times more prone to happen than scission reactions so the authors assumed that pentadecyl radicals do not undergo scission reactions [119]. Direct evidence for covalent sidewall functionalization has been found by Raman spectroscopy [123-124]. G band is a characteristic feature of the graphitic layers and corresponds to the tangential vibration of the carbon atoms. The second characteristic mode is a typical sign for defective graphitic structures (D band). The A_D/A_G ratio, which was defined as the intensity ratio of the D-band to G-band of CNTs, directly indicates the structural changes in nanotubes. Some authors used D to G area ratios (A_D/A_G) rather than intensity [125] which is a better indicator. The relatively high area ratio of the G band relative to D band for penta-g-MWCNT (A_D/A_G = 1.51) in comparison with that of p-MWCNT (i.e. (A_D/A_G =1.20) could be designated as an indicator of grafting species. The ratio between the G band and D band is a good indicator of the changes in chemistry of CNTs. Interestingly, Raman spectra of p-MWCNTs and penta-g-MWCNTs (Figure11a and 11b, respectively) showed two main peaks around 1350cm^{-1} (D band) and 1586cm^{-1} (G band). The relatively high intensity of the G band relative to D band (I_G/I_D=1.25) for penta-g-MWCNT sample in comparison with that of p-MWCNT (i.e. I_G/I_D=0.95) was designated as an indicator of grafting species [119].

It is important to determine whether the results of a CNT surface modification process agree qualitatively with expectations, and equally important is the need for a quantitative assessment of the extent and nature of surface modifications. The course of the generated radical species and the extent of the grafting reaction in regards to the DCP concentration and temperature can be studied through gas chromatography and thermogravimetric analysis (TGA) [119, 126]. TGA permits measurement of the total weight fraction of surface modifiers introduced onto the surfaces of CNTs if the surface-modified CNTs are free of impurities. Indeed, it is well known that heating functionalized nanotubes in an inert atmosphere removes the organic moieties and restores the pristine nanotubes structure. TGA can indicate the degree of surface modification because the type and quantity of surface modifier is identified. It was found that the higher grafting density, as high as 1.46 mmol/g, was obtained at 150°C. At higher temperatures, the grafting density decreases because the β-scission reaction of cumyloxyl radical accelerates as the temperature increases, leading to the formation of methyl radicals. These latter preferentially react by combination whereas cumyloxyl radicals are more prone to hydrogen abstraction from pentadecane. At 150°C, for initiator concentration higher than 3wt%, the grafting density decreases from 1.464mmol/g to 0.371mmol/g upon increasing DCP concentration up to 5%. Thus, to get high grafting efficiency, one should opt for optimal initiator concentration, i.e. 3wt%, and choose the most favourable reaction temperature, i.e. 150°C. Incorporation of TEMPO as radical scavenger in the grafting reaction of pentadecane onto MWCNTs serves two purposes: firstly, it actively suppresses radical combination reactions and hence promotes pentadecyl radicals' addition to nanotubes (~16% increase in grafting density); and secondly, it effectively changes the polarity balance of the grafted species, making pentadecane and TEMPO functionalised nanotubes soluble in solvents such as THF and chloroform [126].

3.1.2. Synthesis of PE grafted carbon nanotubes via peroxide

To make full use of the strength of carbon nanotubes in a composite, it is important to have a high-stress transfer at the matrix-nanotube interface via strong chemical bonding, as discussed by Mylvaganam et al [127]. They have investigated the possible polyethylene-nanotube bonding with the aid of a quantum mechanics analysis with the polyethylene chains represented by alkyl segments, and the nanotubes modeled by nanotube segments with H atoms added to the dangling bonds of the perimeter carbons. Their study has predicted (i) covalent bond formation between alkyl radicals and carbon nanotubes is energetically favourable; and, (ii) this reaction may take place at multiple sites of nanotubes [126]. Hence one way to improve the load transfer of carbon nanotubes/PE composite via chemical bonds at the interface is to use free-radical generators such as peroxide or incorporate nanotubes by means of in situ polymerisation.

Figure 12 sums up main reactive pathways of free radical grafting of PE onto MWCNTs with dicumyl peroxide as initiator and TEMPO as radical scavenger.

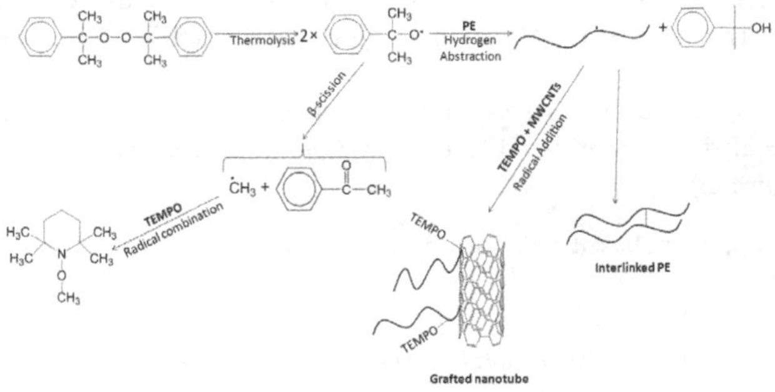

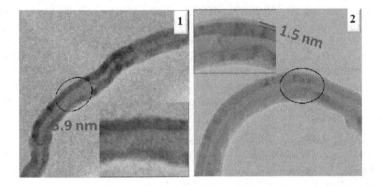

Figure 12. Reaction scheme for PE grafting onto MWCNTs with TEMPO as a radical scavenger. Reproduced from [128] with permission of John Wiley and Sons.

Figure 13. TEM images of p-MWCNTs (1) and PE-g-MWCNTs (2). Reproduced from [128] with permission of John Wiley and Sons.

The formed PE-based radicals are able to react with MWCNTs by radical addition onto sp^2 carbon of the MWCNTs (Figure 12) and with other radical species via the common radical coupling reactions. As discussed using a model compound approach (section 2.1.1), the presence of TEMPO radicals creates competitive combinations reactions (that are actually reversible reactions) which may favour the addition of PE-based radicals to MWCNTs (Figure 12). Before carrying out thermogravimetric analysis to gain a quantitative picture of the extent of nanotubes' functionalisation, the adsorbed (non-covalently attached) PE chains were removed from the grafted ones (covalently attached) by extensive washings with dichlorobenzene (DCB). PE-grafted onto MWCNTs are well known to degrade at 300-540°C, which are nearly the same temperatures as pure PE reactants and the weight of grafted PE is estimated to be in the range 20-24% depending on the experimental procedure [128]. The corresponding grafting densities, calculated using a specific surface area (SSA) of 225m^2/g for MWCNTs [115, 129] are varying from 1.1mg.m^{-2} to 1.4mg.m^{-2}. LDPE grafting density on nanotubes is $1.1mg.m^{-2}$ while incorporation of TEMPO raises the grafting density to $1.4mg.m^{-2}$ [128]. Then, it is usual to examine the morphological structures of p-MWCNTs and PE-grafted MWCNTs by transmission electron microscopy (TEM). In these experiments, a few drops of dilute solutions of PE-grafted nanotubes in hot DCB are initially deposited onto a carbon-coated copper grid and further observed in a dried state after evaporation of the solvent (Figure 13).

A "grafting to" approach based on a radical process can also involve a polymer with reactive end groups.

3.1.3. Synthesis of PE grafted carbon nanotubes via end functionalised PE

Recently, D'Agosto and Boisson [130-134] developed new strategies that rely on a one step *in situ* functionalisation reaction within an ethylene polymerisation reactor to introduce a variety of functional groups including alkoxyamine [130-132] and thiol [132,134] functions at the end of polyethylene chains. Di-polyethylenyl magnesium compound ($MgPE_2$) were prepared using a neodymium metallocene complex which catalysed polyethylene chain growth on magnesium compounds. $MgPE_2$ was in situ reacted with 2,2,6,6-tetramethylpiperidin-yl-1-oxy (TEMPO) radical or elemental sulphur to provide a macroalcoxyamine (PE-TEMPO) and polysulphur based product (PE-S_n-PE) respectively. PE-SH was obtained by simple reduction of PE-S_n-PE. According to these results, a strategy based on the use of those polyethylenes was investigated to generate radical-terminated chains formed either by thermal loss of a nitroxide (PE-TEMPO) or H-abstraction onto a thiol (PE-SH) and graft them onto CNTs (Figure 14) [128].

Indeed, Lou [87] showed an efficient attachment of poly(2-vinylpyridine) (P2VP) end-capped by TEMPO to CNT sidewalls by heating of TEMPO-terminated P2VP. Using the same strategy Liu [137] functionalized shortened CNTs with PS and poly[(*tert*-butyl acrylate)-*b*-styrene] and Wang [138] grafted poly(4-vinylpyridine-*b*-styrene) onto CNTs. In figure 14a, the homolytic cleavage of TEMPO-terminated PE leads to the formation of stable nitroxyl radicals and PE radicals. The reversible termination of the polyethylene chain is the key step for reducing the overall concentration of the radical chain end. The extremely low concentration of reactive chain ends is expected to minimize irreversible termination reactions, such as combination or disproportionation [135] (Figure 14a). Thiol-terminated polyethylene has been also grafted onto CNTs using a similar procedure. The thiol based compounds are widely used for controlling molar mass in free radical polymerizations via a chain transfer process. The chain transfer process displays two contiguous steps: transfer of the thiyl hydrogen to the growing polymer chain followed by re-initiation, whereby a thiyl radical adds to a monomeric double bond. In the presence of DCP-derived radicals and MWCNTs, thiyl radicals are formed and are expected to react by radical addition onto sp^2 carbon of the MWCNTs. For both samples PE-TEMPO and PE-SH, the weight loss observed by thermogravimetric analysis (TGA) has been increased to 36% and 34%, respectively despite their low molar masses (e.g. 1400g/mol and 980g/mol) in comparison with that of LDPE (weight loss = 20-24% ; Mw = 90000g/mol ; see section 2.1.2.). These results indicated that the use of short end-functionalized PE chains has permitted a significant increase of the grafting density (e.g. 1.78 and 2.80μmol.m^{-2}, respectively). These values are approximately increased by two

orders of magnitude in comparison with the LDPE grafting density (e.g. 0.012μmol.m^{-2}) sug-

gesting that longer polymer chains cover a larger surface decreasing the grafting density, as

previously described by Jerome et al [87,136] for the attachment of poly(2-vinylpyridine)

(P2VP) and polystyrene (PS) onto MWCNTs. Indeed, they observed that PS grafting density

decreases from 0.045μmol.m^{-2} to 0.01μmol.m^{-2} by increasing the molecular weight of PS-

TEMPO from 3000g/mol to 30000g/mol [138]. The TEM observations (Figure 15) were consis-

tent with the TGA results : the grafted polymer contents can be highered by using end-

functionalized PE [129].

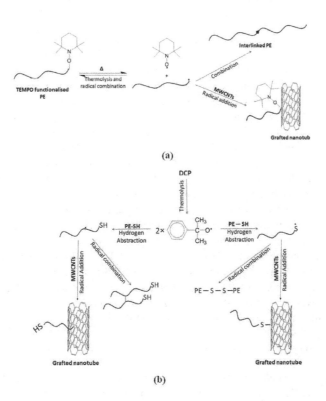

Figure 14. Reaction scheme for end functionalised PE grafting onto MWCNTs: (a) via PE-TEMPO; (b) via PE-SH. Reproduced from [128] with permission of John Wiley and Sons.

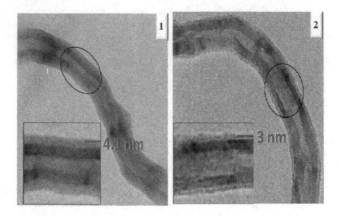

Figure 15. TEM images of PE$_{f\text{-}TEMPO}$-g-MWCNTs (1) and PE$_{f\text{-}SH}$-g-MWCNTs (2). Reproduced from [128] with permission of John Wiley and Sons.

3.2. Radical grafting of polypropylene on carbon nanotubes

Polypropylene (PP) is a widely used commercial polymer due to the excellent combination of mechanical resistance, chemical stability and excellent moisture barrier properties [104]. Although physical blending with CNTs is an economic way to modify polypropylene performance, compatibilizing agents are necessary for creating strong interface between filler particles and the polymer phase. Maleic anhydride grafted polypropylene (MA-*g*-PP) is often used as a compatibilizer which can improve the PP/CNTs composite properties by strong hydrogen bonding between hydroxyl groups located on the acidic-treated CNTs surface and anhydride groups of MA-g-PP [139, 140]. Recently, an original and simple method for promoting mobility sensitivity of carbon nanotubes (CNTs) to an external stress field in polypropylene (PP) matrix was developed [141]. In particular, an interfacial melt reaction initiated by free radicals were used as a tool to prepare PP/CNTs nanocomposites. The presence of tetrakis(phenylmethyl)thioperoxydi(carbothioamide) (TBzTD) increased the interfacial reaction between the PP chains and the CNTs. In addition, the grafted TBzTD to PP backbone could form a physical interaction with CNTs *via* a π–π interaction [141]. According to their previous results [119, 126] based on a study of the radical grafting of polyethylene derivatives onto MWCNTs, Farzi et al. investigated MWCNTs' sidewall functionalization by tetramethylpentadecane and PP in the presence of 1wt% DCP at 160°C.

3.2.2. A model compound approach through the use of 2,6,10,14-tetramethylpentadecane

Simlilarly to the pentadecane grafting procedure (see section 2.1.1.), 2,6,10,14-tetramethylpentadecane (TMP, $C_{19}H_{40}$) has been used as model for the grafting reaction of PP onto MWCNTs [129]. Thermolysis of dicumyl peroxide initiator performed in TMP and in presence of MWCNTs is depicted in Figure 16.

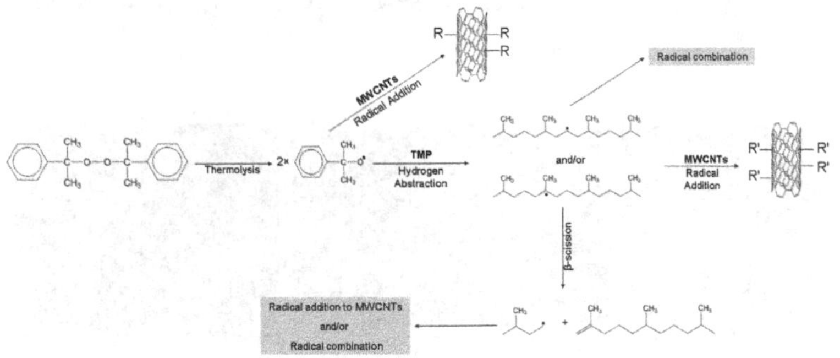

Figure 16. Reaction scheme for the addition of TMP onto CNT in the presence of DCP. Reproduced from [129] with permission of Elsevier.

As shown in Figure 16, the formed peroxide radicals are prone to hydrogen abstraction from hydrocarbon substrates and it is expected that the active species generated onto the hydrocarbon backbone react with unsaturated bonds located on the MWCNTs surface keeping in mind that side reactions such as chain scission for PP derivatives may occur at high temperatures (Figure16) [115, 118]. For experiments conducted in dichlorobenzene (DCB) as solvent at 160°C with 1.5wt% DCP, TMP grafting density was as high as 0.92 mg/m^2.

3.2.2. Synthesis of PP grafted carbon nanotubes via peroxide

Farzi et al. [129] have successfully grafted PP onto MWCNTs through a radical grafting reaction, carried out under similar experimental conditions to PE [128] and TMP [129] (1.5wt% DCP, 160°C) and using 1,2-dichlorobenzene (DCB) as solvent able to solubilize PP partially at elevated temperature. The corresponding PP-grafted nanotubes were analysed by TGA after a purification by soxhlet extraction in DCB. However, the authors were not able to obtain reproductible results with weight losses varying from 50% to 80% for the above-mentioned experiment. This behaviour was attributed to the purification procedure which did not permit to remove all the free PP chains. The authors have also speculated on the degradation behaviour of PP through the well-known β-scission reaction occurring in the presence of radical species therefore the authors were not able to give a PP grafting density. Then, the aforementioned PP coated MWCNTs have been dispersed within a commercially available PP matrix using a contra-rotating Haake Rheomixer and the amount of nanofiller in the final composites has been fixed to 3wt%. The evaluation of MWCNTs dispersion has been examined by using scanning electron microscopy (SEM) (Figure 17).

SEM images of the PP/PP-*g*-MWCNTs composites MWCNTs containing of 3wt% (Figure 17) demonstrated that there were still some areas where PP-*g*-MWCNTs were not found which was obviously connected with improper filler distribution. For a simple melt blend of PP with untreated MWCNTs, SEM images of the resulting material only showed clusters of a few tens micrometers of diameter evidencing a poor interfacial adhesion in the material

(Figure 18), as reported by Lee [140] for untreated MWCNT/PP composites MWCNTs containing of 2wt%.

Figure 17. SEM micrographs of PP/PP-g-MWCNT composites with MWCNTs loading of 3wt%. Reproduced from [129] with permission of Elsevier.

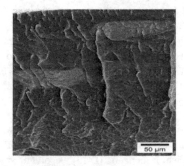

Figure 18. SEM micrographs of PP/MWCNTs composites with MWCNTs loading of 3wt%. Reproduced from [129] with permission of Elsevier.

It was concluded from these results that the grafting of PP onto MWCNTs provided a low steric barrier against the strong intermolecular Van der Waals interactions among nanotubes within the PP matrix.

In a similar approach, isotactic polyrpopylene (iPP) was successfully grafted onto multiwalled carbon nanotubes (MWCNTs) by direct macroradical addition by sonication in hot xylene with BPO as an initiator [142]. It was found that both iPP macromolecular radicals and small-molecular benzoic acid free radicals were grafted onto MWNTs. iPP-g-MWNTs dispersed more uniformly in iPP than pristine MWNTs.

3.3. Rheological behaviour of polyolefin based carbon nanotubes nanocomposites

It has been well known for a century that the addition of fillers, mostly carbon black, to rubber compounds has a strong impact on the viscoelastic properties of materials. In recent years, polymer nanocomposites have been developed as a new class of composites [143]. Actually from a rheological point of view, a direct consequence of incorporation of fillers in molten polymers is the significant change in the steady shear viscosity behavior and the viscoelastic properties. The level of filler dispersion is expected to play a major role in determining the filler effects on non linear responses of nanocomposites. Generally speaking, thermoplastic polymers filled with nano-particles show a solid-like behavior response which includes a non–terminal zone of relaxation, apparent yield stress and a shear-thinning dependence of viscosity on particle concentration, aspect ratio and dispersion.

Since the melt rheological properties of filled polymers are sensitive to the structure, concentration, particle size, shape and surface characteristics of the fillers, rheology offers original means to assess the state of the dispersion in nanocomposites and to investigate the influence of flow conditions upon nano-filler dispersion itself [144]. As discussed previously, one of the most important challenges in filled polymer developments and applications is to obtain a homogeneous dispersion of CNT in polymer matrix by overcoming the van der Waals interaction between elementary tubes. As a result, it can be expected that the rheological percolation, and subsequently the non-linearity effect, depend on nanotube dispersion and aspect ratio. As matter of fact, a great level of activity in the domain of the rheology of polymers filled with CNT is reported in the more recent scientific literature. The rheological behavior of melt thermoplastic polymer filled with NTC was reported to depend on nanotube dispersion, aspect ratio and alignment under flow. However, among the different studies on liquid systems filled with carbon nanotubes two types of relaxation mechanisms of CNT must be differentiated according to the matrix viscosity: Do the carbon nanotubes behave as Brownian particles? The Doi-Edwards theory for dilute regime (the nantotubes are free to rotate without any contacts) allows the rotary diffusivity D_{r0} of a rod (Length: L and diameter: d) in an isotropic suspension to be calculated by equation (1), in which k_B is the Boltzmann constant and η_m is the viscosity of the suspending medium

$$D_{r0} = \frac{3k_B T (\ln(L/d) - 0.8)}{\pi \eta_m L^3} \tag{1}$$

In semi-dilute regime, the rotary diffusivity D_r can be written as equation (2), where A is a dimensionless constant whose value is generally large (A~1000).

$$D_r = AD_{r0}(\nu L^3)^{-2} \tag{2}$$

Consequently, the rotary diffusion of CNT varies according to matrix viscosity:

$$D_r \alpha 1/\eta_m \tag{3}$$

Actually, the rheological behavior of CNT suspension is observed close to the Doi-Edwards theory on the Brownian dynamics of rigid rods. However, it was observed that low shear deformations induced an aggregation mechanism, but these aggregates broke down at high shear, forming small aggregates with less entanglements [145]. The shear rheology of such carbon nanotube suspensions was reviewed by Hobbie [146] from the perspective of colloid and polymer science.

According to the Doi-Edwards theory, Marceau et al [147] have shown that the suspensions of CNT, at low concentration (ϕ=0.2%) and in low fluid matrix (η_m=5 Pa.s), behave as Brownian entities (D_r=5.0x10^{-5} s^{-1}). The diffusion time of these CNTs is then $\tau_r = 1/2D_r \approx 10^4 s$. If we imagine that these same CNT are dispersed in high viscous polymer matrix such as molten PP (η_m~1x10^3 Pa.s), their relaxation time will be then: $\tau_r \approx 2x10^6 s$. The order magnitude of the relaxation time is then one month! Consequently, the carbon nanotubes cannot be considered anymore as Brownian entities in most of the papers that have been addressed to the viscoelastic behavior of carbon nanotubes dispersed in high viscous molten polymers. The main challenge in such nanocomposite systems is to control the dispersion of the nanotubes in high viscous fluid in order to have the lowest percolation threshold regarding the electrical properties. For example, by improving the CNT dispersion using functionalized single wall nanotubes, Mitchell et al [148] observed that the percolation threshold dropped from 3wt% to 1.5% in PS nanocomposites.

Actually, the dispersion of CNT in polymer matrix is strongly difficult mainly due to the nanotube-nanotube interactions higher than the nanotube-polymer interaction. However, optimal dispersion of CNTs can be achieved in polar polymers such as polyamides, polyesters or polycarbonate. This optimal dispersion is generally measured, at least from a qualitative point of view, from the electrical and/or rheological percolation threshold. Nevertheless, the dispersion of CNT in polyolefin (PP, PE or copolymer of ethylene) is most of the time a real challenge due to unfavorable and low nanotube-polymer interactions. On the other hand, the fact that CNT have a high aspect ratio and are not Brownian in polymer matrix leads to the conclusion that the different works of the literature are difficult to compare. The samples, studied in rheology or electric conductivity, have generally undergone different processing conditions. As a result, CNT nanocomposites are totally out of isotropic dispersion and the isotropic equilibrium of CNTs can never be achieved. However, general trends in CNT nancomposite can be described from the open literature.

From a sample preparation point of view, dispersion of CNTs in polyolefins were generally prepared via conventional melt processing, i.e melt blending in batch mixer or in twin screw extruder). Marginal methods may also be used as for example solid-state shear pulverization [149] or in situ lubrication methods [150]. Numerous studies [151-156] have been devoted to the linear viscoelasticity of PP nanocomposites based on CNT dispersion.

All of these papers reported an increase in shear viscosity and storage and loss moduli of the nanocomposites with increasing the CNT concentration as shown in Figure 19.

Furthermore, a general rheological trend for nanocomposites studied in most of these papers is the appearance of a transition from a liquid-like behaviour to a solid-like behaviour, i.e. the apparition of a plateau (second plateau modulus, $G_0 = \lim_{\omega \to 0} G'(\omega)$) of the storage modulus at low frequency which is obviously higher than the loss modulus. Obviously, it is admitted that the increase of the CNT concentration is driving this transition. Above this critical transition, generally associated with the percolation threshold, these nanocomposites show a solid-like behavior response, which includes a non-terminal zone of relaxation leading to apparent yield stress and a shear-thinning dependence on viscosity $|\eta*(\omega)| \alpha \omega^{-1}$.

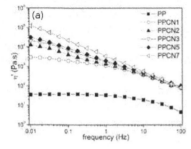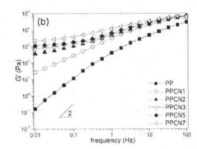

Figure 19. Variation versus frequency of the storage shear modulus G'(ω) and absolute complex viscosity η*(ω) at different concentrations of CNT (1 to 7%). Reproduced from [151] with permission of John Wiley and Sons.

This non-terminal frequency behavior is generally attributed to the formation of an interconnected nanotube network in the polymer matrix. Therefore, the solid-like behavior is associated nanotube–nanotube interactions which increase with the CNT content. Eventually, these interactions lead to percolation and the formation of an interconnected structure of nanotubes in the matrix. Due to the high aspect ratio of CNT (generally, L/d>150), the existence of this percolation threshold is expected at low concentration. For example, from Fig 19, the percolation threshold can be estimated to be less than 2% of CNT. This percolation threshold is generally observed in the range 0.5%-5% depending on CNT nature (aspect ratio, surface chemical modification) and on the processing methods for nanocomposite preparation.

If a lot of works have been devoted to the linear viscoelasticity of CNT nanocomposites whereas a few works have been reported on non-linear properties such as for example the melt flow instabilities. Interestingly, Palza et al [157] showed that carbon nanotubes modify the main characteristics of the spurt instability developed by the linear polyethylene. Furthermore, the sharkskin instability, developed in short chain branched polyethylene, is reduced at low amounts of MWCNT. Furthermore, the critical shear rate for the on-set of the spurt and the sharkskin instabilities decreases in the nanocomposites probably due to the physical interactions between the polymer and the nanofiller. Finally, at high shear rates, the gross melt fracture instability is completely erased in the nanocomposites based on the linear polymer whereas in short chain branched polyethylene the amplitude of this bulk distortion is rather moderated. Clearly, the carbon nanotubes have a drastic effect on the main flow instabilities observed in polyethylene. Consequently, the processing of CNT nanocomposites, i.e under non-linear deformation, is an open investigation domain.

4. Conclusion

The most frequent method for preparing polymer nanotubes composites has been mixing nanotubes and polymer in a suitable solvent and to evaporate the solvent to form composite film. But to increase the advantages at its best, one needs: (i) high interfacial area between nanotubes and polymer; and, (ii) strong interfacial interaction. Unfortunately this solvent technique does not help much in achieving these targets; and as a result a nanocomposite having properties much inferior to theoretical expectations are obtained. In order to obtain higher contact area between nanotubes and polymer, the issue of dispersion needs to be addressed. Uniform dispersion of these nanotubes produces immense internal interfacial area, which is the key to enhancement of properties of interest. On the other hand, modification of nanotubes surface through functionalisation is required for creating an effective interaction with the host matrix and to make nanotubes soluble and dispersible.

The idea of grafting PE or PP with the help of peroxide during extrusion is exciting. It was shown that cumlyoxly radical generated by thermolysis of DCP can abstract hydrogen from polyolefin chains, thus creating polyolefin macroradicals. Then, these macroradicals add to the unsaturated carbon bonds on the surface of the nanotubes. The upside of this strategy is that radicals have short lifetimes which make the procedure possible in an extruder where the residence time is generally low. On the contrary, the downside is the low selectivity of radicals leading to a competition between radical combination reactions and radical addition reactions. Alkanes can be used as model for PE to perform the grafting of PE onto nanotubes. The results were interesting however the degree of PE grafting remained lower than the model (weight loss by TGA = 22% as compared to a weight loss of 30% in case of pentadecane). LDPE grafting density on nanotubes was 1.1mg.m^{-2} while incorporation of TEMPO raised the grafting density to 1.4mg.m^{-2}. End functionalised PE can also be used for PE grafting onto nanotubes with dichlorobenzene as solvent. As emphasized by TEM images, a layer of considerable thickness has been grafted around the periphery of the nanotubes.

In order to follow the same strategy for nanotubes functionalisation with PP, tetramethylpentadecane has been selected as a model compound for PP. It was successfully grafted onto carbon nanotubes with a grafting density of 0.92 mg/m^2. However, the grafting of PP onto nanotubes did not permit to obtain reproductible results. SEM images of the PP-g-MWCNTs nanocomposites with filler loadings of 3wt% in PP matrix did not show a significant improvement in MWCNTs dispersion within the PP matrix although sizes of the aggregates were slightly reduced.

In addition, it can be expected that the rheological percolation, and subsequently the non-linearity effect, depend on nanotube dispersion and aspect ratio. Low shear deformations induced an aggregation mechanism, but these aggregates broke down at high shear, forming small aggregates with less entanglements. In a high viscous polymer media, it was shown that carbon nanotubes could not be considered anymore as Brownian entities. A general rheological trend for CNTs-based nanocomposites is the appearance of a transition from a liquid-like behaviour to a solid-like behaviour increasing with the CNT content because it is

Polymer Nanocomposites Containing Functionalised Multiwalled Carbon NanoTubes:
a Particular Attention to Polyolefin Based Materials

51

associated to nanotube–nanotube interactions. Due to the high aspect ratio of CNT the percolation threshold can be expected to be less than 2% of CNT.

Author details

Emmanuel Beyou, Sohaib Akbar, Philippe Chaumont and Philippe Cassagnau

Université de Lyon, Université Lyon 1, CNRS UMR5223, Ingénierie des Matériaux Polymères: IMP@UCBL, 15 boulevard Latarget, F-69622 Villeurbanne, France

References

[1] Iijima, S. (1991). Helical microtubules of graphitic carbon. *Nature*, 354, 56-58.

[2] Baughman, R. H., Zakhidov, A.A., & de Heer, W. A. (2002). Carbon Nanotubes--the Route Toward Applications. *Science*, 297(5582), 787-792.

[3] Moniruzzaman, M., & Winey, K. I. (2006). Polymer Nanocomposites Containing Carbon Nanotubes. *Macromolecules*, 39(16), 5194-5205.

[4] Grossiord, N., Loos, J., Regev, O., & Koning, C. E. (2006). Toolbox for dispersing carbon nanotubes into polymers to get conductive nanocomposites. *Chemistry of Materials*.

[5] Ciardelli, F., Coiai, S., Passaglia, E., Pucci, A., & Ruggeri, G. (2008). Nanocomposites based on polyolefins and functional thermoplastic materials. *Polymer International*, 57(6), 805-836.

[6] Judeinstein, P., & Sanchez, C. (1996). Hybrid organic-inorganic materials:. *A land of multi-disciplinarity Journal of Materials Chemistry*, 6(4), 511-525.

[7] Connell, M.O. Carbon Nanotubes. *Properties and Applications; CRC Press, 2006.*, 360, 9780849327483.

[8] Gogotsi, Y. *Carbon NanomaterialsCRC Press, 2006.*, 344, 9780849393860.

[9] Spitalsky, Z., Tasis, D., Papagelis, K., & Galiotis, C. (2010). Carbon nanotube-polymer composites: Chemistry, processing, mechanical and electrical properties. *Progress in Polymer Science*, 35(3), 357-401.

[10] Osswald, S., Flahaut, E., & Gogotsi, Y. (2006). In situ Raman spectroscopy study of oxidation of double- and single-wall carbon nanotubes. *Chemistry of Materials*, 18(6), 1525-1533.

[11] Yang, Q., Wang, L., Xiang, W., Zhu, J., & Li, J. (2007). Grafting polymers onto carbon black surface by trapping polymer radicals. *Polymer*, 48(10), 2866-2873.

[12] Deng, X., Jia, G., Wang, H., Sun, H., Wang, X., Yang, S., Wang, T., & Liu, Y. (2007). Translocation and fate of multi-walled carbon nanotubes in vivo. *Carbon*, 45(7), 1419-1424.

[13] Shaffer, M. S. P., & Windle, A. H. (1999). Fabrication and Characterization of Carbon Nanotube/Poly(vinyl alcohol) Composites. *Advanced Materials*, 11(11), 937-941.

[14] Baskaran, D., Mays, J. W., & Bratcher, M.S. (2005). Noncovalent and nonspecific molecular interactions of polymers with multiwalled carbon nanotubes. *Chemistry of Materials*, 17(13), 3389-3397.

[15] Haggenmuller, R., Gommans, H. H., Rinzler, A. G., Fischer, J. E., & Winey, K. I. (2000). Aligned single-wall carbon nanotubes in composites by melt processing methods. *Chemical Physics. Letters*.

[16] Kang, Y. J., & Taton, T. A. (2003). Micelle-encapsulated carbon nanotubes: A route to nanotube composites. *Journal of the American Chemical Society*, 125(19), 5650-5651.

[17] Star, A., & Stoddart, J. F. (2002). Dispersion and solubilization of single-walled carbon nanotubes with a hyperbranched polymer. *Macromolecules*, 35(19), 7516-7520.

[18] Barraza, H. J., Pompeo, F., O'Rear, E. A., & Resasco, D. E. (2002). SWNT-filled thermoplastic and elastomeric composites prepared by miniemulsion polymerization. *Nano Letters*, 2(8), 797-802.

[19] Bonduel, D., Mainil, M., Alexandre, M., Monteverde, F., & Dubois, P. (2005). Supported coordination polymerization: a unique way to potent polyolefin carbon nanotube nanocomposites. *Chemical Communications*, 6, 781-783.

[20] Park, S., Yoon, W., Lee, B., Kim, J., Jung, Y. H., Do, Y., Paik, H. J., & Choi, I. S. (2006). Carbon Nanotubes as a Ligand in Cp2ZrCl2-Based Ethylene Polymerization. *Macromolecular Rapid Communication*, 27(1), 47-50.

[21] Liu, T. X., Phang, I. Y., Shen, L., Chow, S. Y., & Zhang, W. D. (2004). Morphology and mechanical properties of multiwalled carbon nanotubes reinforced nylon-6 composites. *Macromolecules*, 37(19), 7214-7222.

[22] Siochi, E. J., Working, D. C., Park, C., Lillehei, P. T., Rouse, J. H., & Topping, C. C. Melt processing of SWCNT-polyimide nanocomposite fibers. Composites Part B. (2004)., 35(5), 439-446.

[23] Potschke, P., Fornes, T. D., & Paul, D. R. (2002). Rheological behavior of multiwalled carbon nanotube/polycarbonate composites. *Polymer*, 43(11), 3247-3255.

[24] Li, C., Zhao, Q., Deng, H., Chen, C., Wang, K., Zhang, Q., Chena, F., & Fua, Q. (2011). Preparation, structure and properties of thermoplastic olefin nanocomposites containing functionalized carbon nanotubes. *Polymer international*, 60(11), 1629-1637.

[25] Zhang, Z. N., Zhang, J., Chen, P., Zhang, B. Q., He, J. S., & Hu, G. H. (2006). Enhanced interactions between multi-walled carbon nanotubes and polystyrene induced by melt mixing. *Carbon*, 44(4), 692-698.

[26] Shofner, M. L., Khabashesku, V. N., & Barrera, E. V. (2006). Chemistry of Materials. *Processing and mechanical properties of fluorinated single-wall carbon nanotube-polyethylene composites.*, 18(4), 906-913.

[27] Mahfuz, H., Adnan, A., Rangari, V. K., & Jeelani, S. (2005). Manufacturing and characterization of carbon nanotube/polyethylene composites. *International Journal of Nanoscience*, 4(1), 55-72.

[28] Ruan, S. L., Gao, P., & Yu, T. X. (2006). Ultra-strong gel-spun UHMWPE fibers reinforced using multiwalled carbon nanotubes. *Polymer*, 47(5), 1604-1611.

[29] Shofner, M. L., Khabashesku, V. N., & Barrera, E. V. (2006). *Chemistry of Materials*, 18, 906-913.

[30] Bahr, J. L., & Tour, J. M. (2001). Processing and mechanical properties of fluorinated single-wall carbon nanotube-polyethylene composites. *Chemistry of Materials*, 13(4), 3823-3824.

[31] Georgakilas, V., Kordatos, K., Prato, M., Guldi, D. M., Holzinger, M., & Hirsch, A. J. Organic functionalization of carbon nanotubes. . Journal of the American Chemical Society (2002). , 124(5), 760-761.

[32] Peng, H., Reverdy, P., Khabashesku, V. N., & Margrave, J. L. (2003). Sidewall functionalization of single-walled carbon nanotubes with organic peroxides. *Chemical Communications*, 362-363.

[33] Qin, S., Qin, D., Ford, W. T., Resasco, D. E., & Herrera, J. E. (2004). Functionalization of single-walled carbon nanotubes with polystyrene via grafting to and grafting from methods. *Macromolecules*.

[34] Riggs, J. E., Guo, Z., Carroll, D. L., & Sun-P, Y. (2000). Strong luminescence of solubilized carbon nanotubes. *Journal of the American Chemical Society*, 122(24), 5879-5884.

[35] Umek, P., Seo, J. W., Hernadi, K., Mrzel, A., Pechy, P., Mihailovic, D. D., & Forro, L. (2003). Addition of carbon radicals generated from organic peroxides to single wall carbon nanotubes. *Chemistry of Materials*, 15(25), 4751-4755.

[36] Liu, Y. K., Yao, Z. L., & Adronov, A. (2005). Functionalization of single-walled carbon nanotubes with well-defined polymers by radical coupling. *Macromolecules*, 38(4), 1172-1179.

[37] Shaffer, M. S. P., & Koziol, K. (2002). Polystyrene grafted multi-walled carbon nanotubes. *Chemical Communications*.

[38] Kong, H., Gao, C., & Yan, D. Y. (2004). Controlled functionalization of multiwalled carbon nanotubes by in situ atom transfer radical polymerization. *Journal of the American Chemical Society*, 126(2), 412-413.

[39] Baskaran, D., Mays, J. W., & Bratcher, M.S. (2004). Polymer-grafted multiwalled carbon nanotubes through surface-initiated polymerization. *Angewandte Chemie International Edition*.

[40] Breuer, O., & Sundararaj, U. (2004). Big returns from small fibers. *A review of polymer/carbon nanotube composites. Polymer Composites,* 25(6), 630-645.

[41] Tiwari, I., Singh, K. P., & Singh, M. (2009). An insight review on the application of polymer-carbon nanotubes based composite material in sensor technology. *Russian Journal of General Chemistry.*

[42] Martínez- Hernández, A.L., Velasco-Santos, C., & Castaño, V.M. (2010). Carbon Nanotubes Composites: Processing, Grafting and Mechanical and Thermal Properties. *Current Nanoscience,* 6(1), 12-39.

[43] Jin, L., Bower, C. L., & Zhou, O. (1998). *Applied Physics Letters,* 73(9), 1197-1199.

[44] Coleman, J. N., Cadek, M., Blake, R., Nicolosi, V., Ryan, K. P., & Belton, C. (2004). High-performance nanotube-reinforced plastics: Understanding the mechanism of strength increase. *Advanced Functional Materials,* 14(8), 791-798.

[45] Cadek, M., Coleman, J. N., Ryan, K. P., Nicolosi, V., Bister, G., & Fonseca, A. (2004). Reinforcement of polymers with carbon nanotubes: The role of nanotube surface area. *Nano Letters,* 4(2), 353-356.

[46] Cadek, M., Coleman, J. N., Barron, V., Hedicke, K., & Blau, W. J. (2002). Morphological and mechanical properties of carbon-nanotube-reinforced semicrystalline and amorphous polymer composites. *Applied Physics Letters,* 81(27), 5123-5125.

[47] Andrews, R., Jacques, D., Minot, M., & Rantell, T. (2002). Fabrication of carbon multi-wall nanotube/polymer composites by shear mixing. Macromolecular. *Materials and Engineering,* 287(6), 395-403.

[48] Breuer, O., & Sundararaj, U. (2004). Big returns from small fibers. *A review of polymer/carbon nanotube composites. Polymer Composites,* 25(6), 630-645.

[49] Potschke, P., Bhattacharyya, A. R., Janke, A., & Goering, H. (2003). Melt mixing of polycarbonate/multi-wall carbon nanotube composites. *Composite Interfaces,* 10(4), 389-404.

[50] Andrews, R., Jacques, D., Qian, D. L., & Rantell, T. (2002). Multiwall carbon nanotubes: Synthesis and application. *Accounts of Chemical Research,* 35(12), 1008-1017.

[51] Tang, W., Santare, M. H., & Advani, S. G. (2003). Melt processing and mechanical property characterization of multi-walled carbon nanotube/high density polyethylene (MWNT/HDPE) composite films. *Carbon,* 41(14), 2779-2785.

[52] Jia, Z., Wang, Z., Xu, C., Liang, J, Wei, B., & Wu, D. (1999). Study on poly(methyl methacrylate)/carbon nanotube composites. *Materials Science and Engineering A,* 271(1), 395-400.

[53] Bonduel, D., Mainil, M., Alexandre, M., Monteverde, F., & Dubois, P. (2005). Supported coordination polymerization: a unique way to potent polyolefin carbon nanotube nanocomposites. *Chemical Communications,* 41(6), 781-783.

[54] Bryning, M. B., Milkie, D. E., Islam, M. F., Kikkawa, J. M., & Yodh, A. G. (2005). Thermal conductivity and interfacial resistance in single-wall carbon nanotube epoxy composites. *Applied Physics Letters*, 87(87), 161909-161911.

[55] Moniruzzaman, M., Du, F., Romero, N., & Winey, K. I. (2006). Increased flexural modulus and strength in SWNT/epoxy composites by a new fabrication method. *Polymer*, 47(1), 293-298.

[56] Zhu, J., Kim, J., Peng, H., Margrave, J. L., Khabashesku, V. N., & Barrera, E. V. (2003). Improving the dispersion and integration of single-walled carbon nanotubes in epoxy composites through functionalization. *Nano Letters*, 3(8), 1107-1113.

[57] Xia, H., Wang, Q., Li, K., & Hu, G. H. (2004). Preparation of polypropylene/carbon nanotube composite powder with a solid-state mechanochemical pulverization process. *Journal of Applied Polymer Science*, 93(1), 378-386.

[58] Kasimatis, K. G., Nowell, J. A., Dykes, L. M., Burghardt, W. R., Thillalyan, R., Brinson, L. C., Andrews, R., & Torkelson, J. M. (2005). Morphology, thermal, and rheological behavior of nylon 11/multi-walled carbon nanotube nanocomposites prepared by melt compounding. *Programme Making and Special Events Preprint*, 92, 255-256.

[59] Regev, O., El Kati, P. N. B., Loos, J., & Koning, C.E. (2004). Preparation of conductive nanotube-polymer composites using latex technology. *Advanced Materials*, 16(3), 248-251.

[60] Lee, J. U., Huh, J., Kim, K. H., Park, C., & Jo, W. H. (2007). Aqueous suspension of carbon nanotubes via non-covalent functionalization with oligothiophene-terminated poly(ethylene glycol). *Carbon*, 45(5), 1051-1057.

[61] Mcandrew, T., Roger, C., Bressand, E., & Laurent, P. W. (2008). *WO/2008/106572*.

[62] Vigolo, B., Penicaud, A., Coulon, C., Sauder, C., Pailler, R., Journet, C., Bernier, P., & Poulin, P. (2000). Macroscopic fibers and ribbons of oriented carbon nanotubes. *Science*, 290(5495), 1331-1334.

[63] Mamedov, A.A., Kotov, N.A., Prato, M., Guldi, D. M., Wicksted, J. P., & Hirsch, A. (2002). Molecular design of strong single-wall carbon nanotube/polyelectrolyte multilayer composites. *Nature Materials*, 1(3), 190-194.

[64] Calvert, P. (1999). Nanotube composites- A recipe for strength. *Nature*, 399(6733), 210-211.

[65] Hersam, M.C. (2008). Progress towards monodisperse single-walled carbon nanotubes. *Nature Nanotechnology*.

[66] Green, M.J. (2010). Analysis and measurement of carbon nanotube dispersions: nano-dispersion versus macrodispersion. *Polymer Internantional*, 59(10), 1319-22.

[67] Kim, S. W., Kim, T., Kim, Y. S., Choi, H. S., Lim, H. J., Yang, S. J., & Park, C. R. (2012). Surface modifications for the effective dispersion of carbon nanotubes in solvents and polymers. *Carbon*, 50(8), 3-33.

[68] O'Connell, M.J., Boul, P., Ericson, L. M., Huffman, C., Wang, Y., & Haroz, E. (2001). Reversible water-solubilization of single-walled carbon nanotubes by polymer wrapping. *Chemical Physics Letters*, 342(3), 265-271.

[69] Star, A., Stoddart, J. F., Steuerman, D., Diehl, M., Boukai, A., & Wong, E. W. (2001). Preparation and properties of polymer-wrapped single-walled carbon nanotubes. *Angewandte Chemie International Edition*, 40(9), 1721-1725.

[70] Steuerman, D. W., Star, A., Narizzano, R., Choi, H., Ries, R. S., & Nicolini, C. Interactions between conjugated polymers and single-walled carbon nanotubes. Journal of Physical Chemistry B. (2002). , 106(12), 3124-3130.

[71] Star, A., Liu, Y., Grant, K., Ridvan, L., Stoddart, J. F., & Steuerman, D. W. (2003). Noncovalent side-wall functionalization of single-walled carbon nanotubes. *Macromolecules*, 36(3), 553-560.

[72] Star, A., & Stoddart, J. F. (2002). *Macromolecules*, 35(19), 7516-7520.

[73] Nativ-Roth, E., Levi-Kalisman, Y., Regev, O., & Yerushalmi-Rozen, R. (2002). On the route to compatibilization of carbon nanotubes. *Journal of Polymer Engineering*, 22(5), 353-368.

[74] Xia, H. S., Wang, Q., & Qiu, G. H. (2003). Polymer-encapsulated carbon nanotubes prepared through ultrasonically initiated in situ emulsion polymerization. *Chemistry of Materials*, 15(20), 3879-3886.

[75] Li, X., Qin, Y., Picraux, S. T., & Guo-X, Z. (2011). Noncovalent assembly of carbon nanotube-inorganic hybrids. *Journal of Material Chemistry*, 21(21), 25-47.

[76] Sahoo, N. G., Rana, S., Cho, J. W. L. S. H., & Chan, Li. (2010). Polymer nanocomposites based on functionalized carbon nanotubes. *Progress in Polymer Science*, 35(7), 837-867.

[77] Kitano, H., Tachimoto, K., & Anraku, Y. J. (2007). Functionalization of single-walled carbon nanotube by the covalent modification with polymer chains. *Journal of Colloid and Interface Science*, 306(1), 28-33.

[78] Park, H., Zhao, J., & Lu, J. P. (2006). Effects of sidewall functionalization on conducting properties of single wall carbon nanotubes. *Nano Letters*, 6(5), 916-9.

[79] Han, S. J., Kim, B., & Suh, K. D. (2007). Electrical properties of a composite film of poly(acrylonitrile) nanoparticles coated with carbon nanotubes. *Macromolecular Chemistry and Physics*, 208(4), 377-383.

[80] Zhou, C., Wang, S., Zhang, Y., Zhuang, Q., & Han, Z. (2008). situ preparation and continuous fiber spinning of poly(p-phenylene benzobisoxazole) composites with oligo-hydroxyamide-functionalized multi-walled carbon nanotubes. *Polymer*, 49(10), 2520-2530.

[81] Baskaran, D., Mays, J. W., & Bratcher, M.S. (2005). Polymer adsorption in the grafting reactions of hydroxyl terminal polymers with multi-walled carbon nanotubes. *Polymer*, 46(14), 5050-5057.

[82] Zhao, B., Hu, H., Yu, A., Perea, D., & Haddon, R. C. (2005). Synthesis and characterization of water soluble single-walled carbon nanotube graft copolymers. *Journal of American Chemical Society*, 127(22), 8197-8203.

[83] Zhang, N., Xie, J., Guers, M., & Varadan, V. K. *Chemical bonding of multiwalled carbon nanotubes to polydimethylsiloxanes and modification of the photoinitiator system for microstereolithography processing.*

[84] Qin, S., Qin, D., Ford, W. T., Resasco, D. E., & Herrera, J. E. (2004). Functionalization of Single-Walled Carbon Nanotubes with Polystyrene via Grafting to and Grafting from Methods. *Macromolecules*, 37(3), 752-757.

[85] Li, H., Cheng, F., Duft, A. M., & Adronov, A. (2005). Functionalization of single-walled carbon nanotubes with well-defined polystyrene by "click" coupling. *Journal of American Chemical Society*, 127(41), 14518-24.

[86] Huang, H. M., Liu, I. C., Chang, C. Y., Tsai, H. C., Hsu, C. H., & Tsiang, R. C. C. (2004). Preparing a polystyrene-functionalized multiple-walled carbon nanotubes via covalently linking acyl chloride functionalities with living polystyryllithium. *Journal of Polymer Science Part A: Polymer Chemistry*, 42(22), 5802-5810.

[87] Lou, X., Detrembleur, C., Pagnoulle, C., Sciannamea, V., & Jerome, R. (2004). Grafting of alkoxyamine end-capped (co)polymers onto multi-walled carbon nanotubes. *Polymer*, 45(18), 6097-6102.

[88] Viswanathan, G., Chakrapani, N., Yang, H., Wie, B., Chung, H., & Cho, K. (2003). Single-step in situ synthesis of polymer-grafted single-wall nanotube composites. *Journal of American Chemical Society*, 9258-9259.

[89] Liu, Y., Wu, D. C., Zhang, W. D., Jiang, X., He, C. B., & Chung, T. S. (2005). Polyethylenimine-grafted multiwalled carbon nanotubes for secure noncovalent immobilization and efficient delivery of DNA. *Angewandte Chemie-International Edition*, 44(30), 4782-4785.

[90] Bonduel, D., Mainil, M., Alexandre, M., Monteverde, F., & Dubois, P. (2005). Supported coordination polymerization: a unique way to potent polyolefin carbon nanotube nanocomposites. *Chemical Communication*, 6, 781-783.

[91] Park, J. J., Park, D. M., Youk, J. H., Yu, W. R., & Lee, J. (2010). Functionalization of multi-walled carbon nanotubes by free radical graft polymerization initiated from photoinduced surface groups. *Carbon*, 48(10), 2899-2905.

[92] Yao, Z., Braidy, N., Botton, G. A., & Adronov, A. (2003). Polymerization from the surface of single-walled carbon nanotubes- Preparation and characterization of nanocomposites. *Journal of American Chemical Society*, 125(51), 16015-16024.

[93]   Kong, H., Gao, C., & Yan, D. (2004). Controlled functionalization of multiwalled carbon nanotubes by in situ atom transfer radical polymerization. *Journal of American Chemical Society*, 126(2), 412-413.

[94]   Wang, M., Pramoda, K. P., & Goh, S. H. (2005). Enhancement of the mechanical properties of poly (styrene-co-acrylonitrile) with poly(methyl methacrylate)-grafted multiwalled carbon nanotubes. *Polymer*, 46(25), 11510-11516.

[95]   Baskaran, D., Mays, J. W., & Bratcher, MS. (2004). Polymer-grafted multiwalled carbon nanotubes through surface-initiated polymerization. *Angewandte Chemie-International Edition*, 43(16), 2138-2142.

[96]   Shanmugharah, A. M., Bae, J. H., Nayak, R. R., & Ryu, S. H. (2007). Preparation of poly(styrene-co-acrylonitrile)-grafted multiwalled carbon nanotubes via surface-initiated atom transfer radical polymerization. *Journal of Polymer Science Part A-Polymer Chemistry*, 45(3), 460-470.

[97]   Cui, J., Wang, W. P., You, Y. Z., Liu, C., & Wang, P. (2004). Functionalization of multiwalled carbon nanotubes by reversible addition fragmentation chain-transfer polymerization. *Polymer*, 45(26), 8717-8721.

[98]   Hong, C. Y., You, Y. Z., & Pan, C. Y. (2005). Synthesis of water-soluble multiwalled carbon nanotubes with grafted temperature-responsive shells by surface RAFT polymerization. Chemistry of Materials , 17(9), 2247-2254.

[99]   Wang, G. J., Huang, S. Z., Wang, Y., Liu, L., Qiu, J., & Li, Y. (2007). Synthesis of water-soluble single-walled carbon nanotubes by RAFT polymerization. *Polymer*, 48(3), 728-733.

[100] Datsyuk, V., Guerret-Piecourt, C., Dagreou, S., Billon, L., Dupin, J. C., & Flahaut, E. (2005). Double walled carbon nanotube/polymer composites via in-situ nitroxide mediated polymerisation of amphiphilic block copolymers. *Carbon*, 43(4), 873-876.

[101] Zhao, X. D., Fan, X. H., Chen, X. F., Chai, C. P., & Zhou, Q. F. (2006). Surface modification of multiwalled carbon nanotubes via nitroxide-mediated radical polymerization. *Journal of Polymer Science Part A-Polymer Chemistry*, 44(15), 4656-4667.

[102] Dehonor, M., Masenelli-Varlot, K., Gonzalez-Montiel, A., Gauthier, C., Cavaille, J. Y., & Terrones, H. (2005). Nanotube brushes: Polystyrene grafted covalently on CNx nanotubes by nitroxide-mediated radical polymerization. *Chemical Communications*, 42, 5349-5351.

[103] Fan, D. Q., He, J. P., Tang, W., Xu, J. T., & Yang, Y. L. (2007). Synthesis of polymer grafted carbon nanotubes by nitroxide mediated radical polymerization in the presence of spin-labeled carbon nanotubes. *European Polymer Journal*, 43(1), 26-34.

[104] Kaminsky, W. (2008). Trends in polyolefin chemistry. *Macromolecular Chemistry and Physics*, 209(5), 459-466.

[105] Mokashi, V. V., Mokashi, V. V., Qian, D., & Liu, Y. J. A. (2007). study on the tensile response and fracture in carbon nanotube-based composites using molecular mechanics. *Composites science and technology*, 67(3-4), 530-540.

[106] Tang, W. Z., Santare, M. H., & Advani, S. G. (2003). Melt processing and mechanical property characterization of multi-walled carbon nanotube/high density polyethylene (MWNT/HDPE) composite films. *Carbon*, 41(14), 2779-2785.

[107] Yang, B. X., Pramoda, K. P., Xu, G. Q., & Goh, S. H. (2007). Mechanical reinforcement of polyethylene using polyethylene-grafted multiwalled carbon nanotubes. *Advanced Functional Materials*, 17(13), 2062-2069.

[108] Shofner, M. L., Khabashesku, V. N., & Barrera, E. V. (2006). Processing and mechanical properties of fluorinated single-wall carbon nanotube-polyethylene composites. *Chemistry of Materials*, 18(4), 906-913.

[109] Mahfuz, H., Adnan, A., Rangari, V. K., & Jeelani, S. (2005). Manufacturing and Characterization of Carbon Nanotube/Polyethylene Composites. *International Journal of Nanoscience*, 4(1), 55-72.

[110] Ruan, S., Gao, P., & Yu, T. X. (2006). Ultra-strong gel-spun UHMWPE fibers reinforced using multiwalled carbon nanotubes. *Polymer*, 47(5), 1604-1611.

[111] Tong, X., Liu, C., Cheng, H. M., Zhao, H., Yang, F., & Zhang, X. (2004). Surface Modification of Single-walled Carbon Nanotubes with Polyethylene via in situ Ziegler-Natta Polymerization. *Journal of Applied Polymer Science*, 92(6), 3697-3700.

[112] Funk, A., & Kaminsky, W. (2007). Polypropylene carbon nanotube composites by in situ polymerization. *Composites Science and Technology*, 67(5), 906-915.

[113] Bonduel, D., Bredeau, S., Alexandre, M., Monteverde, F., & Dubois, P. (2007). Supported metallocene catalysis as an efficient tool for the preparation of polyethylene/carbon nanotube nanocomposites: effect of the catalytic system on the coating morphology. *Journal of Materials Chemistry*, 17(22), 2359-2366.

[114] Dong, J. Y., & Hu, Y. (2006). Design and synthesis of structurally well-defined functional polyolefins via transition metal-mediated olefin polymerization chemistry. *Coordination Chemistry Reviews*.

[115] Russell, K.E. (2002). Free radical graft polymerization and copolymerization at higher temperatures. *Progress in Polymer Science*, 27(6), 1007-1038.

[116] Chung, T.C. (2002). Synthesis of functional polyolefin copolymers with graft and block structures. *Progress in Polymer Science*, 27(1), 39-85.

[117] Moad, G. (1999). The synthesis of polyolefin graft copolymers by reactive extrusion. *Progress in Polymer Science*, 24(1), 81-142.

[118] Hettema, R., Van Tol, J., & Janssen, L. P. B. M. (1999). In-situ reactive blending of polyethylene and polypropylene in co-rotating and counter-rotating extruders. *Polymer Engineering and Science*, 39(9), 1628-1641.

[119] Akbar, S., Beyou, E., Cassagnau, P., Chaumont, P., & Farzi, G. (2009). Radical grafting
       of polyethylene onto MWCNTs: A model compound approach. *Polymer*, 50(12),
       2535-2543.

[120] Badel, T., Beyou, E., Bounor-Legare, V., Chaumont, P., Flat, J. J., & Michel, A. (2007).
       Melt grafting of polymethyl methacrylate onto poly(ethylene-co-1-octene) by reactive
       extrusion: Model compound approach. *Journal of Polymer Science Part A-Polymer
       Chemistry*, 45(22), 5215-5226.

[121] Johnston, R.T. (2003). Monte Carlo simulation of the peroxide curing of ethylene elas-
       tomers. *Rubber Chemistry and Technology*, 76(1), 174-201.

[122] Johnston, R.T. (2003). Modelling peroxide crosslinking in polyolefins. *Sealing Technol-
       ogy*, 76(2), 6-9.

[123] Osswald, S., Havel, M., & Gogotsi, Y. (2007). Monitoring oxidation of multiwalled
       carbon nanotubes by Raman spectroscopy. *Journal of Raman Spectroscopy*, 38(6),
       728-736.

[124] Pastine, S. J., Okawa, D., Kessler, B., Rolandi, M., Llorente, M., Zettl, A., & Frechet, J.
       M. J. (2008). A facile and patternable method for the surface modification of carbon
       nanotube forests using perfluoroarylazides. *Journal of the American Chemical Society*,
       130(13), 4238-4239.

[125] Park, M.J., Lee, J. K., Lee, B.S., Lee-W, Y., Choi, I. S., & Lee-G, S. (2006). Covalent
       Modification of Multiwalled Carbon Nanotubes with Imidazolium-Based Ionic Liq-
       uids. *Effect of Anions on Solubility. Chemistry of Materials*, 18, 1546-51.

[126] Akbar, S., Beyou, E., Chaumont, P., & Mélis, F. (2010). Effect of a Nitroxyle-Based
       Radical Scavenger on Nanotube Functionalisation with Pentadecane. *A Model Com-
       pound Study for Polyethylene Grafting onto MWCNTs, Macromolecular Chemistry and
       Physics*, 211(22), 2396-2406.

[127] Mylvaganam, K., & Zhang, L. C. (2004). Nanotube functionalization and polymer
       grafting: An ab initio study. *Journal of Physical Chemistry B*, 108(39), 5217-5220.

[128] Akbar, S., Beyou, E., Chaumont, P., Mazzolini, J., Espinosa, E., D'Agosto, F., & Bois-
       son, C. (2011). Synthesis of Polyethylene-Grafted Multiwalled Carbon Nanotubes via
       a Peroxide-Initiating Radical Coupling Reaction and by Using Well-Defined TEMPO
       and Thiol End-Functionalized Polyethylenes. Journal of Polymer Science Part A:Pol-
       ymer Chemistry , 49(4), 957-965.

[129] Farzi, G., Akbar, S., Beyou, E., Cassagnau, P., & Mélis, F. (2009). Effect of radical
       grafting of tetramethylpentadecane and polypropylene on carbon nanotubes' disper-
       sibility in various solvents and polypropylene matrix. *Polymer*, 50(25), 5901-5908.

[130] Lopez, R. G., Boisson, C., D'agosto, F., Spitz, R., Boisson, F., Bertin, D., & Tordo, P.
       (2004). Synthesis and characterization of macroalkoxyamines based on polyethylene.
       Macromolecules, 37(10), 3540-3542.

[131] Lopez, R. G., Boisson, C., D'agosto, F., Spitz, R., Boisson, F., Gigmes, D., & Bertin, D. (2007). Catalyzed chain growth of polyethylene on magnesium for the synthesis of macroalkoxyamines Application to the production of block copolymers using controlled radical polymerization . *Journal of Polymer Science, Part A: Polymer Chemistry*, 45(13), 2705-2718.

[132] Mazzolini, J., Espinosa, E., D'Agosto, F., & Boisson, C. (2010). Catalyzed chain growth (CCG) on a main group metal: an efficient tool to functionalize polyethylene. Polymer Chemistry , 1(6), 793-800.

[133] D'Agosto, F., & Boisson, C. (2010). A RAFT Analogue Olefin Polymerization Technique Using Coordination Chemistry. *Australian Journal of Chemistry*, 63(8), 1155-1158.

[134] Mazzolini, J., Mokthari, I., Briquet, R., Boyron, O., Delolme, F., Monteil, V., Gigmes, D., Bertin, D., D'Agosto, F., & Boisson, C. (2010). Thiol-End-Functionalized Polyethylenes. *Macromolecules*, 43(18), 7495-7503.

[135] Hawker, C. J., Bosman, A. W., & Harth, E. (2001). New polymer synthesis by nitroxide mediated living radical polymerizations. *Chemical Review*, 101(12), 3661-3688.

[136] Lou, X., Detrembleur, C., Sciannamea, V., Pagnoulle, C., & Jerome, R. (2004). Grafting of alkoxyamine end-capped (co)polymers onto multi-walled carbon nanotubes. *Polymer*, 45(18), 6097-6102.

[137] Liu, Y., Yao, Z., & Adronov, A. (2005). Functionalization of single-walled carbon nanotubes with well-defined polymers by radical coupling. *Macromolecules*, 38(4), 1172-1179.

[138] Wang, H. C., Li, Y., & Yang, M. (2007). Sensors for organic vapor detection based on composites of carbon nonotubes functionalized with polymers. *Sensors and Actuators B-Chemical*, 124(2), 360-367.

[139] Wang, Y., Wu, J., & Wei, F. (2003). A treatment method to give separated multi-walled carbon nanotubes with high purity, high crystallization and a large aspect ratio. *Carbon*, 41(15), 2939-2948.

[140] Lee, S. H., Cho, E., Jeon, S. H., & Youn, J. R. (2007). Rheological and electrical properties of polypropylene composites containing functionalized multi-walled carbon nanotubes and compatibilizers. *Carbon*, 45(14), 2810-2822.

[141] Wang, Y., Wen, X., Wan, D., Zhang, Z., & Tang, T. (2012). Promoting the responsive ability of carbon nanotubes to an external stress field in a polypropylene matrix: A synergistic effect of the physical interaction and chemical linking. *Journal Material Chemistry*, 22, 3930-3938.

[142] Zheng, J., Zhu, Z., Qi, j., Zhou, Z., Li, P., & Peng, M. (2011). Preparation of isotactic polypropylene-grafted multiwalled carbon nanotubes (iPP-g-MWNTs) by macroradical addition in solution and the properties of iPP-g-MWNTs/iPP composites. *Journal Material Science*, 46, 648-658.

[143] Jancar, J., Douglas, J. F., Starr, F. W., Kumar, S. K., Cassagnau, P., Lesser, A. J., Stern-
       stein, S. S., & Buehler, M. J. (2010). Current issues in research on structure-property
       relationships in polymer nanocomposites. *Polymer*, 51(15), 3321-3343.

[144] Cassagnau, P. (2008). Melt rheology of organoclay and fumed silica nanocomposites.
       Polymer.

[145] Moreira, L., Fulchiron, R., Seytre, G., Dubois, P., & Cassagnau, P. (2010). Aggregation
       of Carbon Nanotubes in Semidilute Suspension. *Macromolecules*, 43(3), 1467-1472.

[146] Hobbie, E.K. (2010). Shear rheology of carbon nanotube suspensions. *Rheologica Acta*.

[147] Marceau, S., Dubois, P., Fulchiron, R., & Cassagnau, P. (2009). Viscoelasticity of
       Brownian Carbon Nanotubes in PDMS Semidilute Regime. *Macromolecules*, 42(5),
       1433-1438.

[148] Mitchell, C. A., Bahr, J. L., Arepalli, S., Tour, J. M., & Krishnamoorti, R. (2002). Dis-
       persion of functionalized carbon nanotubes in polystyrene. *Macromolecules*, 35(23),
       8825-8830.

[149] Pujari, S., Ramanathan, T., Kasimatis, K., Masuda, J., Andrews, R., Torkelson, J. M.,
       Brinson, L. C., & Burghardt, W. R. (2009). Preparation and Characterization of Multi-
       walled Carbon Nanotube Dispersions in Polypropylene: Melts Mixing versus Soli-
       State Shear Pulverization. *Journal of Polymer Science, Part B: Polymer Physics*, 47(14),
       1426-1436.

[150] Hong, J. S., Hong, I. G., Lim, H. T., Ahn, K. H., & Lee, S. J. (2012). In situ lubrication
       dispersion of Multi-walled carbon nanotubes in Polypropylene melts. *Macromolecular
       Material Engineering*, 297(3), 279-287.

[151] Wu, D., Sun, Y., Wu, L., & Zhang, M. (2008). Linear Viscoelasticity Properties and
       Crystallisation Behavior of Multi-Walled Carbon Nanotube/Polypropylene Compo-
       sites. *Journal of Applied Polymer Science*, 108(3), 1506-1513.

[152] Wu, D., Sun, Y., & Zhang, M. (2009). Kinetics Study on Melt Compounding of Car-
       bon Nanotube/Polypropylene Nanocomposites. *Journal of Polymer Science, Part B: Pol-
       ymer Physics*, 47(6), 608-618.

[153] Jin, S. H., Kang, C. H., Yoon, K. H., Bang, D. S., & Park, Y. B. (2009). Effect of Compa-
       tibilizer on Morphology, Thermal and Rheological Properties of Polypropylene/
       Functionalized Multi-walled Carbon Nanotubes composite. *Journal of Applied Polymer
       Science*, 111(2), 1028-1033.

[154] Menzer, K., Hrausee, B., Boldt, R., Kretzschmar, B., Weidisch, R., & Potschke, P.
       (2011). Percolation Behaviour of multiwalled carbon nanotubes of altered length and
       primary agglomerate morphology in melt mixed isotactic polypropylene-based com-
       posites. *Composite Science and Technology*, 71(16), 1936-1943.

[155] Hemmati, M., Rahimi, G. H., Kaganj, A. B., Sepehri, S., & Rashidi, A. M. (2008). Rheo-
       logical and Mechanical Characterization of Multi-walled Carbon Nanotubes/Poly-

propylene Nanocomposites. *Journal of Macromolecular Science, Part B: Physics*, 47(6), 1176-1187.

[156] Prashantha, K., Soulestin, J., Lacrampe, M. F., Krawczak, P., Dupin, G., & Claes, M. (2009). Masterbatch-based multi-walled carbon nanotube filled polypropylene nano-composites: Assessment of rheological and mechanical properties. *Composite Science and Technology*, 69(11-12), 1756-1763.

[157] Palza, H., Reznik, B., Kappes, M., Hennrich, F., Naue, I. F. C., & Wilhelm, M. (2010). Characterization of melt flow instabilities in polyethylene/carbon nanotube. *Polymer*, 51(16), 3753-3761.

Chapter 3

Production of Carbon Nanotubes and Carbon Nanoclusters by the JxB Arc-Jet Discharge Method

Tetsu Mieno and Naoki Matsumoto

Additional information is available at the end of the chapter

1. Introduction

Since the discovery of methods for the mass production of single-walled carbon nanotubes (SWNTs) [1, 2], applications of SWNTs such as transistor devices, biosensing devices, double-layer-type capacitors, transparent electrode films, radio wave absorbents and material hardeners have been studied [3-5]. Large-scale production and improvement of purity of SWNTs by the electric-arc techniques have been developed [6, 7]. However, the efficient production of high-quality and defect-free SWNTs, and metal/semiconductor selected or diameter-controlled production of SWNTs have not yet been achieved. Therefore, basic study of the various methods of producing SWNTs is still important, by which new high-performance routes to producing desired SWNTs are expected to be found.

Here, the production of SWNTs and carbon nanoclusters by the arc discharge method utilizing a magnetic field, known as the JxB arc-jet discharge method, has been studied [8-10]. Although the application of a steady-state magnetic field to arc discharge is not such a popular method, electromagnetic force can change the flow of hot gas in the arc region and thus control the production process of carbon clusters. To realize the large-scale production of carbon clusters by the arc discharge method, a revolver-injection-type JxB arc-jet producer was successfully developed by our group, by which the continuous mass production of SWNTs and other carbon clusters can be carried out.

As a result, the more efficient production of SWNTs and other carbon clusters compared with conventional arc discharge methods has been achieved. Here, the development of the JxB arc-jet discharge method and results obtained using the method are described.

2. Theoretical model of the *JxB* arc jet discharge method

By applying a steady-state weak magnetic field ($B_0 = 1 - 5$ mT) perpendicular to the discharge current in the arc discharge, the Lorentz force (*JxB* force) causes the ejection of the arc plasma and surrounding gas in the *JxB* direction as shown in Fig. 1 [8, 11]. In the 1960s, this force in a pulsed discharge was actively studied in relation to the electric propulsion engine of rockets [12].

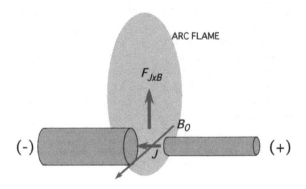

Figure 1. Schematic diagram of the *JxB* arc-jet discharge.

Here, this effect is used to eject sublimated carbon atoms in a selected direction. By controlling the magnetic field, control of the hot gas including the carbon material is possible, and suitable conditions to do hot gas reactions for the production of SWNTs and other carbon clusters can be selected. This method can also reduce the influence of the electrode direction and chamber configuration.

When the discharge current density and applied magnetic field are 50 A/cm^2 and 5 mT, respectively, the Lorentz force causing acceleration of electrons and ions is 0.25 N/cm^3. When the gas pressure and the gas temperature around the arc are 30 kPa and 5000 K, respectively, the mean free path and collision frequency of electrons are about 0.01 mm and 10 GHz, respectively. Because of this high collision frequency, electrons frequently collide with neutral gas atoms and accelerate them in the *JxB* direction, resulting in the ejection of hot gas from the arc region. The acceleration time is related to the electron lifetime in the plasma.

3. Production of carbon clusters by the *JxB* arc-jet discharge method

3.1. Heat flux

To investigate the *JxB* arc-jet discharge reaction, several types of arc reactors are used. Figure 2 shows a schematic of the reactor used to measure the heat flux of the arc plasma [11]. The reactor is made of stainless steel (18 cm diameter, 20 cm height) and has a carbon anode

(8.0 mm), a carbon cathode (15 mm), a viewing port and a movable calorimetric probe. The reactor is evacuated by a rotary pump to a pressure of less than 10 Pa and then closed. After introducing He gas with $p(\mathrm{He}) = 10 - 80$ kPa, discharge starts, where the discharge current is $I_d = 20 - 80$ A and voltage between the electrodes is $V_{rod} = 20 - 35$ V. At the front and back of the reactor, solenoid coils (20 cm inner diameter) are installed to produce a steady state magnetic field of $B_0 = 0 - 5$ mT.

When a magnetic field is applied during the discharge, the shape of the arc flame dramatically changes, and a strong plasma flow in the JxB direction can be observed. Figures 3(a) and (b) respectively show side views of the arc flames for $B_0 = 0$ and $B_0 = 2.0$ mT, where $p(\mathrm{He}) = 40$ kPa and $I_d = 80$ A. The upper direction is the JxB direction. By applying a magnetic field, the plasma and the hot gas are ejected in the vertically upward direction. The upward flow of carbon particles can sometimes be clearly observed. By developing a calorimeter [11] in which flowing water absorbs the heat flux, the local heat flux is measured and the results are shown in Figs. 4 (a) and (b) [11]. By increasing the magnetic field, the heat flux is localized in the upper part of the arc plasma (FWHM value of about 50 mm). Above the arc plasma, the heat flux monotonically increases.

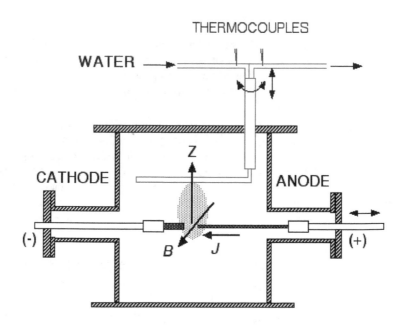

Figure 2. Schematic side view of the arc reactor with a calorimetric probe installed.

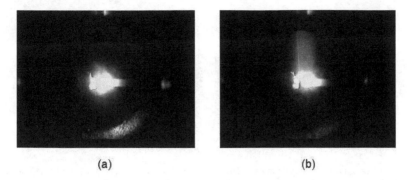

(a)   (b)

Figure 3. Arc flames for B_0= 0 (a) and B_0= 2.0 mT (b) (side views), where p(He)= 40 kPa and I_d= 80 A.

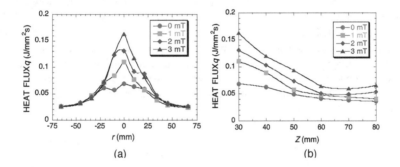

(a)   (b)

Figure 4. Radial profiles (a) and vertical profiles (b) of heat flux above the arc plasma for B_0= 0, 1.0, 2.0 and 3.0 mT. p(He)= 40 kPa and I_d= 60 A.

The heat flux above the arc plasma as a function of He gas pressure is measured and shown in Fig. 5(a), where I_d= 60 A, d_G= 5 mm, and the distance from the arc center is 40 mm. The heat flux increases monotonically with the pressure, which is particularly in the case of B_0= 2.0 mT. Figure 5(b) shows the heat flux above the arc plasma as a function of the gap distance between the two electrodes d_G, where p= 40 kPa, I_d= 60 A and z= 40 mm. By changing d_G, the effect of the arc jet changes, which can be observed from the viewing port. The heat flux gradually increases with the gap distance, and this effect is greatly enhanced when B_0= 2.0 mT.

To summarize these results, that the JxB arc jet is enhanced by increasing the applied magnetic field (B_0= 0 - 3.0 mT), the He pressure and the gap distance. However, under a stronger magnetic field of B_0> 4 mT, the discharge tends to be extinguished easily by fluctuation in the discharge.

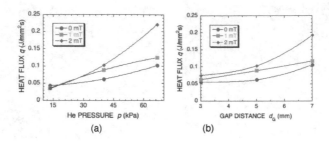

Figure 5. (a) He pressure dependence of the heat flux, where I_d= 60 A, d_G= 5 mm and z= 40 mm from the arc center. (b) Gap distance dependence of the heat flux, where p= 40 kPa, I_d= 60 A and z= 40 mm.

3.2. Relations among directions of the discharge current, magnetic field and gravity

In the case of gas arc discharge, gravity induces strong heat convection. Therefore, by changing the current direction relative to that of gravity, different production characteristics of carbon can be expected [13]. Direction of the *JxB* force compared with that of gravity should also be an important parameter. To examine the relations among the directions of the discharge current, magnetic field and gravity for the production of fullerenes, five experimental configurations are prepared and a discharge experiment is carried out. The 5 configurations are shown in Fig. 6. Here, the carbon anode is 6.5 mm in diameter and the carbon cathode is 15 mm in diameter.

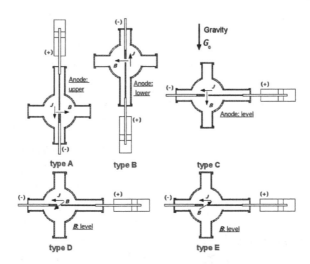

Figure 6. Schematic of five experimental configurations. The directions of the discharge current *J* and magnetic field *B* relative to that of gravity *G₀* are changed.

The production rates of carbon soot W_{soot} (g/h) as a function of B_0 for configurations (types A – E) are obtained and the results are shown in Fig. 7, where p(He) = 40 kPa, I_d = 70 A and d_G ~ 5 mm [13]. Generally, the soot production rate increases steadily with the magnetic field. However, for type A, W_{soot} is very low for B_0 = 0 and it rapidly increases with increasing magnetic field. When B_0 = 4.0 mT, the differences in W_{soot} are very small among the five configurations.

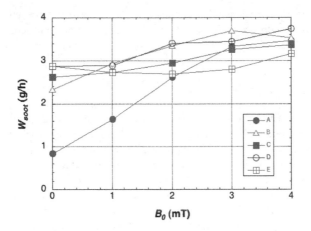

Figure 7. Production rate of carbon soot versus B_0 for the five configurations. p= 40 kPa, I_d= 70 A and discharge time T_d= 60 min

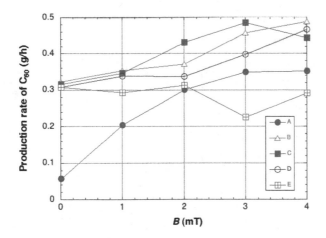

Figure 8. Production rate of C_{60} versus B_0 for the five configurations. p= 40 kPa, I_d= 70 A and discharge time T_d= 60 min.

The production rate of C_{60} as a function of B_0 for configurations (types A – E) is obtained and the results are shown in Fig. 8, where the conditions are the same us those of Fig. 7. Similarly to in Fig. 7, the C_{60} production rate generally increases with B_0, except for type A. C_{60} production rate of type A is very low at $B_0 = 0$. Moreover, for type E, the C_{60} production rate does not increase monotonically with B_0 and the magnetic field does not have a positive effect on the production rate.

From these results, it can be concluded that the directions of the discharge current and magnetic field compared with that of gravity affect the production of carbon soot and fullerenes. The JxB force tends to reduce the effect of gravity when B_0 is sufficiently large. The type A and the Type E are less suitable for the production of fullerenes.

3.3. Production of SWNTs

The production of high-quality SWNTs is one of the most important targets in advancing nanomaterial development. The growth model of SWNTs in the arc-discharge reactors has not been confirmed. Several models show importance of catalyst particles in the hot gas, carbon density and catalyst temperature. [14, 15] Here, the production of SWNTs is examined using the JxB arc-jet method, which could modify the growth reactions in the hot gas. In this case, a Ni/Y catalyst is included in the carbon material rods (6.0 x 6.0 mm, rectangular type, 4.2 W% of Ni and 0.9 W% of Y included), and $p(He)= 60$ kPa and $I_d = 50$ A. The soot production rate as a function of applied magnetic field is shown in Fig. 9(a) [16, 17]. By increasing the magnetic field, the production rate monotonically increases. However, further increasing B_0 to above 3.5 mT makes the discharge unstable. Figure 9(b) shows the pressure dependence of the soot production rate for $B_0 = 0$ and $B_0 = 2.2$ mT. As the pressure increases and the collisional effect of He gas increases, the JxB force clearly affects soot production.

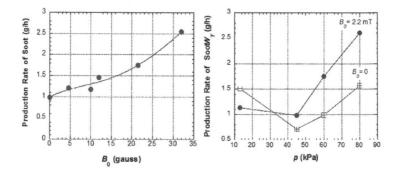

Figure 9. Production rate of soot including SWNTs versus B_0 (a), where $p(He)= 60$ kPa and $I_d = 50$ A, and pressure dependence (b), where $I_d = 50$ A.

Using this JxB arc-jet method, a large amount of SWNTs is produced from carbon rods including a Ni/Y catalyst. Figure 10 shows a typical TEM image of the produced soot, in which many bundles of SWNTs are included. There are also carbon nanoparticles and catalyst nanoparticles in the soot, which should be removed during the purification of the SWNT products. The quality of the products is measured by a Raman spectrometer (Jasco Co., NR-1800. An Ar ion laser of λ= 488.0 nm is used.), and the results are shown in Figs. 11 (a) and 11(b) for B_0= 0 and B_0= 3.2 mT. In both cases, there are very small D(disorder) band peaks and large G (graphite) band peaks, from which we can estimate the content and quality of SWNTs in the produced carbon soot. These figures show that the JxB arc discharge does not degrade the quality of the SWNTs. From the signals of the radial breathing mode in Fig. 11(a), we can evaluate the SWNT diameters [18]. The major diameter is 1.40 nm, and SWNTs with a diameter of 1.26 nm also exist in the case of B_0= 3.2 mT. SWNTs with diameters of 1.70 nm, 1.16 nm and 1.0 nm also exist in the case of B_0= 0.

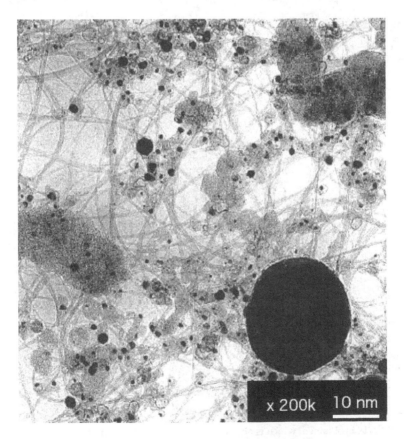

Figure 10. Typical TEM image of SWNTs produced by the JxB arc-jet discharge method.

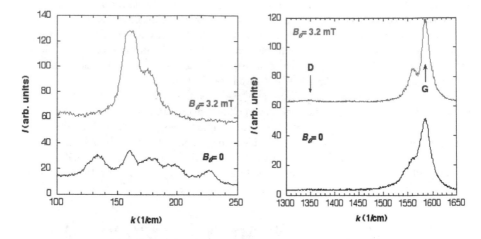

Figure 11. Raman spectra of the produced samples for magnetic fields of B_0= 3.2 mT and 0 T. p(He)= 60 kPa and I_d= 50 A. λ = 488.0 nm.

Figure 12. Photograph of SWNTs dispersed in pure water.

In the arc production of SWNTs, effect of gas species is examined. Ar, Ne or N_2 gas is used instead of He gas, all of which degrade production of SWNTs. When amount of H_2 gas is included in He gas, the SWNT production rate considerably decreases, which is not consistent with the previous report [19]. It is conjectured that He atom has high ionization potential, and it causes almost no chemical reactions and less emission loss. When Co or Fe particles are used as catalyst material instead of Ni/Y particles, the SWNT production rate considerably decreases. However, in case of of Co catalyst, a little amount of very long bundles of SWNTs is obtained, which is about 5 cm long. Improvement of the long-SWNT yield by this method is one of our study targets.

Usually SWNTs have poor dispersibility in water, which limits their potential applications. Therefore, the development of water-dispersible SWNTs is very important. Here we attempt to dissolve a SWNT sample in pure water. First, a small amount of SWNTs is placed in 20 mL of pure water, which is then mixed using a supersonic homogenizer (Sonics Co., VC-130, 25 W) for about 40 min. Then, a small amount of surfactant is added, which is one of bio-polymers [20, 21]. And it is mixed by sonication again for about 40 min. As a result, the SWNTs are well dispersed in water, and the dispersion remains very stable for more than 1 month. Figure 12 shows a photograph of SWNTs dispersed in water after 50 times dilution. The study of SWNTs dispersed in water is continuing with the aim of realizing biological applications.

3.4. Production of endohedral metallofullerenes

Using the above arc-jet discharge methods, endohedral metallofullerenes (such as $Gd@C_{82}$ and $La@C_{82}$) [22] and carbon nanocapsules are efficiently produced. Applications of these materials are expected.

By performing arc discharge using a Gd_2O_3-containing carbon rod (6.0 x 6.0 mm, rectangular), metallofullerenes are produced, where $p(He)= 50$ kPa and $I_d= 58$ A. The production rate of soot is about 2.5 g/h. Figure 13(a) shows a typical TEM image of the sample obtained, in which gadolinium nanoparticles with a diameter on the order of 10 nm are covered with carbon atoms, resulting in the formation of carbon capsules with a size of 10 – 50 nm, which are very stable in air. By refluxing the sample with toluene, fullerenes can be extracted from the soot. After 4 h of reflux, the liquid is filtered and a reddish liquid is obtained. A mass spectrum of this sample obtained using a laser-desorption time-of-flight mass spectrometer (LD-TOF-MS) (Bruker Co., Autoflex, + ion mode, 50 shots averaged) is shown in Fig. 13(b). We can confirm the existence of not only C_{60} and higher fullerenes but also endohedral metallofullerenes $Gd@C_{82}$. Although the peak intensities are not quantitative, the relative yield of $Gd@C_{82}$ compared with that of C_{60} is several mol%. The $Gd@C_{82}$ is expected to be used as a contrast agent in MRI [23].

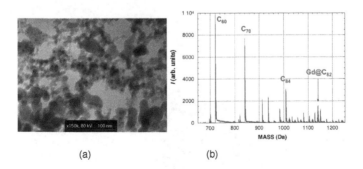

(a)   (b)

Figure 13. (a) TEM image of gadolinium-continuing carbon nanoparticles produced by the arc-jet discharge method. (b) LD-TOF-MS spectrum of a carbon sample extracted from soot using toluene.

3.5. Production of magnetic nanoparticles

Using the $J\times B$ arc-jet discharge method, many types of metal-particle-encapsulated carbon nanoparticles [24, 25] can be easily produced. As examples, iron-encapsulated carbon nano-particles and cobalt-encapsulated carbon nanoparticles have been produced. Both are ferro-magnetic nanoparticles with a size of 10 - 100 nm, and are very stable.

Using iron-containing carbon rods (6.0 x 6.0 mm, rectangular), iron-encapsulated carbon nanoparticles are produced, where p(He)= 50 kPa and I_d= 50 A. The soot production rate is about 0.43 g/h. Figure 14 (a) shows a photograph of the produced soot suspended by a mag-net, demonstrating the good ferromagnetic property. A typical TEM image of the sample is shown in Fig. 14 (b). Iron particles with a size of 1 nm to 20 nm are covered with carbon atoms, resulting in the formation of carbon particles with a size of 10 – 100 nm. These parti-cles are very stable in air and inactive in hydrochloric acid.

Figure 14. (a) Photograph of iron-containing carbon nanoparticles suspended by a magnet. (b) TEM image of iron-containing carbon nanoparticles.

Figure 15. Photograph of iron-containing carbon nanoparticles dispersed in water.

Cobalt-encapsulated carbon nanoparticles, which also have ferromagnetic properties, are produced by the arc-jet discharge method. They are dispersed in pure water with a small amount of surfactant (gelatin *etc.*) and mixed using a supersonic homogenizer (Sonic Co., VX-130) for 1 h. Finally, a black inklike liquid is obtained. The dispersion is homogeneous and stable, and most of the particles do not precipitate even after one month. These water-soluble magnetic nanoparticles potentially have many applications in the fields of liquid sealing, medical diagnostics and medical treatment [26]. Figure 15 shows a photograph of the stable iron-containing carbon nanoparticles dispersed in water.

4. Development of automatic *JxB* arc-jet producer

To produce SWNTs and carbon nanoclusters at a commercial scale by the *JxB* arc-jet discharge method, a revolver-injection-type arc jet producer (RIT-AJP) has been developed by collaboration with Daiavac Ltd. (Japan) [9].

A schematic and photograph of RIT-AJP are shown in Fig. 16. The left side of the machine is an arc discharge chamber, which consists of a vacuum vessel made of stainless steel 25 cm in diameter and 70 cm high that is uniformly cooled by water jackets. About 5 L of water is stored in the jackets and cooling water is slowly supplied to the jackets. In the central part of the chamber, a cathode electrode (30 mm in diameter), an anode electrode, an exhaust port, a viewing port and an electrode-cleaning hand are mounted. The top and bottom parts of the chamber are soot collectors, with an inner diameter of about 25 cm and a height of 24 cm, in which produced soot is deposited. Using these collectors, as much as 25 L of soot can be easily collected after a single operation.

The right side of the apparatus is a revolver-type carbon rod magazine. In the cylindrical metal vacuum vessel, which is 34 cm in diameter and 49 cm long, there is a rotatable cylindrical magazine, in which as many as 50 carbon rods of 6 - 10 mm diameter and 300 mm length can be loaded. A schematic figure of the material-feeding mechanism of the magazine is shown in Fig. 17(a), and a photograph of a rotatable carbon rod magazine (for 50 carbon rods) is shown in Fig. 17(b). Under the vacuum chamber, there is a vacuum pump, an electrical controller and a microcomputer. Discharge power is supplied by an inverter-type DC power supply (Daihen Co., ARGO-300P).

The production sequence is as follows. First, up to 50 carbon rods are loaded in the magazine, and the chamber is evacuated by the vacuum pump. After evacuation to a pressure of less than 10 Pa, pure helium gas is introduced and the chamber is closed. Upon turning on the electrical controller, a metal striker pushes one of the carbon rods towards the cathode, and the discharge starts upon electrical contact. As the discharge conditions are determined by the discharge voltage and discharge current, the carbon rod is automatically moved until both parameters match the set values. Once the carbon rod is consumed, the cylindrical carbon magazine rotates 1/50 of a turn and the next carbon rod is inserted by the striker. A magnetic field can be applied by a block-type ferrite magnet located outside the chamber, by which a magnetic field of about 2 mT is applied horizontally to the discharge space. Carbon deposited on the cathode is removed by a cathode-cleaning hand. After the discharge, produced soot that has been deposited is carefully collected.

As an example of the continuous production of carbon clusters, fullerenes are produced. Using 134 carbon rods of 8 mm diameter, continuous JxB arc-jet discharge is carried out, where p(He)= 40 kPa, the discharge current is I_d= 120 A, the voltage between electrodes is $V_{rod}\sim$ 33 V and the gap distance is $d_G\sim$ 5 mm. The insertion speed of the carbon rods is about 30 cm/h. After the discharge, carbon soot from three parts (the top collector, central chamber and bottom collector) is collected separately and their masses are measured. The amount of soot deposited on the top wall is considerably increased by applying the magnetic field, because the carbon molecules are blown upward onto the top wall. After sufficient mixing, the C_{60} content in the soot is measured by a UV/visible spectrometer (Shimadzu Co. UV-1200). At the top collector, the C_{60} content is the highest and about 14 W% of C_{60} is present, whereas, 4.2 W% is present on the center wall and 2.9 W% is present on the bottom wall. In total, about 105 g of soot containing about 7 g of C_{60} is produced in 12 h.

The contents of higher fullerenes in the soot are measured using a high-pressure liquid chromatograph (HPLC) (Jasco Co., Gulliver Series, PU980) [27]. The collection rates of C_{60}, C_{70}, C_{76}, C_{78} and C_{84} for two different magnetic fields are shown in Fig. 18. White rectangles in the graphs show the measurement errors. By applying a magnetic field, the collection rates of these fullerenes considerably increase.

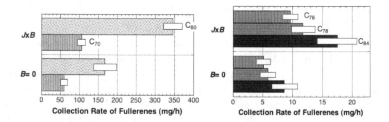

Figure 18. Collection rates of C_{60} and C_{70} (a), and C_{76}, C_{78} and C_{84} (b) for two different magnetic fields.

5. Summary

1. By applying a steady state magnetic field perpendicular to the discharge current, JxB arc-jet discharge is successfully produced. The flow of hot gas and the heat flux are concentrated in the JxB direction.

2. Carbon atoms sublimated from the anode are continuously ejected from the arc plasma in the JxB direction. As a result of the JxB force, the effect of gravity (heat convection) can be reduced.

3. By increasing the magnetic field, the production rate of carbon soot including SWNTs considerably increases, in which the quality of the SWNTs remains high. Water-soluble SWNTs can be obtained by additional processing.

4. Using the *JxB* arc-jet discharge, endohedral metallofullerenes, Gd@C_{82} and magnetic nano-particles (iron-encapsulateted carbon nanoparticles and cobalt-encapsulated carbon nano-particles) are sucessfully produced.

5. For the continuous and large-scale production of carbon clusters, a revolver-injection-type arc-jet producer (RIT-AJP) has been developed. Using this machine, the automatic mass production of SWNTs and carbon clusters is realized. We are currently attempting to fabricate many new types of carbon clusters using this machine.

Acknowledgments

We thank H. Inoue of Daiavac Co. (Chiba, Japan) for his technical support during the development of the RIT-AJP machine. We also thank W. Tomoda, Md. K. H. Bhuiyan and S. Aoyama of Shizuoka University for their technical assistance.

Author details

Tetsu Mieno and Naoki Matsumoto

Department of Physics, Shizuoka University, Japan

References

[1] Iijima, S. Single-Shell Carbon Nanotubes of 1-nm Diameter, Nature 1993; 363, 603-605.

[2] Bethune, D. S., Klang, M. S., de Vries, M. S., Gorman, G., Savoy, R., Vazquez, J., Beyers, R., Cobalt-catalysed growth of carbon nanotubes with single-atomic-layer walls, Nature 1993; 363, 605-607.

[3] Dresselhaus, M. S., Dresselhaus, G., Avouris, Ph., (eds.) Carbon Nanotubes, Springer, Berlin, 2000, ISBN: 3-540-41086-4.

[4] Jorio. A., Dresselhaus, M. S., Dresselhaus, G., (eds.) Carbon Nanotubes, Springer, Berlin, 2008, ISBN : 978-3-540-72864-1.

[5] Harris, P. J. Carbon Nanotubes Science, Cambridge University Press, Cambridge, 2009, ISBN: 978-0-521-53585-4.

[6] Journet, C., Maser, W. K., Bernier, P., Loiseau, A. Lamy de la Chapele, M., Lefrant, S., Deniard, P., Lee, R., Fischer, J. E., Large-scale production of single-walled carbon nanotubes by the electric-arc technique, Nature, 388, pp. 756-758.

[7] Mansour, A., Razafinimanana, M., Monthoux, M., Pachco, M., Gleizes, A., A siginificant improvement of both yield and purity during SWNT synthesis via the electric arc process, Carbon, 45, 2007, pp. 1651-1661.

[8] Mieno, T. Automatic Production of Fullerenes by a JxB Arc Jet Discharge Preventing Carbon Vapor from Depositing on the Cathode, Fullerene Science & Technology 1995; 3 (4), 429-435.

[9] Mieno, T. JxB Arc Jet Fullerene Producer with a Revolver Type Automatic Material Injector, Fullerene Science & Technology 1996; 4 (5), 913-923.

[10] Mieno, T. Production of Fullerenes from Plant Materials and Used Carbon Materials by Means of a Chip-Injection-Type JxB Arc Reactor, Fullerene Science & Technology 2000; 8 (3), 179-186.

[11] Matsumoto, N., Mieno, T. Characteristics of Heat Flux of JxB Gas-Arc Discharge for the Production of Fullerenes, Vacuum 2003; 69, 557-562.

[12] Jahn, R. G. Physics of Electric Propulsion, McGraw-hill, New York, 1968, 257-316, (NC)ID: BA18312459.

[13] Aoyama, S., Mieno, T. Effects of Gravity and Magnetic Field in Production of C_{60} by a DC Arc Discharge, Japanese Journal of Applied Physics 1999; 38 (3A), L267- L269.

[14] Kanzow, H., Ding, A, Formation mechanism of single-wall carbon nanotubes on liquid-metal particles, Physical Review B 1999 ; 60 (15) 11180-11186.

[15] Gavillet, J. *et al.*, Microscopic mechanisms for the catalyst assisted growth of single-wall carbon nanotubes, Carbon 2002 ; 40, 1649-1663.

[16] Mieno, T., Tan, G.-D. Effect of Gravity and Magnetic Field on Production of Single-Walled Carbon Nanotubes by Arc-Discharge Method, In: Yellampalli, S. (ed.) Carbon Nanotubes- Synthesis, Characterization, Application; InTech; 2011, 61-76, ISBN: 978-953-307-497-9.

[17] Mieno, T., Matsumoto N., Takeguchi, M. Efficient Production of Single-Walled Carbon Nanotubes by JxB Gas-Arc method, Japanese Journal of Applied Physics 2006; 43 (12A), L1527- L1529.

[18] Kataura, H. *et al.* Optical Properties of Single-Walled Carbon Nanotubes, Synthetic Metals 1999; 103, 2555-2558.

[19] Zhao, X., Inoue, S., Jinno, M., Suzuki T., Ando Y., Macroscopic oriented web of single-wall carbon nanotubes, Chemical Physics Letters 2003; 373, 266-271.

[20] Takahashi, T., Tsunoda, K., Yajima, H., Ishii, T., Purification of Single Wall Carbon Nanotubes Using Gelatin, Japanese Journal of Applied Physics 2004; 43 (3), 1227-1230.

[21] Takahashi, T., Luculescu, C. R., Uchida, K., Tsunoda, K., Yajima, H., Ishii, T., Dispersion Behavior and Spectroscopic Properties of Single-Walled Carbon Nanotubes in Chitosan Acidic Aqueous Solutions, Chemistry Letters 2005; 34 (11), 1516-1517.

[22] Shinohara, H., Endohedral metallofullerenes, Reports on Progress in Physics 2000; 63, 843-892.

[23] Mikawa, M. et al. Paramagnetic Water-Soluble Metallofullerenes Having the Highest Relaxivity for MRI Contrast Agents, Bioconjugate Chemistry 2001; 377, 510-514.

[24] Saito, Y., Nanoparticles and Filled Nanocapsules, Carbon 1995; 33 (7) 979-988.

[25] Saito, Y. et al. Iron particles nesting in carbon cages grown by arc discharge, Chemical Physics Letters 1993; 212 (3-4) 379-383.

[26] Pankhurst, Q. A., Connolly, J., Jones, S. K., Dobson, J. Applications of Magnetic Nanoparticles in Biomedicine 2003; 36, R167-R181.

[27] Bhuiyan, Md. K. H., Mieno, T. Production Characteristics of Fullerenes by Means of the JxB Arc Discharge Method, Japanese Journal of Applied Physics 2002; 41 (1), 314-318.

Chapter 4

Classification of Mass-Produced Carbon Nanotubes and Their Physico-Chemical Properties

Heon Sang Lee

Additional information is available at the end of the chapter

1. Introduction

Carbon nanotube (CNT) may be classified into single walled (SWCNT), double walled (DWCNT), and multiwalled carbon nanotube (MWCNT) according to the number of graphene layers. In some cases, bamboo-shaped multiwalled carbon nanotubes were also synthesized. Among these carbon nanotubes, multiwalled carbon nanotubes have been mass-produced in hundreds metric tons level. Many researchs on multiwalled carbon nanotubes point to an electrode, polymer composites, coating, and others. The number of graphene layers, purity, and crystallinity are the main features of multiwalled carbon nanotubes, which need to be characterized. We have proposed one important characteristic of multiwalled carbon nanotubes, the mesoscopic shape of MWCNT, of which many industrial applications may be comprised. According to our suggestion, one can determine the degree of tortuousness of MWCNT, quantitatively. (*see ref.1-5 and sections 1-6 in this chapter*)In this chapter, we will describe the mesoscopic shape factor of MWCNT in detail. Various physical properties as well as toxicity may strongly depend on the mesoscopic shape factor of MWCNT. Our suggestion has also been published as an international standard ISO/TS11888 by international organization for standardization (ISO) in 2011.

I hope readers enjoy the concepts and expressions shown in this chapter. Especially, this chapter shall be helpful to whom may want to develop a commercial application by selecting a proper CNT.

2. Static bending persistence length (SBPL, l_{sp})

If MWCNTs have no defect along their axis, their appearance would be straight to several hundred micro meter. Persistence length is the maximum straight length that is not bent by thermal energy. The persistence length of MWCNT is expected to be several hundred micro meter due to its exceptional high modulus. Static bending persistence length (SBPL) has been proposed in our earlier work to quantify the mesoscopic shape of MWCNT. SBPL is the maximum straight length that is not bent by permanent deformation. Fig. 1 shows the concept of SBPL. When a length considered is longer than SBPL, the shape of MWCNT looks tortuous. On the contrary, the shape of MWCNT looks straight as a length considered is shorter than SBPL.

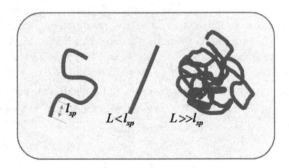

Figure 1. The concept of static bending persistence length of MWCNT (l_{sp}).

If length considered is longer than SBPL, the shape of MWCNT looks tortuous. On the contrary, the shape of MWCNT looks straight as a length considered is shorter than SBPL.

3. Mathematical expression of SBPL (l_{sp})

The end-to-end vector can be obtained such as eq 1 when the distribution of bending points ($\{\varphi\} \equiv (\varphi_1, \varphi_2,...,\varphi_k)$) is given.

$$\mathbf{R} = N\sum_{i=1}^{k}\varphi_i\mathbf{r}_i \tag{1}$$

The spatial average of end-to-end distance $\langle R \rangle$ should be zero, since probability to bend to one direction is the same as that to the opposite direction. Then spatial average of square end-to-end vector is obtained as eq 2

$$\left\langle \mathbf{R}^2 \right\rangle = N^2 \sum_{i=1}^{k} \sum_{j=1}^{k} \left(\varphi_i \mathbf{r_i} \right) \cdot \left(\varphi_j \mathbf{r_j} \right) = N^2 b^2 D_b \tag{2}$$

$$D_b \equiv \left\langle \mathbf{R}^2 \right\rangle / L^2 \cong \sum_{i=1}^{k} \varphi_i^2 \tag{3}$$

where D_b is a bending ratio, $\varphi_i = N_i / N$, N_i is the number of unit segment in i-direction seg-ment, N is the total number of unit segment, $k = m + 1$, m is the number of static bending points on a coil, and r_i is i-direction segment vector with the length of b. The expression shown in eq 3 is significant. This indicates that we can obtain the distribution function when we have enough data. This is often called as ill-posed problem. Regularization method in applied mathematics gives us the solution for solving the problems. Equation 3 holds only if a probability of the fold-back conformation is the same as that of the straight conforma-tion.By using the scaling law, the coil expressed in eq 2 and 3 can be renormalized into the coil that has constant segment length, $2l_{p0}$. Then we can obtain eq 3 with $\varphi_i = 2l_{p0} / L$ and $k = L / 2l_{p0}$. We can also consider a case where the bent angle (θ) between the *ith* and (*i+1*)th segments is a fixed small angle. The spatial average of the square end-to-end vector is ob-tained as following

$$\left\langle \mathbf{r}_n \cdot \mathbf{r}_m \right\rangle = N \left(\sum_{i=1}^{k} \varphi_i^2 \right) b^2 \left(\cos\theta \right)^{|n-m|} \tag{4}$$

$$\left\langle \mathbf{R}^2 \right\rangle = \sum_{n=1}^{k} \sum_{m=1}^{k} \left\langle \mathbf{r}_n \cdot \mathbf{r}_m \right\rangle = \sum_{n=1}^{k} \sum_{p=-n+1}^{k-n} \left\langle \mathbf{r}_n \cdot \mathbf{r}_{n+p} \right\rangle \cong \sum_{n=1}^{k} \sum_{p=-\infty}^{\infty} \left\langle \mathbf{r}_n \cdot \mathbf{r}_{n+p} \right\rangle \tag{5}$$

$$\sum_{p=-\infty}^{\infty} \left\langle \mathbf{r}_n \cdot \mathbf{r}_{n+p} \right\rangle = N \left(\sum_{i=1}^{k} \varphi_i^2 \right) b^2 \left(1 + 2 \sum_{p=1}^{\infty} \cos^p \theta \right) = N \left(\sum_{i=1}^{k} \varphi_i^2 \right) b^2 \left(\frac{1 + \cos\theta}{1 - \cos\theta} \right) \tag{6}$$

$$\left\langle \mathbf{R}^2 \right\rangle = (N^2 b^2)(\sum_{i=1}^{k} \phi_i^2)(\frac{1 + \cos(\theta)}{1 - \cos(\theta)}) = L^2 D_b \tag{7}$$

$$D_b \equiv \frac{\left\langle \mathbf{R}^2 \right\rangle}{L^2} \cong \left(\sum_{i=1}^{k} \varphi_i^2 \right) \left(\frac{1 + \cos(\theta)}{1 - \cos(\theta)} \right) \tag{8}$$

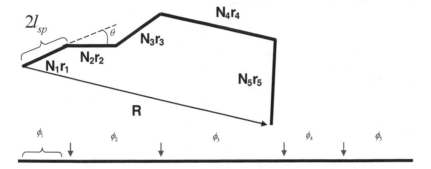

Figure 2. Tortuous MWCNT; bent points are distributed randomly along MWCNT axis.

Equation 7 can also be renormalized into the coil that has a constant segment length, $2l_{sp}$. The bending ratio (D_b) is expressed as eq 9

$$D_b \equiv \frac{\langle \mathbf{R}^2 \rangle}{L^2} \cong \left(\frac{2l_{p0}}{L} \right) \left(\frac{1 + \cos(\theta)}{1 - \cos(\theta)} \right) = C \left(\frac{2l_{p0}}{L} \right) = \frac{2l_{sp}}{L} \tag{9}$$

where $l_{sp} = Cl_{p0}$ is the static bending persistence length and C should be a constant for a fixed bent angle. The static bending persistence length is a statistical quantity, representing the maximum straight length that is not bent by static bending. In the case of continuous curvature, a more accurate statement is that the static bending persistence length is the mean radius of curvature of the rigid random-coil due to static bending. The same quantity arising from dynamic bending instead of static bending is dynamic bending persistence length (l_p). The dynamic bending persistence length represents the stiffness of the molecules as determined by the effective bending modulus against thermal energy in Brownian motion. Equation 5 is valid when $L \gg l_{sp}$, the coil limit. $D_b = \langle \mathbf{R}^2 \rangle / L^2 = 1$ when $L < l_{sp}$, the rod limit. If we know the values of end-to-end distance and contour length, the bending ratio can be obtained from the mean-squared end-to-end distance divided by the mean-squared contour length. The end-to-end distance of RRC varies with the change of bending angle. The difference can be compromised by using an arbitrary unit segment length which is similar to the scaling of polymer chain. The mean-squared end-to-end distance by the Kratky-Porod (KP) expression is given by eq 10 when the dynamic bending persistence length (l_p) is replaced by the static bending persistence length (l_{sp}) and the twice l_{sp} equals to Kuhn length.

$$\langle \mathbf{R}^2 \rangle = 2l_{sp}L + 2l_{sp}^2 \left(e^{-L/l_{sp}} - 1 \right) \tag{10}$$

4. Measurement methods for SBPL

The plot of eq 10 is presented in Fig. 3. Given data, the SBPL can be obtained by eq 11.

$$l_{sp} = \lim_{L \to \infty} \frac{1}{2} \frac{dD_b}{d \ln L} \tag{11}$$

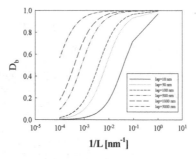

Figure 3. Bending ratio (D_b) with respect to reciprocal contour length.

In this method, one need to have experimental data for $\langle R^2 \rangle$ and L. In order to obtain these data, one have to cut MWCNTs into pieces with various L. Acid cutting or mechanical cutting method may be applied to obtain pieces of MWCNT. It is worth to note that $\langle R^2 \rangle$ are Gaussian, given contour length (L). That is, various end-to-end distances may be measured for a constant L. This method is exact, but hard to obtain the experimental data.

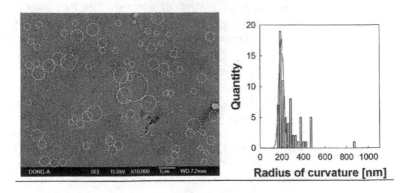

Figure 4. Approximation method to determine SBPL; the mean radius of curvature approximate SBPL.

The mean radius of curvature approximates the SBPL. One can easily obtain the mean value of the radius of curvatures of MWCNTs from any SEM images as seen in Fig. 4. The approximation method is convenient because SEM images of as-synthesized or as-received MWCNT can be directly used. The SBPL obtained by the approximation could have an error up to 200% compared to those obtained by exact method. However, the approximated value of SBPL still has physical significant in many applications, since many applied properties depend on the order of magnitude of SBPL.

5. Intrinsic viscosity of MWCNTs

From the molecular weight, the contour length, and the persistence length, the intrinsic viscosity of MWCNTs can be calculated. If we apply the intrinsic viscosity model of a worm-like coil to the rigid random-coil, the following expressions are obtained,

$$[\eta] = 2.20 x 10^{21} \frac{\langle R^2 \rangle^{3/2}}{M} f \tag{12}$$

$$f \cong \left[1 + 0.926\theta \left(D_b \right)^{1/2} \right]^1 \tag{13}$$

$$\theta = \ln \left(\frac{2l_{sp}}{e} \right) - 2.431 + \left(e / a \right) \tag{14}$$

where M is molecular weight, e is spacing between frictional elements along the contour, $a = \zeta / 3\pi \eta_s$, ζ is the friction factor for a single frictional element, and η_s is the solvent viscosity. For the non-draining limit for the random coil, $f = 1$, giving the maximum value of intrinsic viscosity in the model. When we take the static bending persistence length (l_{sp}) as the length of a single frictional element, the friction factor of the element in eq 14 may follow the rigid-rod model such that $\zeta_T = 3\pi \eta_s l_{sp} / (\ln(l_{sp}/d) + 0.3)$ for the translational motion and may be $\zeta_r = \pi \eta_s l_{sp}^3 / (3(\ln(l_{sp}/d) - 0.8))$ for the end-over-end rotational motion. Translational-rotational coupling and hydrodynamic shielding may also be considered for the evaluation of friction factor in eq 14. In this case, we can surmise that friction factor in eq 14 is scaled with l_{sp}^s, where s is larger than unit value. We can reasonably neglect e/a in eq 14. The measurement of intrinsic viscosity assumes the deformation rate is slow enough. The intrinsic viscosity is determined by the competition of tendency of orientation toward flow direction and tendency to random orientation due to thermal motion (Brownian motion). The measurement often performed at shear rate of several hundreds reciprocal second. At this regime, CNTs may be extended to the static shape by shear force where peclet number

$(Pe = \dot{\gamma}(2R_h)^2/D_T)$ is over 10.It is worth noting that the static bending persistence length determined from intrinsic viscosity is consistent with that determined from 3-D SEM analysis in dried state.

6. Diffusions of MWCNTs

Not only the toxicological issues but also researches on novel hybrid materials or nano-scale devices points to the need for the understanding of overall shape and mobility of carbon nanotube particles in a solution or in atmosphere.The degree of flexibility of carbon nanotubes is the major ingredient for the shape and mobility, however it is also puzzling.The persistence lengths of single-walled carbon nanotubes are expected to be in the order of tens to hundreds of micrometers due to their exceptionally large modulusand to have longer persistence lengths for muliwalled nanotubes, indicating currently prepared several-micrometer long nanotubes behave like rigid rods. Elastic fluctuations of semi-rigid particles by thermal energy have been described exactly by the worm-like coil model proposed more than 50 years ago by Kratky and Porod. The model describes the stiffness of molecules by dynamic bending persistence lengths (mean radius of curvatures) which are determined by effective bending modulus (E_{eff}) against thermal energy (kT) in a solution. Theoretical calculationshave shown that the dynamic bending persistence lengths (l_ps) of carbon nanotubes (CNTs) are up to several millimetersdue to their exceptionally large Young's modulus of about 1.5 TPa. Real-time visualization technique revealed that l_p s of singlewalled carbon nanotubes (SWCNTs) are between 32 and 174μm, indicating SWCNTs shorter than $l_p(=32\mu m)$ may be rigid around room temperature in a solution. However, rippling developed on the compressive side of the tube leading to a remarkable reduction of the effective bending modulus, which is more pronounced for multiwalled carbon nanotubes (MWCNTs). Theoretical calculationshave shown that the effective bending moduli of MWCNTs are around 0.5 $nN\ nm^2$ when the radii of curvatures are around 150 ~ 500nm. This indicates MWCNTs longer than 0.5 μmmight be flexible in a solution around room temperature, since thermal energy is about 4.1 x $10^{-3}nN\ nm$. It seems not likely that van der Waals interaction between graphene layers is the only reason that makes the effective bending moduli of MWCNTs more than 100 times smaller than SWCNT.

Both MWCNTs and SWCNTs discussed above are no more than worm-like coils (WLCs) where ensemble average of overall size (end-to-end distance) scales with the square root of molecular weight (contour length) in asymptotic limit. Our recent work has revealed that the spatial average of overall size of MWCNTs also follows the same scaling as WLCs in spite of their static bent points. We designated these MWCNTs as rigid random-coils (RRCs).The only difference between RRCs and WLCs is whether the bending points are static or dynamic by thermal energy.The relationship between the shape and size of RRCs has been characterized by static bending persistence lengths (l_{sp}s). Because both RRCs and WLCs are Gaussian, the models for the mobility of WLCs may also work for RRCs.

Translational diffusion coefficient is defined by the mobility of particle against thermal energy as Einstein relation, eq 15.

$$D = \frac{kT}{\zeta}$$ (15)

where k is Boltzman constant, T is temperature, and $1/\zeta$ is the mobility. By analogy to macromolecule, a MWCNT with static bend points can also be considered to be made up N identical structural elements with a frictional factor ζ_e per unit element and a spacing e between elements along the contour of the coil.In this case, the mobility in eq 15 may be expressed as the sum of free-draining contribution $(1/N\zeta_e)$ and hydrodynamic interaction contribution which is called non-draining term.

$$\frac{1}{\zeta} = \frac{1}{N\zeta_e} + \frac{1}{2!N^2}\sum_i\sum_j\frac{1}{\zeta_{ij}} + \cdots$$ (16)

where ζ_{ij} is the frictional factor by interaction between ith and jth element and $i \neq j$.

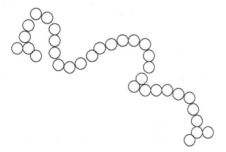

Figure 5. CNT made up N identical frictional element.

When we choose a spherical bead having diameter of a as a frictional element, the frictional factor of each element follows Stokes-Einstein relation, $\zeta_e = 3a\pi\eta_s$ where η_s is the viscosity of solvent. The frictional factor by interaction between ith and jth element may also follow Stokes-Einstein relation since a mean value of distance between elements i and $j$is small. Then, Kirkwood expression is obtained such as eq 17.

$$\frac{1}{\zeta} = \frac{1}{N3a\pi\eta_s}\left[1 + \frac{a}{2Ne}\sum_i\sum_j(1-\delta_{ij})\langle r_{ij}^{-1}\rangle\right]$$ (17)

$$\frac{1}{\overline{R}_{ij}} = \frac{1}{e\overline{r}_{ij}} = \left(\frac{1}{e}\right)\sum_i\sum_j\left\langle\left(1-\delta_{ij}\right)\left\langle r_{ij}^{-1}\right\rangle\right\rangle \tag{18}$$

where $r_{ij} \equiv R_{ij}/e$ and R_{ij} is the distance between element i and j,e is solvated diameter of molecule, and N is number of frictional element, $N = L / e$. Here, we can see that mathematical expression for the mobility of RRCs is similar to that of WLCs. Equation 17 is widely used for the estimation of translational diffusion coefficient of macromolecules. Equation 18 has been solved by Hearst and Stockmayer using Riseman-Kirkwood theory and Daniel distribution. We notice that $\overline{r}_{ij}$ is no more than a mean value of distance between the frictional element i and j. Then, $\overline{r}_{ij}$ must depend on the conformation of the carbon nanotubes. Hearst and Stockmayer obtained $\overline{r}_{ij}$ by the Kirkwood-Riseman theory as eq 19.

$$\frac{1}{2}\sum_x\sum_n\left\langle\left(1-\delta_{xn}\right)\left\langle r_n^{-1}\right\rangle_x\right\rangle = \int_1^N dl\int_1^l dn\int_0^\infty F(r,n)rdr = \int_1^N dl\int_1^l dn\int_0^\infty\left[F(r,n)-f(r,n)\right]rdr + \int_1^N dl\int_1^l dn\int_0^\infty f(r,n)rdr \tag{19}$$

where $F(r, n)$ is the unknown distribution for all n, $f(r, n)$ is the known distribution, x is the contour distance of the point of interest from one end of the carbon nanotube, n is the contour distance from the point of interest to the frictional element n, and r is the displacement of frictional element n from the point of interest. Hearst and Stockmayer chose the Daniels distribution which includes a first-order correction to a Gaussian distribution as $f(r, n)$, and obtained $\overline{r}_{ij}$ as eq 20.

$$\overline{r}_{ij} = \left[2N\left(\ln\left(2L_p\right)-2.431+1.843\left(N/2L_p\right)^{1/2}+0.138\left(N/2L_p\right)^{-1/2}-0.305\left(N/2L_p\right)^{-1}\right)\right]^{-1} \tag{20}$$

where $N = L / e$, L is contour length, e is spacing between frictional elements along the contour, $L_p = l_p/e$, and l_p is persistence length. The translational diffusion coefficient of worm-like coil can be estimated by eqs 15, 17, and 20. Rotational diffusion coefficient is expressed as eq 21.

$$D_r = \left(\frac{kT}{\eta_s}\right)\left(\frac{2}{l_pL^2}\right)\left[0.253\left(\frac{L}{4l_p}\right)^{1/2}+0.159\ln\left(2L_p\right)-0.387+0.160\right] \tag{21}$$

Equations 20 and 21 are valid for a semi-flexible rod when the contour length of rod is much longer than its persistence length such that the mean squared end-to-end distance follows random-coil scaling, $\langle R^2\rangle = Nb^2$ where $b = 2l_p$. The semi-flexible rods in this coil limit, $L >> l_p$, are so-called worm-like coils (WLCs). We see that the mobility is determined solely by the average conformation of particle with a given solvent viscosity and a contour length in eqs 13 and 14. We can reasonably surmise that the diffusion coefficients of RRCs are similar to those of WLCs with a given contour length, if the values of static bending persistence lengths (l_{sp}s) of RRCs are the same as those of the dynamic bending persistence lengths (l_ps) of WLCs.When hydrodynamic shielding effect is taken into account, the diffusion coefficients of RRCs might be slightly larger than those of WLCs due to the static bent points. The root mean-squared end-to-end distance of RRCs are given by eq 22.

$$\left\langle \mathbf{R}^2 \right\rangle = \left(N^2 b^2\right)\left(\sum_{i=1}^{k} \varphi_i^2\right)\left(\frac{1+\cos(\theta)}{1-\cos(\theta)}\right) = L^2 D_b \tag{22}$$

$$D_b \equiv \frac{\left\langle \mathbf{R}^2 \right\rangle}{L^2} \cong \left(\sum_{i=1}^{k} \varphi_i^2\right)\left(\frac{1+\cos(\theta)}{1-\cos(\theta)}\right) \cong \left(\frac{2l_{p0}}{L}\right)\left(\frac{1+\cos(\theta)}{1-\cos(\theta)}\right) = C\left(\frac{2l_{p0}}{L}\right) = \frac{2l_{sp}}{L} \tag{23}$$

where D_b is bending ratio, l_{sp} is static bending persistence length, l_{p0} is an arbitrary constant segment length, θ is static bent angle from the MWCNT axis, $\varphi_i = N_i / N$, N_i is the number of unit segment in i-direction segment, N is the total number of unit segment, $k = m + 1$, m is the number of static bending points on a coil. When RRCs have semi-flexibility by thermal energy, the ensemble average of bent angle (θ_2) always becomes larger in amount of $\Delta\theta$ than the static bent angle θ; $\theta_2 = \theta + \Delta\theta$. This is due to the fact that the effective bending modulus toward the bent direction is smaller than that toward the opposite direction.This indicates that the overall size of RRCs may be decreased when they are fluctuated by thermal energy. Because frictional elements of RRCs have Gaussian distribution by the definition of RRCs, those of semi-flexible RRCs also have Gaussian distrbution. Therefore, eqs 20 and 21 are also valid for semi-flexible RRCs when the persistence length is replaced by an apparent persistence length. The apparent persistence length (l_{ap}) is determined by the static bent angle (θ) and dynamic bent angle due to thermal energy ($\Delta\theta$) as following.

$$l_{ap} = l_{sp}\left(\frac{1+\cos(\theta + \Delta\theta)}{1-\cos(\theta + \Delta\theta)}\right)\left(\frac{1-\cos(\theta)}{1+\cos(\theta)}\right) \tag{24}$$

Expression for the translational diffusion coefficient of RRCs can be obtained from eqs 15, 17,18, and 24.

$$D_T = \frac{kT}{3\pi\eta_s L}\left[1+\ln\left(2L_{ap}\right)-2.431+1.843\left(N/2L_{ap}\right)^{1/2}+0.138\left(N/2L_{ap}\right)^{-1/2}-0.305\left(N/2L_{ap}\right)^{-1}\right] \tag{25}$$

Similary, expression for the rotational diffusion coefficient of RRCs can be obtained as following.

$$D_r = \left(\frac{kT}{\eta_s}\right)\left(\frac{2}{l_{ap}L^2}\right)\left[0.253\left(\frac{L}{4l_{ap}}\right)^{1/2} + 0.159\ln\left(2L_{ap}\right) - 0.387 + 0.160\right] \tag{26}$$

where $L_{ap} = l_{ap}/e$. Equations 25 and 26 are promising for the estimation of diffusion coefficients of MWCNTs synthesized by a CVD method. In other words, eqs 25 and 26 give us the information of the shape and size of MWCNTs if we have the measured values of diffusion coefficients.

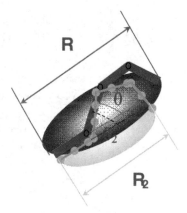

Figure 6. Size of carbon nanotube (R) is decreased to R2 at the elevated temperature.

7. Dynamic light scattering

The translational and rotational Brownian motions lead to fluctuation in the intensity of scattered light. The velocity of particles in Brownian motion can be directly measured by using dynamic light scattering (DLS) method, since time is correlated to obtain the intensity autocorrelation function ($g^{(2)}(t)$). The intensity autocorrelation function is connected to the electric field autocorrelation function ($g^{(1)}(t)$) which is given by eq 27 for a monodisperse solution.

$$\left|g^{(1)}(t)\right| = \exp(-\Gamma t) \tag{27}$$

$$\int_0^\infty G(\Gamma)d\Gamma = 1$$

where $\Gamma = Dq^2$ with D, the translational diffusion coefficient of the molecules, and q, the scattering vector magnitude ($q = 4\pi n \sin(\theta/2)/\lambda_0$ where n is the solution refractive index, θ is the scattering angle, and λ_0 is the incident light wavelength *in vacuo*). For polydisperse solutions, the electric field correlation function is given by a sum or distribution of exponentials,

$$\left|g^{(1)}(t)\right| = \int_0^\infty G(\Gamma)\exp(-\Gamma t)d\Gamma \tag{28}$$

where

In many cases, a single exponent obtained from an average translational diffusion coefficient value fits the decay rate of the electric field correlation function such as eq 29

$$K_1 = \langle\Gamma\rangle_{av} = \int_0^\infty \Gamma G(\Gamma)d\Gamma \tag{29}$$

where K_1 is the first cumulant. When a solution is dilute enough neglecting the interaction between particles, the effective diffusion coefficient can be obtained by the intensity average. The intensity of light scattered by macromolecule species i is often proportional to the molecular weight (M_i) times the weight concentration (c_i). In this case, the intensity average diffusion coefficient equals to z average diffusion coefficient.

The average decay rate (Γ) of the electric field autocorrelation function can be obtained by using conventional DLS. The first cumulant generally fits the data well for carbon nanotube solutions. When the incident light and detector are both vertical, V_v, translational diffusion are characterized by eq 30.

$$\Gamma_{Vv} = q^2 D_T \tag{30}$$

When the incident light vertical and detector horizontal, H_v, the diffusion of anisotropic particle are characterized by eq 31

$$\Gamma_{Hv} = q^2 D_T + 6D_R \tag{31}$$

where D_T is translational diffusion coefficient and D_R is rotational diffusion coefficient. This equation is valid if the particle rotates many times while diffusing a distance comparable to q^{-1} or if there is little anisotropy in particle dimension. MWCNTs solution meets with the former case in this work. Now we have three independent mathematical model eqs 25, 26, and 12. And two equations for DLS measurement.Three unknown shape factors of static bending persistence length, contour length, and thermal fluctuation angle can be determined from the measured diffusion coefficients and intrinsic viscosity using eqs 25, 26, and 12.This is uniqueness of carbon nanotubes compared to macromolecules, since macromolecules have only two unknown shape factors of persistence length and contour length.

8. Micro rheology

The terminology of "microrheology" is used, to distinguish the technique from conventional (macro) rheology. In the microrheology, colloidal particles are used for probing the rheology of material of interest. The starting point is the Stokes-Einstein relation.

$$D_T = \frac{kT}{6a\pi\eta_s} \tag{32}$$

If we have measured values of translational diffusion coefficient, the viscosity of material of interest can be easily obtained by

$$\eta_s = \frac{kT}{6a\pi D_T} \tag{33}$$

where a is the radius of spherical colloidal particle; colloidal particles have usually average diameter between 1 nm and 1000 nm. It is comparable the ISO definition of nanoparticles those have average diameter between 1 nm and 100 nm. In this sense, nanoparticles are just some kinds of colloidal particles. When we have measured value of mean-squared displacement of probe particle, the following equation can be utilized.

$$\left\langle r^2\left(t\right)\right\rangle = 6D_T t = \frac{kT}{\pi a\eta_s} \tag{34}$$

This seemingly simple idea has done a great impact on various research fields, indeed. One example is the nanoparticles dispersed in a polymer melt. It is often reported that nanoparticles seems diffuse faster than expected. The origin of this phenomenon lies in the "Nano" size. The viscosity of polymer melt is well described by integral constitutive equations such as reptation model. In this model, the viscosity is determined by the stress relaxation time of polymer chain from the constraint of entanglement. When the observation time is much shorter than any relaxation time of polymer in rheometry of frequency sweep, the polymers behave like a crosslinked rubber, exhibiting a plateau modulus. The plateau modulus of polymers is determined from the entanglement lengths of polymer such as

$$G_o^N = \frac{\rho RT}{M_e} \tag{35}$$

The plateau modulus of polymer is usually reported in the order of 10^{6-7}Pa. Entanglement molecular weight of polymer is about 1000~2000 g/mole. The entanglement length is about 10~100 nm.The particles having comparable size to the entanglement length of a polymer would feel less frictional force than expected from the melt viscosity in macrorheology. Therefore, viscosity of polymer melt is much lower for the nanoparticles. This may lead to the faster thermal motion of nanoparticle compared to a larger particles.

9. Applications

In our previous works, we demonstrated that MWCNT having shorter SBPL have a certain merit in a polymer composite for electrical conductive application. When MWCNTs are needle-like, polymer composites comprised of them exhibit higher electrical conductivity compared to those comprised of tortuous MWCNTs. The situation changed drastically when the composites were molded into a specimen by injection molding machine. The needle like CNTs aligned to the flow direction, which broke the electrical conductive networks, then the composites lose the electrical conductivity. However, this problem was not observed when the composites contained tortuous MWCNTs which have a short SBPL. We also showed that the electrical percolation threshold depends on the length of MWCNT when MWCNT are needle-like. But, the electrical percolation threshold depends on the SBPL for a tortuous MWCNT. Thermal conductivity and linear thermal expansivity are also strongly dependant properties on the SBPL of MWCNT. Especially, thermal shrinkable material can be fabricated as well as thermal expansive material by controlling SBPL of MWCNT.

Author details

Heon Sang Lee

Address all correspondence to: heonlee@dau.ac.kr

Dong-A University, Department of Chemical Engineering, Sahagu, Busan, South Korea

References

[1] Lee, H. S.; Yun, C. H.; Kim, S. K.; Choi, J. H.; Lee, C. J.; Jin, H. J.; Lee, H.; Park, S. J.; Park, M. Appl. Phys. Lett. 2009, 95, 134104.

[2] Lee, H. S.; Yun, C. H. J. Phys. Chem. C 2008, 112, 10653-10658.

[3] Lee, H. S.; Yun, C. H.; Kim, H. M.; Lee, C. J. J. Phys. Chem. C 2007, 111, 18882- 18887.

[4] Han, M. S.; Lee, Y. K.; Yun, C. H.; Lee, H. S.; Lee, C. J.; Kim, W. N. Synthetic Metals 2011, 161, 1629-1634.

[5] Heo, Y. J.; Yun, C. H.; Kim, W. N.; Lee, H. S. Curr. Appl. Phys. 2011, 11, 1144-1148.

[6] Hearst, J. E.; Stockmayer, W. H. J. Chem. Phys. 1962, 37, 1425-1433.

[7] Squires, T. M.; Mason, T. G. Ann. Rev. Fluid Mech. 2010, 42, 413-438.

[8] Hearst, J. E. J. Chem. Phys. 1963, 38, 1062-1065.

[9] Hearst, J. E. J. Chem. Phys. 1964, 40, 1506-1509.

[10] Kratky, O; Porod, G. Recl. Trav. Chim. Pays-Bas 1949, 68, 1106-1122.

[11] Doi, M.; Edwards, S. F. The theory of polymer dynamics; Oxford University Press: New York, U.S., 1986.

[12] Berne, B. &Pecora, R. Dynamic Light Scattering; Wiley: New York, 1976.

[13] Tanford, C. Physical Chemistry of Macromolecules; John Wiley & Sons: New York, 1961.

[14] Koppel, D. E. J. Chem. Phys. 1972, 57, 4814-4820.

[15] Cush, R.; Russo, P. S.; Kucukyavuz, Z.; Bu, Z.; Neau, D.; Shih, D.; Kucukyavuz, S.; Ricks, H. Macromolecules1997, 30, 4920-4926.

Chapter 5

Large Arrays and Networks of Carbon Nanotubes: Morphology Control by Process Parameters

I. Levchenko, Z.-J. Han, S. Kumar, S. Yick, J. Fang and
K. Ostrikov

Additional information is available at the end of the chapter

1. Introduction

Large arrays and networks of carbon nanotubes, both single- and multi-walled, feature many superior properties which offer excellent opportunities for various modern applications ranging from nanoelectronics, supercapacitors, photovoltaic cells, energy storage and conversation devices, to gas- and biosensors, nanomechanical and biomedical devices etc. At present, arrays and networks of carbon nanotubes are mainly fabricated from the pre-fabricated separated nanotubes by solution-based techniques. However, the intrinsic structure of the nanotubes (mainly, the level of the structural defects) which are required for the best performance in the nanotube-based applications, are often damaged during the array/network fabrication by sur-factants, chemicals, and sonication involved in the process. As a result, the performance of the functional devices may be significantly degraded. In contrast, directly synthesized nanotube arrays/networks can preclude the adverse effects of the solution-based process and largely pre-serve the excellent properties of the pristine nanotubes. Owing to its advantages of scale-up production and precise positioning of the grown nanotubes, catalytic and catalyst-free chemi-cal vapor depositions (CVD), as well as plasma-enhanced chemical vapor deposition (PECVD) are the methods most promising for the direct synthesis of the nanotubes.

On the other hand, these methods demonstrate poor controllability, which results in the unpre-dictable properties, structure and morphology of the resultant arrays. In our paper we will dis-cuss our recent results obtained by the application of CVD and PECVD methods. Specifically, we will discuss carbon nanotube arrays and networks of very different morphology. The fabri-cation of the arrays of vertically aligned and entangled nanotubes, as well as arrays of arbitrary shapes grown directly on the pre-patterned substrates will be considered with a special atten-tion paid to the fabrication methods and the influence of the process parameters on the array growth morphology (see Figure 1). Besides, the possibility of creating the 3D structures of carbon nanotubes through post-processing of the arrays by liquids will be discussed.

The fabrication methods involved are the conventional CVD utilizing various gases such as methane, ethane, acetylene, argon, and hydrogen; plasma-enhanced CVD based on inductively-coupled low-temperature plasma reactor; microwave PECVD. The advantages of the plasma-based CVD process will be shown and discussed with a special attention. We will also discuss the influence of the process parameters such as process temperature, pressure, gas composition, discharge power etc. on the morphology of the nanotube arrays and networks, and demonstrate that the proper selection of the parameters ensures very high level of the process controllability and as a result, sophisticated control and tailoring of the growth structure and morphology of the carbon nanotube arrays.

Characterization technologies used are scanning and transmission electron microscopy (SEM and TEM), as well as atomic force microscopy (AFM), Raman and X-ray photoelectron spectroscopy techniques. The results of the numerical simulations will also be used to support the growth models and proposed growth mechanisms.

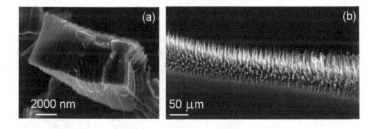

Figure 1. Morphologies of the representative CNT arrays grown by CVD (a) and PECVD (b).

2. CVD versus PECVD: Morphology control issues

2.1. CVD and PECVD: General

The term 'chemical vapor deposition', or 'CVD', is commonly used for describing the processes and chemical reactions which occur in a solid material deposited onto a heated substrate using a gaseous precursor. However, more complicated process than the 'common' CVD takes place [1,2] during the growth of CNTs. In this case, the carbon-containing gaseous precursors (e.g., CH_4, C_2H_4, C_2H_2, CO) firstly dissociate into atomic or molecular carbon species on the surface of catalyst nanoparticles, and then the nucleation occurs as these carbon species diffuse into the catalyst nanoparticles, reach a supersaturated state, and then segregate from the surface of nanoparticles to form a nanotube cap. Subsequently, the growth of nanotubes is sustained by the continuous incorporation of carbon atoms *via* bulk and/or surface diffusion. Figure 2 shows the SEM image of randomly-oriented SWCNTs with a unique 'bridging' morphology in catalytic

CVD. This array was grown by using Ar/H$_2$/CH$_4$ gas mixture on Fe/Al$_2$O$_3$ catalyst. By tuning the growth condition (e.g., temperature and pressure), it was demonstrated that the SWCNTs could be of a high quality (a high I$_G$/I$_D$ in the Raman spectra) and could contain a significantly higher content of metallic nanotubes as compared to the 'standard' metallic nanotube content of 33% (1/3 metallic and 2/3 semiconducting) produced in many CVD processes [3].

On the other hand, PECVD refers to the CVD process that uses plasma environment as an extra dimension to control the growth of CNTs. Plasma by definition contains ionized species and is generally considered as the fourth state of matter along with solid, liquid and gas. Recent advances in the plasma-based nanofabrication offer unprecedented control over the structure and surface functionalities of a range of nanomaterials [4]. One of the major advantages, as compared to the conventional CVD processes, is that nanostructures can grow vertically-aligned due to the electrical field in the vicinity of surface [5]. Another benefit of using plasma is that the temperature required to dissociate carbon feedstock could be greatly reduced [6]. Figure 3 illustrates the isolated CNTs grown in a PECVD system using Ni/SiO$_2$ as the catalyst, C$_2$H$_2$/NH$_3$ as the gas precursors, and a DC glow discharge. It can be seen clearly that these nanotubes are aligned vertically to the substrate surface, due to the plasma sheath-directed growth. These freestanding nanotubes could give many opportunities to custom-design novel functional devices.

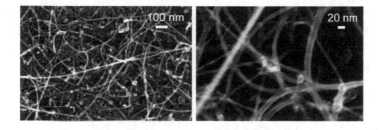

Figure 2. Typical randomly-oriented SWCNT networks with a unique "bridging" morphology grown in catalytic CVD [3].

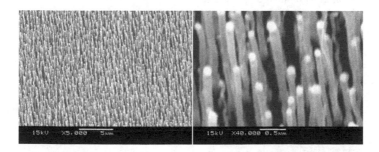

Figure 3. Low- and high-resolution SEM images of the typical arrays of vertically-aligned CNTs grown in PECVD process with a glow discharge. The growth followed the 'tip-growth' mode as the catalyst nanoparticles are noticeable on the top of each nanotube [4].

2.2. Morphologies of nanotube arrays

In general, there are three types of morphologies observed in the directly-grown nanotube arrays: entangled, horizontally aligned, and vertically aligned. Each of these morphologies has their specific functionalities and can be desirable for different applications. In this work, we will briefly describe the first two morphologies and then pay the most attention to the arrays of vertically aligned nanotubes.

The horizontally-aligned CNT arrays were usually grown on the quartz wafers using CVD. The alignment could in such arrays be controlled by two factors: gas flow direction and substrate lattice. These arrays could have a very high density (up to 50 SWNT/μm) over large area. These nanotubes have also a large diameter and good electrical properties desirable for the nanoelectronic applications [7].

On the other hand, the vertically aligned CNTs could be grown using both CVD and PECVD. *Hata et al.* demonstrated that by using Fe/Al_2O_3 as the catalysts, C_2H_4 as the feedstock and a trace amount of water vapor (100 – 300 ppm) as the growth enhancer, high-yield, milli-meter long vertically aligned SWCNTs could be produced [8]. Water in this process was used to etch the possible amorphous carbon phase deposited onto the catalyst during the growth, therefore enhancing the lifetime and activity of the catalyst. The vertical alignment was supported by the collective van der Waals' interactions among the nanotubes [9]. In contrast, the CNTs grown in PECVD process do not require such growth enhancer to align them vertically, since the electrical field in the plasma-surface sheath at the vicinity of the substrate could easily direct the growth.

The third type of CNT arrays is the entangled network consisting of interconnected randomly-oriented nanotubes. In some cases, these networks are not entirely 'random'; instead, they may form certain unique features such as the 'Y-junctions', as well as 'knotted' and 'bridging' structures. *Sun et al.* demonstrated that by using a porous membrane filter to collect the nanotubes at room temperature, a unique 'Y-junction' with high electronic performance could be induced in an aerosol CVD process [10]. Similar to VACNTs, they can be produced in both CVD and PECVD processes. Figure 4 illustrates both the horizontally and vertically aligned morphologies obtained by our group.

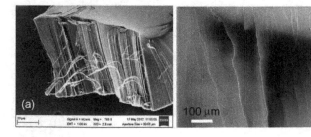

Figure 4. Highly uniform, dense array of vertically aligned single-walled carbon nanotubes (SWCNTs) grown on trilayered $Fe/Al_2O_3/SiO_2$ catalyst (a). Horizontally-aligned nanotubes (b).

2.3. Vertically-aligned arrays of carbon nanotubes

Vertically-aligned CNTs not only preserve the excellent intrinsic properties of individual nanotubes, but also show a high surface-to-mass ratio owing to their three-dimensional microstructure. Moreover, the surface of the vertically-aligned CNTs could be easily function-alized. These advantageous features have placed the vertically-aligned CNT arrays among the most promising materials for a variety of applications ranging from field emitters, heat sinks, nanoelectrochemical systems, gas- and bio-sensors, drug delivery systems, to molecular/particular membranes. For example, Wu et al. used the functionalized vertically-aligned CNTs to deliver nicotine for therapeutic purposes [11]; Han et al. studied the release behaviors of bone morphogenetic protein-2 (BMP-2; a growth factor for human mesenchymal stem cells) on the vertically-aligned CNTs with different surface wettability, in attempting to control the differentiation and proliferation of these stem cells [12].

Growth of the vertically-aligned CNTs can be easily obtained in PECVD. Figure 5 shows SEM images (high and low magnification) of the vertically-aligned nanotubes grown in the low-temperature plasma [13]. These CNTs have a diameter of 50-200 nm, a height of several micrometers, and followed a "tip-growth" mechanism. Interesting, they collapse upon liquid wetting (this will be discussed in more detail in the next section).

Figure 5. Dense array of vertically aligned single-walled carbon nanotubes.

The the vertically-aligned CNTs grown using CVD process are much denser and longer, and have more uniform distribution of diameters. In such arrays, very strong Van der Waals forces are present. The CVD process is therefore suitable for mass production of CNTs, and may contribute to lowering the price of CNTs. We have recently demonstrated that highly uniform and dense arrays of SWCNTs with more than 90% population of thick nanotubes (>3 nm in diameter) could be obtained by tailoring the thickness and microstructure of the catalyst supporting SiO2 layer [14].

2.4. Entangled arrays of carbon nanotubes

Networks of entangled nanotubes consist of randomly-oriented nanostructures. The thickness of the entangled array may vary from sum-monolayer to a few monolayers. Advantages of

such morphology, as compared to individual nanotubes, are scalability, stability, reproducibility, and low cost of the CNT-based devices. They are therefore widely used as thin film
transistors, transparent conductive coatings, solar cells, gas and biosensors. The electrical
resistivity in entangled SWCNTs is determined by the nanotube-nanotube junctions in the
network, and the nanotube-metal junctions at the electrodes (so-called Schottky barrier). The
intrinsic resistance of the nanotubes usually plays a minor role if the array density is not far
away from the percolation threshold [15]. In addition, it is generally perceived that for the
CNT-based device to deliver outstanding performance, chirality-selected growth of CNT is a
pre-requisite. However, for entangled SWCNTs, this stringent requirement may be avoided if
the density is within a certain range (usually 1–3 nanotubes/μm^2) [16,17,18].

There are many parameters of the CVD process that should be controlled to grow entangled
CNTs with some special patterns. For example, the length of the nanotubes could be determined by the exposure time of the carbon feedstocks. Recently, we have demonstrated that
the density of entangled SWCNTs, which is a critical factor in device performance, could be
controlled over 3-order-of-magnitude in acetylene-modulated CVD processes (Figure 6a) [2].
In addition, we also obtained a special 'knotted' morphology of the CNT network by using
porous silica as the catalyst-supporting layer (Figure 6b) [19]. In contrast to this morphology,
a much lower density of nanotubes was observed on flat silica surface.

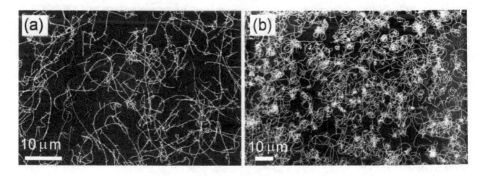

Figure 6. Representative arrays of entangled carbon nanotubes [2,19].

3. Complex catalyst-free arrays by mechanical writing

Plasma-based techniques yet being capable of producing high-quality nanotube arrays, still
require metal catalyst to initialize the control the nanotube growth process. However, there is
a strong demand in metal-free CNTs, i.e. the CNTs not containing a catalyst metal which is
usually incorporated in the nanotube structure (from the nanotube top or bottom, depending
on the process used). Removal of metal catalyst from CNTs implies a complex post-processing
[20] which results in significant disadvantages, such as essential change in electronic properties

or degradation of the nanotube ordering or orientation (in particular, post-processing deteriorates the vertical orientation of the nanotubes), damages the substrate structure in high temperature annealing process, etc. Thus, removal of the metallic catalyst by after-growth post-processing is feasible only for limited small-scale experimental production [21]. Hence, the development of the catalyst-free methods for growing arrays of high quality, dense vertically aligned nanotubes is a pressing demand now. The metal-free nucleation and growth of carbon nanotubes is possible, yet with the use of other catalytic material, and with a low quality outcome. For example, the nucleation and growth on semiconductor nanoparticles in CVD process was recently reported [22,23,24]. In these works, the nanotubes were catalyzed and grown without metal catalyst, but those nanotubes are not vertically aligned but highly tangled, tousled, and the surface density is quite low. Therefore, obtaining high quality arrays of CNTs on a catalyst-free silicon substrate still remains elusive.

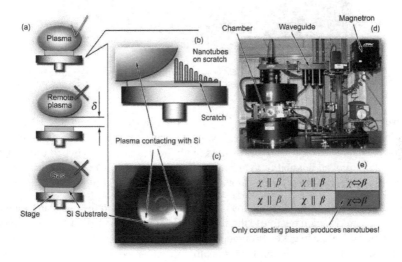

Figure 7. a) Three typical process configurations: localized plasma, remote plasma, gas environment; (b) nanotubes growth on Si substrate contacting with plasma: dense nanotubes as-grown on a doted spot; (c) photo of the plasma above substrate and (d) photo of the microwave reactor; (e) complete experiment matrix, which indicates the substrate condition (for scratched or non-scratched surface), and the process environment condition; remote gas/plasma and contacting gas/plasma. Among all possible 6 variants tested, only localized plasma process have produced nanotubes on substrate [25].

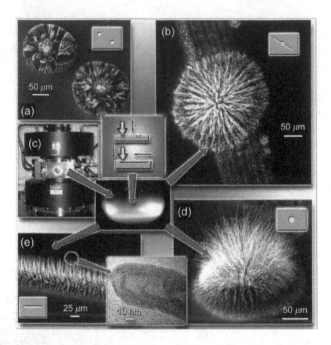

Figure 8. a,b) Top-view of CNTs on doted spots (SEM images); (c) microwave reactor; (d, e) tilted SEM image of CNT arrays showing a high number density of CNTs. Insets illustrate the process of making pattern and TEM image of the carbon nanotube [25].

Here we describe a novel plasma-based catalyst-free growth technique that is capable of producing very dense, strongly aligned arrays of extremely long (up to several hundred μm) CNTs on Si wafer surface in very fast process (with growth rate achieving 50 μm/sec), with experimentally proven possibility to arrange the nanotubes into complex arrays of various shapes such as separate nests and linear strands.

The six different experimental variations were used, with respect to the plasma/gas environments and plasma location relative to the substrate, as shown in Figure 7. We did not observe the nanotube nucleation in gas environment, on both smooth and patterned surfaces; we also did not observe the nucleation on both smooth and patterned surfaces with the remotely located plasma, and only the process conducted in plasma contacting the patterned surface resulted in the nucleation and growth of CNTs. The process starts by applying a special notch pattern (NP) on the prepared Si(100) wafers.

Then, the substrates with a specific NP (we used a linear NP consisting of parallel notches, and spot pattern of small pits) was treated in a chemical vapor deposition (CVD) reactor (Figure 8) where a microwave discharge was ignited in gas mixture of CH_4 and N_2, at pressure of 13 Torr and power density of 1.28 W/cm³, typically for 3 min. The substrates were heated up to ~800 °C only by the plasma. The plasma localization relative to the substrate was varied to study in detail the plasma effect on CNT growth process (Figure 1e); namely, the process conducting with plasma contacting the wafer surface was effective for nucleation and growth of CNTs. More details on making the mechanical pattern are shown in Figure 8.

The scanning electron microscopy (SEM) investigations (Figure 9) show that a fast growth of a high-density, highly aligned CNTs are produced exactly replicating the pattern configuration (a complex pattern configuration which consists of a linear notch and spot applied directly on the notch have been achieved). The SEM images clearly show that the complex pattern of CNTs was perfectly replicated by nanotubes and the longest nanotubes reach ~140 µm in length. The growth sites are very densely occupied, and the rest wafer surface is absolutely free of nanotubes. Notably, these very dense arrays were formed in a very fast process, such unusual growth rates (up to ~48 µm/minute) were not reported previously in the absence of metal catalyst. Figure 9a shows the high-magnification SEM image of the nanotube array.

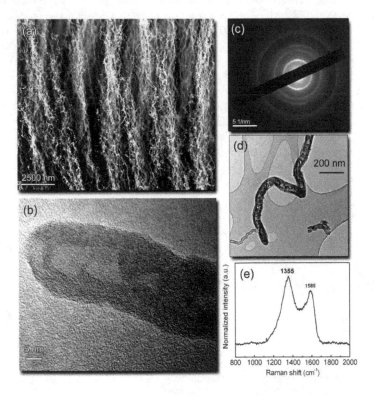

Figure 9. a) High-magnification SEM image of the vertically-oriented CNTs; (b) a high-resolution TEM image showing the planes in CNTs, with the inter-planer distance of ~0.34 nm; (c) the electron diffraction pattern of CNTs; (d) low-resolution TEM image showing the nanotube diameter of about 10 nm; (e) micro-Raman spectrum of the carbon nanotubes [25].

Further characterization of the nanotube structure was done with a high-resolution transmission electron microscopy (TEM) and Raman techniques (Figure 9). The TEM images (inset in Figure 8, Figures 9b and 9d) clearly show the absence of catalyst particle at the closed end tip of the CNTs, this reveals that the nanotubes were following in a "base-growth" mode [25]. As follows from TEM images, the diameters of the nanotubes are in the range of 10-80 nm, with up to 25 walls. Figure S13c shows the electron diffraction pattern of multi-wall nanotubes. Raman spectrum of as-grown nanotubes obtained at a room temperature (Figure 9e) shows a Raman broad-band peak at 1585 cm^{-1}, which is the characteristic of in-plane C-C stretching E2g mode of the hexagonal sheet. The appearance of a broad-band peak at 1355 cm^{-1} indicates the disordered graphitic nature of the nanotubes.

Thus, the nanotubes in our experiments were grown on the features mechanically written on the surface of Si wafer, and no nanotubes were formed on the intact silicon. To explain this, we propose a mechanism based on the key role of nano-elements on Si surface, so-called 'nano-hillocks'. These hillocks are formed on the surface when writing pattern, they establish a strong covalent bond to the Si surface at a temperature of ~800 °C during the process of CNT nucleation, and thus remain on Si surface, and hence at the bottom of nanotube during the whole growth process. Indeed, the solubility of carbon in Si is very low (10^{-3} %) [26] as compared to the conventional metallic catalyst such as Fe, Co, Ni etc., and thus the extremely high (up to 1 μm/s) growth rate observed in these experiments indicates that the nanotubes were grown via a surface diffusion, without involving very slow bulk-Si diffusion. Thus, a vapor-liquid-solid (VLS) mechanism was not involved, and the plasma played a key role in this process. We propose the following mechanism, so-called *reshaping-enhanced surface catalyzed* (RESC)growth. During the first stage, the tip region of a Si nano-hillock was heated up by plasma due to increased current density to the nano-sized tip (Figure 10).

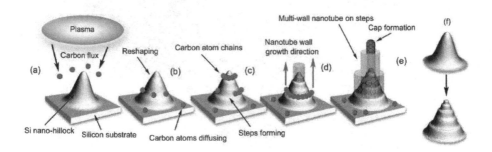

Figure 10. Scheme of the proposed mechanism of carbon nanotube nucleation and growth on silicon nano-hillocks in the plasma environment. (a) Si nano-hillock (with the shape 'as-produced' by mechanical patterning) is locally (mainly at the top) heated by the plasma; (b) heated Si nano-hillock starts reshaping – multiple step-like features are formed due to thermal re-arrangement and carbon saturation of the upper (overheated) Si layer; single carbon atoms incorporate into the steps; (c) reshaping continues, the steps become well-shaped, carbon atoms form chains (nanotube nuclei) along the multiple steps; (d) carbon chains close, nanotube start growing; (e) nanotube grow and close; (f) supposed reshaping of the silicon nano-hillock during plasma heating and nanotube nucleation [25].

Further, the heated silicon nano-hillock starts reshaping [27] by forming multiple step-like features due to thermal re-arrangement of silicon (to minimize the surface energy), and partially due to the possible carbon solution and saturation in the upper overheated Si layer. Then, carbon atoms incorporate into the steps and form closed chains. Simultaneously, the steps become well-shaped and thus carbon atoms assist the nanotube nucleation along the multiple steps. Later, multi-walled nanotubes start growing. Eventually, when the nanotube reaches 100-150 μm in length, the tip of carbon nanotube closes.

Thus, in this process the carbon catalization proceeds by the minimization of surface energy at the nano-hillock steps [28], since the adatom adsorbed in the step can be considered as 'partially dissolved'. As a result, this process leads to the formation of very dense array of very long multiwall nanotubes on the mechanically patterned areas. Thus, the proposed growth mechanism explains all the observed features; it is noteworthy that just the effect of plasma on the patterned surface explains several important characteristics due to plasma-related heating and high rate of material delivery. As a result, the catalyst-free, very dense arrays of long (up to 150 μm) vertically oriented multiwall carbon nanotubes were grown on the mechanically patterned silicon wafers in a low-temperature microwave discharge. These experiments have demonstrated an extremely high (up to 48 μm/min) growth rate.

4. Three-dimensional CNT arrays by post-processing with liquids

The above-described method can be used to produce 'planar', drawing-like arrays of the verti-cally-aligned carbon nanotubes on silicon surfaces. When a need in a complex three-dimen-sional array arises, post-processing of the uniform array (array-precursor) can be used. Among others, the post-treatment with a liquid is the most cheap and convenient [29-38]. Nevertheless, this technique still lacks controllability. In this section we show several possible ways of en-hancing controllability of the fabrication of three-dimensional structures of the vertically-aligned carbon nanotube arrays. Specifically, we show that the array structure can be a key factor of the resultant structure fabricated by immersing the CNT array into liquid.

Figure 11a is an SEM image of the cross-section of the array of vertically-aligned CNTs grown using the CVD technique. This array exhibits super hydrophobic properties and thus, it cannot be wetted by water. After immersing into water, only weakly-collapsed irregular structure was produced (Figure 11b). In contrast, this array does not show super-hydrofobicity to acetone, and thus, highly-regular completely collapsed pattern was produced by immersing this array into acetone (Figures 11c, 11d). As one can see in this figure, this pattern exhibits very high surface area of the 'sponge', produced by carbon nanotubes (and hence, the walls of this sponge can be highly-conductive or semi-conductive). Such structures could be very useful for the fabrication of gas and bio-sensors, gas storage devices, as well as energy-transforming applications requiring very high levels of the light absorbance.

It is apparent that the control over the resultant structure of such patterns is a key issue for the above applications. Using different growth conditions, we have grown a similar CNT array with denser structure (see Figure 12a), which does not exhibit super-hydrophobic properties.

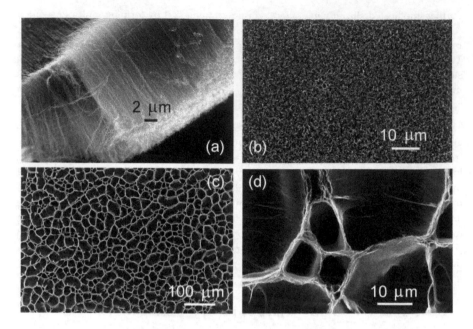

Figure 11. Treatment of super-hydrophobic sample with water and acetone. (a) Cross-section of the arrays of vertical-ly-aligned nanotubes. (b) Irregular structure was produced after treating with water. (c, d) Regular structure (com-pletely collapsed pattern) was produced by acetone, low and high magnifications.

Post-treatment of this array with ethanol and acetone has produced the sponge-like structure consisting of separated fine-porous island (Figures 12b, c, d). This structure is significantly different of that fabricated using super-hydrophobic sample (Figure 11). Moreover, variation of the dosage of liquid (we used 5 and 10 droplets of acetone, applied in sequence after complete drying of the preciously-applied drop) can be used to slightly change the structure. A comparison of the structures produced by 5 and 10 drops (Figures 12c and d) reveals a slight change in the pore sizes.

The use of water to treat this weakly hydrophobic CNT array produces slightly different pattern (Figure 13) consisting of smaller islands, which still demonstrate the sponge-like structure, i.e., each island is not a solid, intact array of the vertically-aligned carbon nanotubes but also consists of collapsed CNTs forming fine pores. One can expect that the fine-sponged structures produced using weakly hydrophobic CNT arrays can be very promised for the gas storage applications, whereas the highly-collapsed patterns may be more promising for sensing and other applications requiring control of the electrical resistivity of the surface. Thus, different internal structure of the vertically-aligned CNT array, together with the type of liquid and dosage, can be control parameters for the production of CNT patterns with a high level of the controllability.

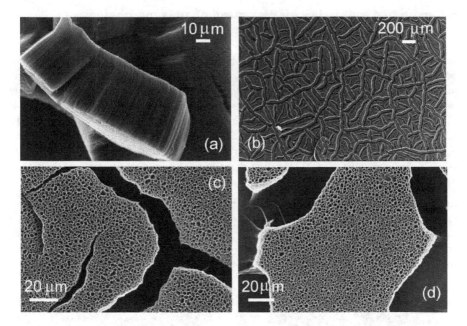

Figure 12. Treatment of weakly hydrophobic CNT array (a) with ethanol (b) and acetone, application of 5 drops (c) and 10 drops (d). Acetone produces a patterned sponge-like structure consisting of fine-porous island.

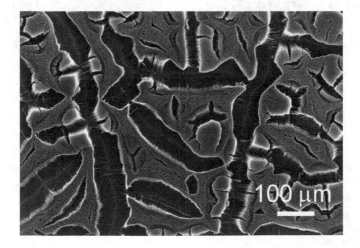

Figure 13. Treatment of weakly hydrophobic sample with water. Smaller islands demonstrate the sponge-like structure.

5. Perspective approaches for the control of array morphology

Dense arrays of highly ordered surface-bound vertically aligned nanotubes on silicon and metal oxides have a great potential for the fabrication of various advanced nano- and micro-devices such as fuel cells, sensors, and field emission element [39,40,41,42] One possible way to integrate the carbon nanotube array in the silicon platform is the use of anodized aluminum oxide (AAO) membranes to grow the pre-structured CNT patterns, bonded to the template surface. Indeed, the use of AAO membranes as growth templates was successful for the fabrication of, i.e., electron emitters [43]. Synthesis of carbon nanotubes on AAO templates allows precise and reproducible control of the dimensions of nanotubes [44, 45]. In this section we will review in short the AAO template characteristics important for growing the carbon nanotube arrays, and discuss the most important control parameters.

An AAO template can be prepared by the anodic oxidation of aluminum in various acid solutions. The thickness, pore size and interpore distance can be easily controlled by varying conditions of anodization such as composition of electrolyte, process temperature, applied voltage, process time and pore widening time [46,47]. Figure 14 shows SEM images of the free standing AAO templates fabricated by the two-step anodization.

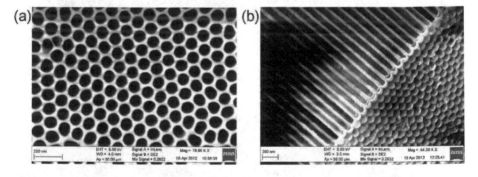

Figure 14. A Free-standing AAO fabricated using a two-step anodization. (a) Top view and (b) side view.

However, the free-standing AAO templates and membranes fabricated on aluminum foil are not be suitable for growing carbon nanotube arrays due to the thermal instability. Under thermal treatment, which is inevitable in the nanotube fabrication process, the AAO templates fabricated on aluminum foil easily crack due to the difference in thermal expansion coefficient of the alumina oxide and underlying aluminum. Moreover, the growth temperature cannot exceed the aluminum melting point e. In addition, a free-standing AAO template easily cracks due to its ceramic nature. Therefore, the conventional approach based on the use of the aluminum foil is not suitable for the CNT growth and fabrication of the carbon nanotube-based electronic devices.

To avoid this problem, it is necessary to fabricate AAO templates on other functional sub-strates. The alternative materials include silicon, quartz and ITO glass, on which the highly ordered structure of very thin AAO templates can be fabricated. For example, AAO templates fabricated on silicon wafers have already been used to fabricate highly ordered carbon nanotubes [48]. The AAO templates fabricated on non-aluminum substrates can be compatible with much higher processing temperatures, well above the 700 °C. Silicon substrates may be also useful for protecting AAO from distortion during the CNT growth.

However, quartz is more advantageous substrate for AAO membranes to be used as templates for the CNT synthesis. Quartz has a very high melting point, allowing for much higher temperatures, and can protect the AAO templates from cracking during the thermal treatment. Another advantage of quartz is transparency enabling the use of AAO templates as photonic crystals, and thus significantly broadening the application of fabricated AAO templates in other optics related applications.

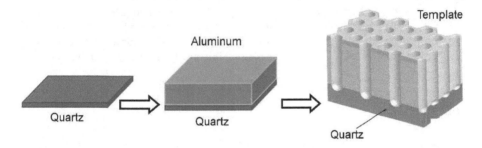

Figure 15. Schematic of the fabrication of AAO on quartz. High purity aluminum is deposited onto cleaned quartz, and the template is fabricated by anodization.

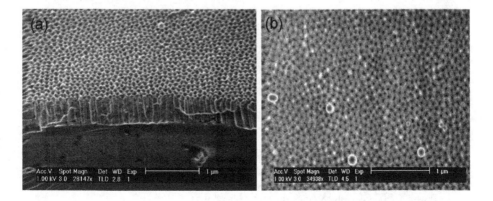

Figure 16. SEM images of the quartz-based AAO template suitable for the fabrication of carbon nanotube arrays. (a) Side view and (b) top view.

The high quality AAO templates were fabricated directly on quartz, using a two-step anodization without using any inter-layers between the deposited aluminum and quartz substrate. Figure 15 illustrates the schematic of this process. Prior to the fabrication of AAO template, quartz samples are cleaned in boiling solution of 30% w.t. of H_2SO_4 and 70% W.t. of H_2O_2. After that, the samples were etched in HF solution (0.1 w.t.% for 30 seconds), placed into an e-beam evaporator and coated with high purity (99.999%) aluminum to a thickness of 1.0 μm at a deposition rate of ≈1.5 nm×s^{-1}. The deposited Al film was then anodized to produce porous alumina templates in an electrolytic cell using a two-step anodization process. As a result, high quality AAO templates on quartz were fabricated. Figure 16 shows the SEM image of the AAO templates.

Above results demonstrate that AAO template technology not only can be used in a piece of aluminum foil, but also can be combined with silicon and other functional substrate technology. AAO template on functional substrate were used in the fabrication of CNT arrays can be realize the field emitters or possible become optical devices when CNT in quartz-AAO. Moreover, since the crystallinity of CNTs increase with the synthesis temperature, the emission current density increases with the synthesis temperature of CNTs. This again demonstrated that only AAO on functional substrate can realize high quality of CNTs array fabrication.

Author details

I. Levchenko[1,2], Z.-J. Han[1,2], S. Kumar[1,2], S. Yick[1,2], J. Fang[1,3] and K. Ostrikov[1,2]

1 Plasma Nanoscience Centre Australia (PNCA), CSIRO Materials Science and Engineering, Lindfield, Australia

2 Plasma Nanoscience, School of Physics, The University of Sydney, Sydney, Australia

3 School of Physics, The University of Melbourne, Parkville, VIC, Australia

References

[1] C. A. Crouse, B. Maruyama, R. Colorado Jr., T. Back, A. R. Barron. Growth, new growth, and amplification of carbon nanotubes as a function of catalyst composition. J. Am. Chem. Soc. 2008;130(25), 7946–7954.

[2] Z.-J. Han, I. Levchenko, S. Yick, K. Ostrikov. 3-Orders-of-magnitude density control of single-walled carbon nanotube networks by maximizing catalyst activation and dosing carbon supply. Nanoscale 2011;3, 4848-4853.

[3] Z.-J. Han, S. Yick, I. Levchenko, E. Tam, M. M. A. Yajadda, S. Kumar, P. J. Martin, S. Furman, K. Ostrikov. Controlled synthesis of a large fraction of metallic single-wal-

led carbon nanotube and semiconducting carbon nanowire networks. Nanoscale 2011;3, 3214-3220.

[4] Z. J. Han, B. K. Tay, M. Shakerzadeh, K. Ostrikov. Superhydrophobic amorphous carbon/carbon nanotube nanocomposites. Appl. Phys. Lett. 2009;94, 223106.

[5] I. Levchenko and K. Ostrikov, M. Keidar, S. Xu. Deterministic nanoassembly: Neutral or plasma route? Appl. Phys. Lett. 2006;89, 033109.

[6] M Meyyappan. A review of plasma enhanced chemical vapour deposition of carbon nanotubes. J. Phys. D: Appl. Phys. 2009;42, 213001.

[7] W. Zhou, L. Ding, S. Yang, J. Liu. Synthesis of high-density, large-diameter, and aligned single-walled carbon nanotubes by multiple-cycle growth methods. ACS Nano 2011;5, 3849–3857.

[8] K. Hata, D. N. Futaba, K. Mizuno, T. Namai, M. Yumura, S. Iijima. Water-assisted highly efficient synthesis of impurity-free single-walled carbon nanotubes. Science 12004;306, 1362-1364.

[9] G. Zhang, D. Mann, L. Zhang, A. Javey, Y. Li, E. Yenilmez, Q. Wang, J. P. McVittie, Y. Nishi, J. Gibbons, H. Dai. Ultra-high-yield growth of vertical single-walled carbon nanotubes: Hidden roles of hydrogen and oxygen. PNAS 2005;102, 16141-16145.

[10] D. Sun, M. Y. Timmermans, Y. Tian, A. G. Nasibulin, E. I. Kauppinen, S. Kishimoto, T. Mizutani, Y. Ohno. Flexible high-performance carbon nanotube integrated circuits. Nature Nanotech. 2011;6, 156–161

[11] J. Wu, K. S. Paudel, C. Strasinger, D. Hammell, A. L. Stinchcomb, and B. J. Hinds, Programmable transdermal drug delivery of nicotine using carbon nanotube membranes. Proc. Natl. Acad. Sci. USA 2010;107, 11698.

[12] Z. J. Han, K. Ostrikov, C. M. Tan, B. K. Tay, S. A. F. Peel. Effect of hydrophilicity of carbon nanotube arrays on the release rate and activity of recombinant human bone morphogenetic protein-2. Nanotechnology 2011;22, 295712.

[13] S. Kumar, I. Levchenko, Q. J. Cheng, J. Shieh, K. Ostrikov. Plasma enables edge-to-center-oriented graphene nanoarrays on Si nanograss. Appl. Phys. Lett. 2012;100, 053115.

[14] Z. J. Han, K. Ostrikov. Uniform, dense arrays of vertically aligned, large-diameter single-walled carbon nanotubes. J. Am. Chem. Soc. 2012;134, 6018–6024.

[15] M. P. Garrett, I. N. Ivanov, R. A. Gerhardt2,3, Alex A. Puretzky2, and David B. Geohegan. Separation of junction and bundle resistance in single wall carbon nanotube percolation networks by impedance spectroscopy Appl. Phys. Lett. 97, 163105 (2010); http://dx.doi.org/10.1063/1.3490650 (3 pages)

[16] Q. Cao, H. Kim, N. Pimparkar, J. P. Kulkarni, C. Wang, M. Shim, K. Roy, M. A. Alam, J. A. Rogers. Medium-scale carbon nanotube thin-film integrated circuits on flexible plastic substrates. Nature 2008;454, 495-500.

[17] M. A. Topinka, M. W. Rowell, D. Goldhaber-Gordon, M. D. McGehee, D. S. Hecht, G. Gruner. Charge transport in interpenetrating networks of semiconducting and metallic carbon nanotubes. Nano Letters 2009;9, 1866–1871.

[18] K. Ostrikov. Reactive plasmas as a versatile nanofabrication tool. Rev. Mod. Phys. 2005;77, 489–511.

[19] Z. J. Han, H. Mehdipour, X. Li, J. Shen, L. Randeniya, H. Y. Yang, K. Ostrikov. SWCNT networks on nanoporous silica catalyst support: morphological and connectivity control for nanoelectronic, gas-sensing, and biosensing devices. ACS Nano 2012;6, 5809–5819.

[20] T. W. Ebbesen, P. M. Ajayan, H. Hiura, K. Tanigaki. Purification of nanotubes. Nature 1994;367 519-page.

[21] X. M. H. Huang, R. Caldwell, L. Huang, S. C. Jun, M. Huang, M. Y. Sfeir, S. P. O'Brien, J. Hone. Controlled Placement of Individual Carbon Nanotubes Nano Letters 2005;5 1515.

[22] D. Takagi, H. Hibino, S. Suzuki, Y. Kobayshi, and Y. Homma. Carbon Nanotube Growth from Semiconductor Nanoparticles. Nano Letters 2007;7(8) 2272-page.

[23] B. Liu, W. Ren, L. Gao, S. Li, S. Pei, C. Liu, C. Jiang and H. M. Cheng. Metal-Catalyst-Free Growth of Single-Walled Carbon Nanotubes. J. Am. Chem. Soc. 2009;131 2082.

[24] S. Huang, Q. Cai, J. Chen, Y. Qian, and L. Zhang. Metal-Catalyst-Free Growth of Single-Walled Carbon Nanotubes on Substrates. J. Am. Chem. Soc. 2009;131, 2094-page.

[25] S. Kumar, I. Levchenko, K. Ostrikov and J. McLaughlin. Plasma-enabled, catalyst-free growth of carbon nanotubes on mechanically-written Si features with arbitrary shape. Carbon 2012, 50, 321.

[26] P. Laveant, P. Werner, G. Gerth, and U.Gosele. Incorporation, diffusion and agglomeration of carbon in silicon. Solid State Phenomeona 2002;82-84, 189.

[27] S. Helveg, C. Lopez-Cartes, J. Sehested, P. L. Hansen, B. S. Clausen, J. R. Rostrup-Nielsen, F. Abild-Pedersen, J. K. Norskov. Atomic-scale imaging of carbon nanofibre growth. Nature 2004;427, 426.

[28] J. Pezoldt, Y. V. Trushin, V. S. Kharlamov, A. A. Schmidt, V. Cimalla, O. Ambacher. Nuclear Instr. Meth. in Phys. Res. B 2006;253, 241.

[29] E.M. Kotsalisa, E. Demosthenousb, J.H. Walthera, c, S.C. Kassinosb, P. Koumoutsakos. Wetting of doped carbon nanotubes by water droplets. Chem. Phys. Lett. 2005;412, 250–254.

[30] M. A. Correa-Duarte, N. Wagner, J. Rojas-Chapana, C. Morsczeck, M. Thie, M. Giersig. Fabrication and Biocompatibility of Carbon Nanotube-Based 3D Networks as Scaffolds for Cell Seeding and Growth. Nano Lett. 2004;4 2233-2236.

[31] B. Q. Wei, R. Vajtai, Y. Jung, J. Ward, R. Zhang, G. Ramanath, P. M. Ajayan. Organized assembly of carbon nanotubes. Nature 2002; 416, 495-496.

[32] C. L. Pint, Y.-Q. Xu, M. Pasquali, R. H. Hauge. Formation of Highly Dense Aligned Ribbons and Transparent Films of Single-Walled Carbon Nanotubes Directly from Carpets. ACS Nano 2008;2, 1871–1878.

[33] B. Pokroy, S. H. Kang, L. Mahadevan, J. Aizenberg. Self-Organization of a Mesoscale Bristle into Ordered, Hierarchical Helical Assemblies. Science 2009; 323, 237-240.

[34] D. N. Futaba, K. Hata, T. Yamada, T. Hiraoka, Y. Hayamizu, Y. Kakudate, O. Tanaike, H. Hatori, M. Yumura, S. Iijima. Shape-engineerable and highly densely packed single-walled carbon nanotubes and their application as super-capacitor electrodes. Nature Materials 2006;5, 987-994.

[35] S. Li, H. Li, X. Wang, Y. Song, Y. Liu, L. Jiang, D. Zhu. Super-hydrophobicity of large-area honeycomb-like aligned carbon nanotubes. J. Phys. Chem. B 2002;106, 9274-9276.

[36] W. Wang, M. Tian, A. Abdulagatov, S. M. George, Y.-C. Lee, R. Yang. Three-dimensional Ni/TiO2 nanowire network for high areal capacity lithium ion microbattery applications. Nano Lett. 2012;12, 655–660.

[37] H. Liu, J. Zhaia, L. Jiang. Wetting and anti-wetting on aligned carbon nanotube films. Soft Matter. 2006;2, 811–821.

[38] C. T. Wirth, S. Hofmann, J. Robertson. Surface properties of vertically aligned carbon nanotube arrays. Diam. Relat. Mater. 2008;17, 1518–1524.

[39] Z. Wang, L. Ci, L. Chen, S. Nayak, P. M. Ajayan and N. Koratkar. Polarity-Dependent Electrochemically Controlled Transport of Water through Carbon Nanotube Membranes. Nano Letters 2007;7, 697.

[40] K. Gong, F. Du, Z. Xia, M. Durstock and L. Dai. Nitrogen-doped carbon nanotube arrays with high electrocatalytic activity for oxygen reduction. Science 2009;323, 760.

[41] I. Levchenko, S.Kumar, M. M. A. Yajadda, Z. J. Han, S. Furman, K. Ostrikov. Self-organization in arrays of surface-grown nanoparticles: characterization, control, driving forces. J. Phys. D: Appl. Phys. 2011;44, 174020.

[42] M. E. Itkis, A. Yu and R. C. Haddon. Single-Walled Carbon Nanotube Thin Film Emitter–Detector Integrated Optoelectronic Device. Nano Letters 2008;8, 2224.

[43] S.-K. Hwang, J. Lee, S.-H. Jeong, P.-S. Leen, K.-H. Lee. Fabrication of carbon nanotube emitters in an anodic aluminum oxide nanotemplate on a Si wafer by multi-step anodization. Nanotechnology 2005;16, 850–858.

[44] K. Liu, M. Burghard and S. Roth. Conductance spikes in single-walled carbon nano-tube field-effect transistor. Appl. Phys. Lett. 1999;75, 2494.

[45] J. S. Suh, J. S. Lee. Highly ordered two-dimensional carbon nanotube arrays. Appl. Phys. Lett. 1999;75 2047.

[46] J. Fang, P. Spizzirri, A. Cimmino, S. Rubanov, and S. Prawer, Nanotechnology, 20, 065706 (2009). Extremely high aspect ratio alumina transmission nanomasks: their fabrication and characterization using electron microscopy.

[47] A. P. Li, F. Müller, A. Birner, K. Nielsch, U. Gösele. Hexagonal pore arrays with a 50–420 nm interpore distance formed by self-organization in anodic alumina. J. Appl. Phys. 1998;84, 6023.

[48] S. Chu, K.Wada, S. Inoue, S. Todoroki. Synthesis and characterization of titania nano-structures on glass by Al anodization and sol-gel process. Chem. Mat. 2002;14(1), 266.

Chapter 6

Characterization and Morphology of Modified Multi-Walled Carbon Nanotubes Filled Thermoplastic Natural Rubber (TPNR) Composite

Mou'ad A. Tarawneh and Sahrim Hj. Ahmad

Additional information is available at the end of the chapter

1. Introduction

Carbon nanotubes describes a specific topic within solid-state physics, but is also of interest in other sciences like chemistry or biology. Actually the topic has floating boundaries, because we are at the molecule level. In the recent years carbon nanotubes have become more and more popular to the scientists. Initially, it was the spectacularly electronic properties, that were the basis for the great interest, but eventually other remarkable properties were also discovered.

The first CNTs were prepared by M. Endo in 1978, as part of his PhD studies at the University of Orleans in France. Although he produced very small diameter filaments (about 7 nm) using a vapour-growth technique, these fibers were not recognized as nanotubes and were not studied systematically. It was only after the discovery of fullerenes, C60, in 1985 that researchers started to explore carbon structures further. In 1991, when the Japanese electron microscopist Sumio Iijima [1] observed CNTs, the field really started to advance. He was studying the material deposited on the cathode during the arc-evaporation synthesis of fullerenes and came across CNTs. A short time later, Thomas Ebbesen and Pulickel Ajayan, from Iijima's lab, showed how nanotubes could be produced in bulk quantities by varying the arc-evaporation conditions. However, the standard arc-evaporation method only produced only multiwall nanotubes. After some research, it was found that the addition of metals such as cobalt to the graphite electrodes resulted in extremely fine single wall nanotubes.

The synthesis in 1993 of single-walled carbon nanotubes (SWNTs) was a major event in the development of CNTs. Although the discovery of CNTs was an accidental event, it opened the way for a flourishing research into the properties of CNTs in labs all over the world, with many scientists demonstrating promising physical, chemical, structural, and optical properties of CNTs.

CNTs exhibit a great range of remarkable properties, including unique mechanical and electrical characteristics. These remarkable high modulus and stiffness properties have led to the use of CNTs to reinforce polymers in the past few years. Both theoretical (e.g. molecular structural mechanics and tight-binding molecular dynamics) and experimental studies have shown SWCNTs to have extremely high elastic modulus (≈1 TPa) [2-3]. The tensile strength of SWCNTs estimated from molecular dynamics simulation is ≈150 MPa [4]. The experimental measurement of 150 MPa was found for the break strength of multi-walled carbon nanotubes (MWCNTs) [5].

The remarkable properties of CNTs offer the potential for improvement of the mechanical properties of polymers at very low concentrations. In practice, MWCNTs are preferred over SWCNTs as the reinforcing fillers for polymers due to their lower production cost. However, slippage between the shells of MWCNTs would undermine the capability of the fillers to bear the external applied load.

Mixed 1 wt.% MWCNTs with polystyrene (PS) in toluene via ultrasonication, achieved about 36–42% increase in the elastic modulus and a 25% increase in the tensile strength of the PS–MWCNT film compared to pure PS [6]. They found that nanotube fracture and pullout are responsible for the failure of the composite. The fracture of MWCNTs in a PS matrix implies that certain load transfer from the PS to the nanotubes has taken place. However, the pullout of MWCNTs from the PS matrix indicating that the PS–nanotube interfacial strength is not strong enough to resist debonding of the fillers from the matrix. It is considered that some physical interactions exist at the PS–MWCNT interface, thereby enabling load transfer from the matrix to the fillers.

The additions of 0.25– 0.75 wt.% SWCNTs to polypropylene (PP) considerably its tensile strength and stiffness as well as storage modulus. The elongation at break reduces from 493 (PP) to 410% with the addition of 0.75 wt.% filler, corresponding to -17% reduction in ductility. At 1 wt.% SWNT, both stiffness and strength are significantly reduced due to the formation of aggregates [7].

The morphology and mechanical properties of the melt-compounded polyamide 6 (PA6)–MWNT nanocomposites were studied by [8]. The MWCNTs were purified by dissolving the catalyst in hydrochloric acid followed by refluxing in 2.6 Mnitric acids to increasing the carboxylic and hydroxyl groups. It was also found that with the addition of only 1 wt.% MWCNTs, the tensile modulus and the tensile strength are greatly improved by ≈115 and 120%, respectively compared to neat PA6. The tensile ductility drops slightly from 150 to 125%. They attributed the improvements of these mechanical properties to a better dispersion of MWCNTs in PA6 matrix, and to a strong interfacial adhesion between the nanofillers and PA6 matrix which leads to favorable stress transfer across the polymer to the MWCNTs.

The influence of SWNT and carbon nanofiber additions on the mechanical performances of silicone rubber was reported by [9]. They reported that SWCNTs are effective reinforcements for silicone rubber due to their large aspect ratio and low density. The initial modulus (measured by fitting a straight line to the data below 10% strain) tends to increase almost linearly with increasing filler content. The effect of SWCNT and carbon fiber additions on the tensile ductility of silicone rubber is shown that the strain to failure drops from 325 to 275% upon loading with 1 wt.% SWCNTs, corresponding to ≈15% reduction.

The carbon nanotube additions to polyurethane (PU) improve the mechanical properties such as increased modulus and yield stress, without loss of the ability to stretch the elastomer above 1000% before final failure; the addition of CNTs increases the modulus and strength of PU without degrading deformabilty. The elongation at break decreases very slightly with CNT loading up to 17 wt.%. At this filler loading, the nanocomposite still maintains a very high value of elongation at rupture, i.e. 1200% [10].

Theoretical prediction showed an extremely high thermal conductivity (6000 W/mK) of an isolated SWCNTs [11]. High thermal conductivity of the CNTs may provide the solution of thermal management for the advanced electronic devices with narrow line width. Revealed the thermal conductivity of epoxy-based composites reinforced with 1.0 wt.% SWCNTs increased over 125% reaching a value of ~0.5 W/mK [12]. The variation of thermal conductivity with the values of 35 and 2.3 W/mK for a densepacked mat and a sintered sample, respectively [13]. High thermal conductivity of 42 and ~18W/mK of the aligned and the random bucky paper mats, respectively. However, the thermal conductivity drops significantly by almost an order of magnitude when the aligned bucky paper mats were loaded with epoxy, the volume fraction of the aligned bucky paper composites is about 50%[14].

Developed an infiltration method to produce CNTs/epoxy composites and showed a 220% increase in thermal conductivity (~0.61 W/mK) at 2.3 wt.% SWCNT loading, and they found that the electrical resistance between SWCNT-polymer is more severe than that of SWCNT–SWCNT [15]. Prepared SWCNT and MWCNT films and reported the thermal conductivity of 1.64 and 1.51 W/mK, respectively; they concluded that the intra-tube spacing affects the thermal transfer more significantly than that of the nanotubes themselves [16].

The thermal conductivities of composites reinforced with 1.0 wt.% SWCNTs and 4.0 wt.% MWCNTs are 2.43 W/mK and 3.44 W/mK, respectively. Composites reinforced with the unpurified CNTs have higher thermal conductivity than that of the purified CNTs reinforced composite. This is attributed to the generation of defects on the CNT surface during acid treatment. Moreover, due to longer phonon propagation length, it is found that thermal conductivity increases with temperatures over the range from 25 to 55°C for both SWCNTs/ Poly (methyl methacrylate) PMMA and MWCNTs/PMMA composites. However, the thermal conductivities of CNT films decrease with increasing temperature, which results from phonon scattering during transfer due to the presence of defects coupled with smaller phonon mean free path at higher temperature [17].

The differences in the composite manufacturing methods, powder-(MWCNTs and ball milled SWCNTs) or liquid- (chemically treated SWCNTs) based approach, can not account for the differences in the properties, since both methods were used for the SWNT-composites and resulted in similar thermal behaviour [18]. Thus, they concluded that in this case, there must be a very large interface resistance to the heat flow associated with poor phonon coupling between the stiff nanotubes and the (relatively) soft polymer matrix. In addition it is possible that the phonon vibrations in the SWCNTs are dampened by the matrix interaction, while in the MWCNTs the phonons can be carried in the inner walls without hindrance.

The precise sectioning of CNTs provides an effective way to shorten carbon nanotubes with controlled length and minimum sidewall damage [19]. For shortened nanotubes they found that they are easily dispersed into polymer matrices, which effectively improved the percolation. The minimum CNT sidewall damage and improved percolation in short SWCNT composites led to an obvious improvement of thermal conductivity. Hence, their research suggests an effective way to improve dispersion of CNTs into polymer matrices and also retain the perfect electronic structure of the CNTs, resulting in desired functional materials.

Accurate measurement of the thermal conductivity of composites and nanocomposites can be done using the transient hot-wire technique which is capable of measuring the thermal conductivity of solid materials in an absolute way. The enhancement in the thermal conductivity was measured as 27% in relation to the thermal conductivity of the epoxy-resin polymer, which is satisfactory taking into account the low volume fraction (28%) of the glass fibres used in the composite [20]. They reported that when 2% by weight C-MWNT were mixed with the epoxy-resin, the enhancement of thermal conductivity was 9% while using both glass fibres and C-MWNT the enhancement was 48%.

For sufficient enhancement of most of the nanocomposites' properties, the dispersion of the CNTs should be very fine in the polymer matrix, which means that the surface of interaction between the filler and the matrix should be optimised. However, this is difficult to achieve since their long length results in them becoming entangled. Moreover, their very large surface-to-volume ratio and strong van der Waals interactions keep them tied together, which in most cases leads to the formation of large agglomerates in polymer matrices. The interfacial adhesion between CNTs and the polymeric matrix is also crucial. In order to increase the interfacial adhesion between the polymer and the CNTs various routes of surface modification of the nanotubes have been considered. One is non-covalent functionalisation of molecules and the other is covalent functionalisation from the walls of the nanotubes. Non-covalent functionalisation is based on weak Van der Waals forces [21]. The advantage of non-covalent functionalisation is that the perfect structure of the nanotubes is not altered while the covalent attachment can greatly improve the load transfer to the matrix; however, it usually introduces structural defects on the nanotubes' surface.

Although both probe style and bath style ultrasonic systems can be used for dispersing CNTs, it is widely believed that the probe style ultrasonic systems work better for dispersing CNTs [22]. It is also widely known that adding a dispersing reagent (surfactant) into the solution will accelerate the dispersion effect.

The most common procedure used for covalent attachment of reactive groups is the treatment with inorganic acids. Usually the nanotubes are refluxed with a nitric acid solution or a mixture of nitric and sulfuric acid, sometimes concurrently with the application of high power sonication [23]. These oxidative treatments usually result in shortening of the CNTs' length and formation of surface reactive groups, such as hydroxyl, carbonyl and mainly carboxylic acid. Oxidation of the nanotubes starts at the tips and gradually moves towards the central part of the tube and the layers are removed successively [24].

The synthesized carbon nanotubes usually exist as agglomerates of the size of several hundred micrometers [25]. Such entanglements make it difficult to disperse nanotubes uniformly in a polymer matrix. To overcome the dispersion problem, it is necessary to tailor the chemical nature of the nanotube surface. One of the most straightforward methods for nanotube dispersion is direct mixing; however, it does not always yield a homogeneous distribution of nanotubes because of the lack of compatibility between the MWCNTs and polymer matrix. Solution processing has been a commonly used method in fabrication of the well-dispersed carbon nanotube composites. However, it is hard to achieve homogeneous dispersion of nanotubes in a polymer matrix because carbon nanotubes are insoluble and bundled.

Chemical functionalization of the MWCNTs surface increases the interfacial interaction between MWCNTs and the polymer matrix. This enhances the adhesion of the MWCNTs in various organic solvents and polymers, reduces the tendency to agglomerate, and improves dispersion. The improved interactions between MWCNTs and the polymer matrix govern the load-transfer from the polymer to the nanotubes and, hence, increase the reinforcement efficiency. Attachment of oxygen containing functional groups (i.e., carboxyl groups, carbonyl groups, hydroxyl groups, etc.) on the surface of the MWCNTs could be achieved by applying several chemical treatments.

The chemically functionalized MWCNTs can be easily mixed with the polymer matrix. Acid treatment of the nanotube is an especially well-known technique to remove catalytic impurities, generate functional groups on open ends or sidewalls of nanotubes, and facilitate good dispersion of MWCNTs in polymeric solutions or melts.

The emergence of thermoplastic elastomers (TPEs) is one of the most important developments in the area of polymer science and technology. TPEs are a new class of material that combines the properties of vulcanized rubber with the ease of processability of thermoplastics [26]. Thermoplastic elastomers can be prepared by blending thermoplastic and elastomers at a high shear rate. Thermoplastics, for example, polypropylene (PP), polyethylene (PE) and polystyrene (PS), and elastomers, such as ethylene propylene diene monomer (EPDM), natural rubber (NR) and butyl rubber (BR), are among the materials used in thermoplastic elastomer blends.

Blends of natural rubber (NR) and polypropylene (PP) have been widely reported by previous researchers [26]. According to them, polypropylene is the best choice for blending with natural rubber due to its high softening temperature (150°C) and low glass transition temperature (-60°C, is Tg for NR), which makes it versatile in a wide range of temperatures. Even though NR and PP are immiscible, their chemical structure is nearly the same. Thus, stable dispersion of NR and PP is possible. Incompatibility between NR and PP can be overcome by the introduction of a compatibiliser that can induce interactions during blending. Compatibility is important as it may affect the morphology, mechanical and thermal properties of the blends. Among the commonly used compatibilisers are dicumyl peroxide (DCP), m-phenylene bismaleimide (HVA-2) and liquid natural rubber (LNR). Apart from compatibility, mixing torque and curing are interrelated in determining the homogeneity of the TPNR blend.

Mechanical blending of PP and NR with the addition of LNR as a compatibiliser has been reported to be optimal at a temperature of 175-185°C and a rotor speed of 30-60rpm. The percentage of LNR used depends on the ratio of NR to PP. For a NR:PP ratio of 30:70 the best physical properties are obtained at 10% LNR [27]. The compatibiliser helps to induce the interaction between the rubber and plastic interphase and thereby increases the homogeneity of the blend.

MWCNTs/TPNR composites with different amounts of MWCNT were prepared and their thermal properties have been investigated by [28]. The higher thermal conductivity was achieved in the samples with 1 and 3wt% of MWCNTs compared to the pristine TPNR. Any sample with MWCNTs content higher than 3wt% caused the conductivity to decrease. In addition, the improvement of thermal diffusivity and specific heat was also achieved at the same percentage. DMA confirmed that the glass transition temperature (Tg) increased with the increase in the amount of MWCNTs.

The tensile strength, tensile modulus, and also the impact strength of TPNR/MWCNTs are improved significantly while sacrificing high elongation at break by incorporating MWCNTs. The reinforcing effect of MWCNTs was also confirmed by DMA where the addition of nanotubes has increased the storage modulus, the loss modulus, and also the glass transition temperature (Tg). Homogeneous dispersion of MWCNTs throughout the TPNR matrix and strong interfacial adhesion between MWCNTs and matrix as confirmed by SEM images are proposed to be responsible for the significant mechanical enhancement [29].

The reinforcing effect of two types of MWCNTs has also confirmed by dynamic mechanical analysis where the addition of nanotubes have increased in the storage modulus E', and the loss modulus E", in the addition the glass transition temperature (Tg) increased with an increase in the amount of MWCNTs. The addition of MWCNTs in the TPNR matrix improved the mechanical properties. The tensile strength and elongation at break of MWCNTs 1 increased by 23%, and 29%, respectively. The Young's modulus had increased by increasing the content of MWCNTs. For MWCNTs 2 the optimum result of tensile strength and Young's modulus was recorded at 3% which increased 39%, and 30%, respectively. The laser flash technique was used to measure the thermal conductivity, thermal diffusivity and specific heat, from the results obtained. The high thermal conductivity was achieved at 1 wt% and 3 wt% of MWCNTs compared with TPNR after 3 wt% it decreased, also the improvement of thermal diffusivity and specific heat was achieved at the same percentage. The MWCNTs 1 and 2/TPNR nanocomposites were fabricated and the tensile and properties were measured [30].

In this chapter, the effect of multi-walled carbon nanotubes with and without acid treatment on the properties of thermoplastic natural rubber (TPNR) was investigated. Two types of MWCNTs were introduced into TPNR, which are untreated multi-walled carbon nanotubes (UTMWCNTs) (without acid treatment) and treated multi-walled carbon nanotubes (TMWCNTs) (with acid treatment). Using this method, MWCNTs are dispersed homogeneously in the TPNR matrix in an attempt to increase the properties of these nanocomposites. The effect of MWCNTs on the mechanical and thermal properties of TPNR nanocomposites is reported in this chapter.

2. Experiment Details

Polypropylene, with a density of 0.905 g cm-3, was supplied by Propilinas (M) Sdn. Bhd, natural rubber was supplied by Guthrie (M) Sdn. Bhd, and polypropylene (PP) with a density of 0.905 g/cm3 was supplied by Polipropilinas (M) Sdn. Bhd were used in this research. Maleic anhydride–grafted–polypropylene (MAPP) with a density of 0.95 g/cm3 was supplied from Aldrich Chemical Co., USA. Liquid natural rubber (LNR) was prepared by the photochemical degradation technique.

A Multi-walled carbon nanotubes (MWCNTs) were provided by Arkema (GraphistrengthTM C100). Table 1 shows the properties of multi-walled carbon nanotubes (MWCNTs).

MWCNTs	Purity	Length	Diameter	Manufactured
MWCNTs	"/90%	0.1-10 µm	10-15 nm.	Catalytic Chemical Vapour Deposition (CVD)

Table 1. Properties of multi-walled carbon nanotubes (MWCNTs).

2.1. Preparation of TPNR-Multi-Walled Carbon Nanotubes (MWCNTs) Composite

Mixing was performed by an internal mixer (Haake Rheomix 600P). The mixing temperature was 180°C, with a rotor speed of 100 rpm and 13 min mixing time. The indirect technique (IDT) was used to prepare nanocomposites, this involved mixing the MWCNTs with LNR separately, before it was melt blended with PP and NR in the internal mixer. TPNR nanocomposits were prepared by melt blending of PP, NR and LNR with MWCNTs in a ratio of 70 wt% PP, 20 wt% NR and 10wt% LNR as a compatibiliser and 1,3,5 and 7% MWCNTs.

2.2. Acid Treatment of MWCNTs

Two types of MWCNTs were introduced to the TPNR which is untreated MWCNTs (MWCNTs 1) and treated MWCNTs (MWCNTs 2), MWCNTs 2 were treated by immersing neat MWCNTs in a mixture of nitric and sulfuric acid with a molar ratio of 1:3, respectively. In a typical experiment, 1g of raw MWCNTs was added to 40ml of the acid mixture. Then, the oxidation reaction was carried out in a two-necked, round-bottomed glass flask equipped with reflux condenser, magnetic stirrer and thermometer. The reaction was carried out for 3 hours at 140°C. After that, this mixture was washed with distilled water on a sintered glass filter until the pH value was around 7, and was dried in a vacuum oven at 70°C for 24hours [31].

2.3. Characterizations

Fourier transform infrared (FTIR) spectroscopy analysis was carried out on the Perkin Elmer spectrum V-2000 spectrometer by the potassium bromide (KBr) method for MWCNTs. The samples were scanned between 700 to 4000 cm-1 wave number. Differences in the peaks as well as the new peaks of MWCNTs and MWCNTs after acid treatment were observed to identify any functional groups on the MWCNTs tubes surface.

The tensile properties were tested using a Testometric universal testing machine model M350-10CT with 5 kN load cell according to ASTM 412 standard procedure using test specimens of 1 mm thickness and a crosshead speed 50 mm min-1. At least five samples were tested for each composition, and the average value was reported.

The impact test was carried out using a Ray Ran Pendulum Impact System according to ASTM D 256-90b. The velocity and weight of the hammer were 3.5m/s and 0.898kg, respectively.

Dynamic mechanical analysis for determining the glass transition temperature, storage and loss modulus was carried out using DMA 8000 (PerkinElmer Instrument), operating in single cantilever mode from -100 to 150°C at a constant frequency of 1 Hz, with a heating rate of 5°C/min. The dimensions of the samples were 30 x 12.5 x 3 mm.

The thermal conductivity was measured by a laser flash method. Disk-type samples (12.7 mm in diameter and 1mm in thickness) were set in an electric furnace. Specific heat capacities were measured with a differential scanning calorimeter DSC. Thermal diffusivity (λ, Wm_1 K_1) was calculated from thermal diffusivity (α, m2 s_1), density (ϱ, g cm_3) and specific heat capacity (Cp, J g_1 K_1) at each temperature using the following:

$$\lambda = \alpha.\,\rho.\,C. \tag{1}$$

The reference used for the heat capacity calculation was a 12.7mm thick specimen of pyroceram. The reference sample was coated with a thin layer of graphite before the measurement was performed. The thermal conductivity of MWCNTs reinforced TPNR matrix composites of all volume fractions was studied from 30°C to 150°C. The morphology of the MWCNTs and the composite were examined using a scanning electron microscope (Philips XL 30). The samples were coated with a thin layer of gold to avoid electrostatic charging during examination.

3. Results and Discussion

3.1. Fourier-Transform Infrared Spectroscopy

The method used to functionalize the pristine MWCNTs in this study was the acid treatment method, which is described in section 2.2. Through this process, MWCNTs were oxidized and purified by eliminating impurities such as amorphous carbons, graphite particles, and metal catalysts [32]; the functional group of the surface of the CNTs are as shown in Figure 1.

Characterization and Morphology of Modified Multi- Walled Carbon Nanotubes Filled
Thermoplastic Natural Rubber (TPNR) Composite

123

The generation of chemical functional groups on MWCNTs was confirmed using Fourier transform infrared spectroscopy (FT-IR) spectra which were recorded between 400 cm_1 and 4000 cm_1.

The FT-IR spectra of pure MWCNTs and the surface treated MWCNTs are shown in Figure. 2 and Figure 3. The characteristic bands due to generated functional groups are observed in the spectrum of each chemically treated MWCNTs. In figure 2 we could not see any band compared with the treated MWCNTs. The acid treated MWCNTs shows new peaks in comparison with the FT-IR spectrum of the untreated MWCNTs, which lack the hydroxyl and carbonyl groups. The peaks around 1580 cm_1 are assigned to the O–H band in C-OH, and the peaks at 674 cm-1 are assigned to COOH, as shown in Figure 3. This demonstrates that hydroxyl and carbonyl groups have been introduced on the nanotube surface [33].

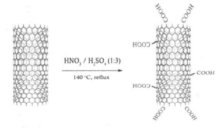

Figure 1. Example of chemical functionalization of carbon nanotubes.

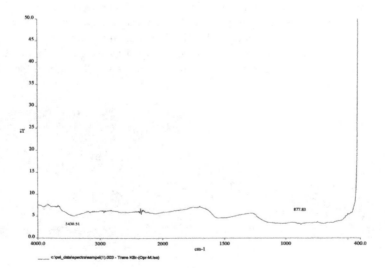

Figure 2. FTIR spectra of MWCNTs (before acid treatment).

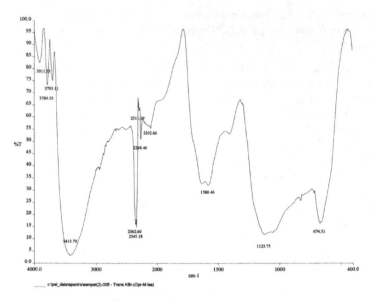

Figure 3. FTIR spectra of MWCNTs (after acid treatment).

3.2. Transmission Electron Microscopy (TEM)

TEM microphotographs of pure MWCNTs are shown in Figure 4 (A and B). The figure presents unmodified MWCNTs containing particles with diameters of 5–12 nm. The nanoparticles may be impurities from amorphous carbon and can be removed by acid treatment. According to the supplier, the unmodified MWCNT contains approximately 5% amorphous carbon. Figure 4 B demonstrates that most of the nanoparticles were deposited on the surface of the carbon nanotubes and some of them were dispersed throughout the solution used to view the MWCNTs by TEM.

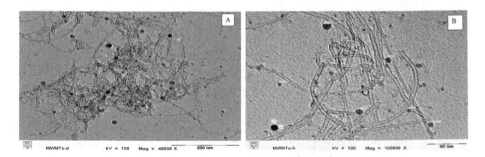

Figure 4. TEM micrograph of Pure MWCNTs before acid treatment with different magnifications (A) 45000 (B) 100000.

Figure 5 (A and B) displayed no nanoparticles in the acid-modified MWCNTs. The particles might have been removed during acid modification. This reveals that the acid-modified MWCNTs were straight and that some of them aggregated in bundles, which were dispersed well in the matrix. The length of the MWCNTs were reduced during acid modification, since the mixed acid corroded the MWCNTs. TEM microphotographs of the unmodified and acid-modified, curled and entangled MWCNTs demonstrate that the MWCNTs are straight.

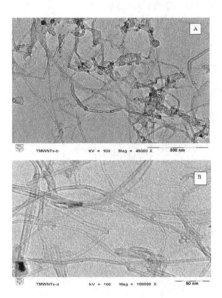

Figure 5. TEM micrograph of MWCNTs after acid treatment with different magnifications (A) 45000 (B) 100000.

3.3. Mechanical Properties

3.3.1. Tensile strength

The tensile strengths of TPNR reinforced with MWCNTs (with and without treatment) of different percentages (1%, 3%, 5% and 7%) are shown in Figure 6. Generally, both MWCNTs exhibited an increasing trend up to 3wt% content. Further increments in MWCNTs content decreased the tensile strength compared to the optimum filler loading.

From Figure 6, TPNR with UTMWCNTs and TMWCNTs have optimum results at 3 wt%, which, compared with TPNR, increased by 23% and 39%, respectively. The tensile strength increased radically as the amount of MWCNTs concentration increased. The mechanical performance, such as tensile properties, strongly depends on several factors such as the properties of the filler reinforcement and matrix, filler content, filler length, filler orientation, and processing method and condition. The improvement in the tensile strength may be caused by the good dispersion of MWCNTs in the TPNR matrix, which leads to a strong interaction between the TPNR matrix and MWCNTs. These well-dispersed MWCNTs may have the effect of physically crosslinking points, thus, increasing the tensile strength.

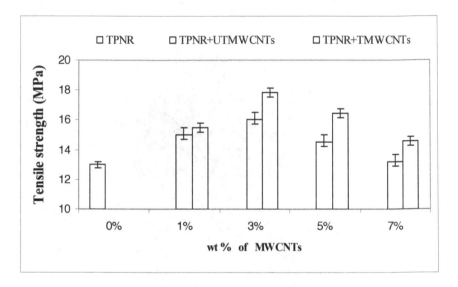

Figure 6. Tensile strength of TPNR reinforced with MWCNTs (with and without treatment).

A good interface between the CNTs and the TPNR is very important for a material to stand the stress. Under load, the matrix distributes the force to the CNTs, which carry most of the applied load. The order of these value is TPNR/TMWCNTs > TPNR/UTMWCNTs > TPNR. The better properties in tensile strength for the TPNR/TMWCNTs nanocomposites could be due to the improved dispersion of the MWCNTs, as well as the response to the opportunities offered by the acid treated MWCNTs. Furthermore, the MWCNTs after acid treatment contain many defects as well as acidic sites on CNTs, such as carboxylic acid, carbonyl and hydroxyl groups. These will greatly enhance the combination of CNTs in a polymer matrix, thus improving the mechanical strength of the nanocomposites [34]. When the content of MWCNTs is higher, the MWCNTs cannot disperse adequately in the TPNR matrix and agglomerate to form a big cluster. This is because of the huge surface energy of MWCNTs as well as the weak interfacial interaction between MWCNTs and TPNR, which leads to inhomogeneous dispersion in the polymer matrix and negative effects on the properties of the resulting composites that cause a decrease in the tensile strength [35].

3.3.2. Young's Modulus

Figure 7 shows the effect of filler content on the tensile modulus of TPNR reinforced by TMWNTs and UTMWCNTs. The same trend as for the tensile strength in Figure 6 was observed for the tensile modulus of TMWCNTs. Figure 6 clearly shows that the presence of MWCNTs has significantly improved the tensile modulus of the TPNR.

The remarkable increase of Young's modulus with TMWCNTs content shows a greater improvement than that seen in the tensile strength at high content, which indicates that the Young's modulus increases with an increase in the amount of the TMWCNTs. At 3 wt% of TMWCNTs the Young's modulus is increased by 34 % compared to TPNR. The Young's modulus of UTMWCNTs increased with the increase in the amount of UTMWCNTs. The maximum result was achieved at 3wt%, with an increase of about 22%, which was due to the good dispersion of nanotubes displaying perfect stress transfer [36].The improvement of modulus is due to the high modulus of MWCNTs [37]. The further addition of TMWCNTs and UTMWCNTs from 5 to 7 wt% increased the Young modulus dropped respectively.

As explained before, a reduction in performance occurred at higher filler contents for both types of MWCNTs, as depicted in Figure 7. Initially it increases with filler content and then decreases when exceeding the filler loading limit due to the diminishing interfacial filler-polymer adhesion. It is assumed that aggregates of nanotube ropes effectively reduce the aspect/ratio (length/diameter) of the reinforcement.

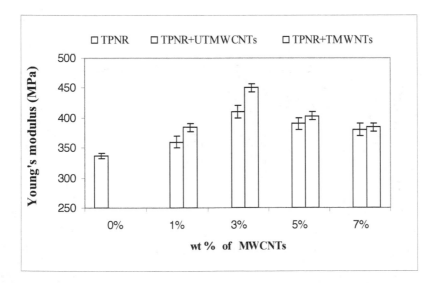

Figure 7. Young's Modulus of TPNR reinforced with MWCNTs (with and without treatment).

3.3.3 Elongation at Break

The elongation at the break of TPNR with TMWCNTs and UTMWCNTs is shown in Figure 8. For TMWCNTs and UTMWCNTs, the elongation at break decreased with the increase in the amount of MWCNTs, compared with TPNR.

It can be deduced that the reinforcing effect of MWCNTs is very marked. As the MWCNTs content in the TPNR increases, the stress level gradually increases, however, the strain of the nanocomposites decreased at the same time. This is because the MWCNTs included in the TPNR matrix behave like physical crosslinking points and restrict the movement of polymer chains. This indicates that, when the amount of CNTs incorporated into the rubber increase it tends to decrease the ductility and the material become stronger and tougher, however, at the same time, it is also

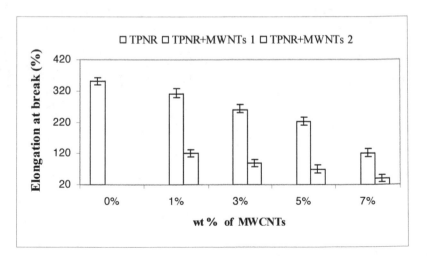

Figure 8. Elongation at Break of TPNR reinforced with MWCNTs (with and without treatment).

3.3.4 Impact Strength

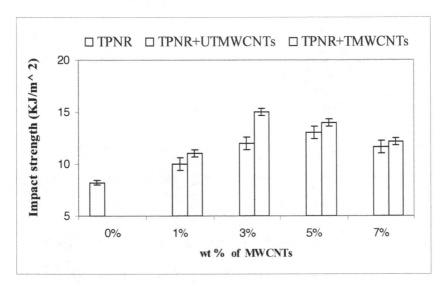

Figure 9. Impact Strength of TPNR reinforced with MWCNTs (with and without treatment).

The effect of filler loading on the impact strength of TPNR/TMWCNTs and TPNR/UTMWCNTs nanocomposites is given in Figure 9. It shows that incorporation of MWCNTs into TPNR considerably affects the impact strength of TPNR nanocomposites.

The results exhibited better impact strength for TMWCNTs and UTMWCNTs at 3 wt% with an increase of about 82% and 46%, respectively. This is due to the better dispersion of carbon nanotubes in the matrix, which generated a significant toughening effect on the TPNR/TMCWNTs nanocomposite compared with TPNR/UTMWCNTs nanocomposites. However, when the load is transferred to the physical network between the matrix and the filler, the debonding of the chain segments from the filler surface facilitates the relaxation of the matrix entanglement structure, leading to higher impact toughness.

The low impact energy was attributed to the filler content being more than 3wt%. This will reduce the ability of reinforced composites to absorb energy during fracture propagation. However, in the case of elastomer-toughened polymer, the presence of the elastomer basically produces stress redistribution in the composite, which causes micro cracking or crazing at many sites, thereby resulting in a more efficient energy dissipation mechanism [38].

Consequently, because of their higher surface energy and large aspect ratio, it will be difficult for the nanotubes to disperse in the TPNR when the TMWCNTs and UTMWCNTs content are higher. This will lead to less energy dissipating in the system due to the poor interfacial bonding and induces micro spaces between the filler and polymer matrix. This causes micro-cracks when impact occurs, which induces easy crack propagation. Therefore, the higher agglomeration of MWCNTs can cause the mechanical properties of the composites to deteriorate [39].

3.4. Thermal Properties

3.4.1. Glass Transition Temperature

The dynamic mechanical data shows that the glass transition temperature of the TPNR/UTMWCNTs and TPNR/TMWCNTs is affected by the addition of the different amounts of MWCNTs, as depicted in Figure 10.

From the figures, the Tg for the TPNR/TMWCNTs nanocomposites is higher than the corresponding temperature for the TPNR and TPNR/UTMWCNTs nanocomposites, usually the Tg of a polymeric matrix tends to increase with the addition of carbon nanotubes. The rise in Tg in any polymeric system is associated with a restriction in molecular motion, reduction in free volume and/or a higher degree of crosslinking (TPNR/TMWCNTs > TPNR/UTMWCNTs) due to the interactions between the polymer chains and the nanoparticles, and the reduction of macromolecular chain mobility.

With the high amount of MWCNTs (after 3wt %) of TMWCNTs and UTMWCNTs the Tg drops. This might be due to the phase separation/agglomeration of MWCNTs, this allows the macromolecules to move easily. When the content of MWCNTs is higher, the MWCNTs congregate, possibly because the intrinsic van der Waals forces occurs, which leads to bubbles and small aggregates. The conglomerations and matrix holes existing in the network of MWCNTs may perform as defects, which make the macromolecules move easily, and the Tg of the matrix is decreased.

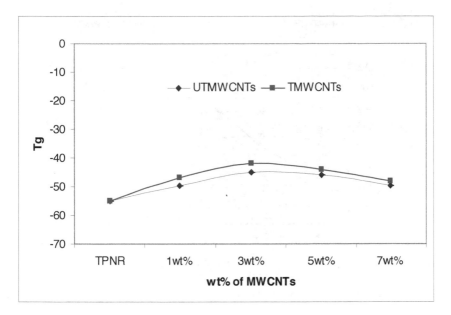

Figure 10. Glass Transition Temperature of TPNR reinforced with MWCNTs (with and without treatment).

3.4.2. Thermal Conductivity

To study the effect of MWCNTs filler on thermal conductivity, the temperature was varied from (30 – 150) °C. The carbon filler loading was from 1wt% to 7wt% for two types of carbon nanotubes (UTMWCNTs and TMWCNTs). Introducing MWCNTs to TPNR can significantly enhance the thermal conductivity of the TPNR matrix, as shown in Figure 11 and Figure 12.

As shown in figure 11 at 30°C the thermal conductivity of TPNR/TMWCNTs composites, Thermal conductivity increased at 3wt% compared to 1wt%, 5wt% and 7wt%, respectively, and for TPNR/UTMWCNTs, the thermal conductivity increased at 3wt%, as compared to TPNR at the same temperature as shown in figure 12. Thermal transport in the CNT composites includes phonon diffusion in the matrix and ballistic transportation in the filler.

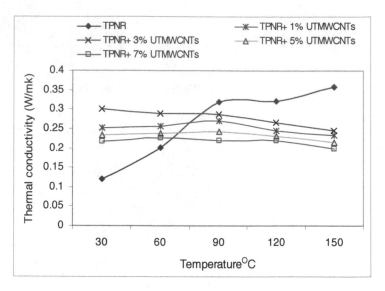

Figure 11. Thermal Conductivity of TPNR reinforced with UTMWCNTs.

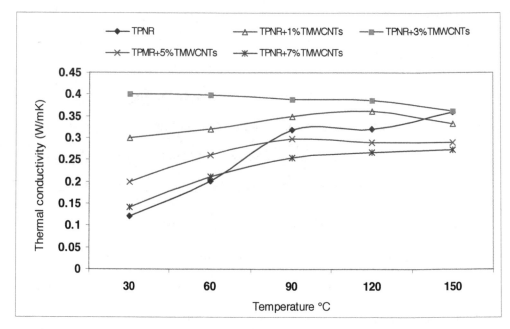

Figure 12. Thermal Conductivity of TPNR reinforced with TMWCNTs.

The improvement in thermal conductivity in TMWCNTs/TPNR may stem from the improved percolation because of the better dispersion and formation of a network [40]. The dispersion of 1wt% and 3wt% TMWCNTs is better than 5wt% and 7wt% in TPNR, at 5% and 7% the TMWCNTs agglomerated inside TPNR. Therefore, the large amounts of junctions among the carbon nanotubes form a single conducting path, which is believed to be the reason why the measured thermal conductivity is low. For the UTMWCNTs the conductivity at 3wt% and 1wt% is better than 5wt% and 7wt%, respectively.

The significant enhancement in the thermal conductivity of CNT nanocomposites is possibly attributed to the kinks or twists of UTMWCNTs. When the phonon travels along the nanotubes, if it meets the kinks or twists it would be blocked at those sites. The existence of such kinks or twists in CNTs would lead to a decrease in the effective aspect ratio of the nanotubes [41] when the amount of UTMWCNTs increases, and, thus, the thermal conductivity of UTMWCNTs-TPNR nanocomposites would be reduced. Therefore, the acid treatment of MWCNTs in TPNR could reduce these kinks or twists of TMWCNTs due to the good dispersion of MWCNTs in TPNR, causing the thermal conductivity of the nanocomposites to increase.

Two factors have been proposed to explain the significant enhancement of thermal conductivity with TMWCNTs (1) the rigid linkage between TMWCNTs and TPNR matrix with provides good interface compatibility which may reduce interface thermal resistance; (2) the good interface compatibility allows TMWCNTs to disperse well in the matrix, consequently, the results of the TEM indicate that TMWCNTs possess good dispersion and good compatibility in the TPNR matrix.

The formation of the UTMWCNTs bundles restrict the phonon transport in composites, which maybe be attributed to two reasons (1) the UTMWCNTs aggregation reduces the aspect ratio, consequently, decreasing the contact area between the UTMWCNTs and the TPNR matrix; (2) the UTMWCNTs bundles cause the phenomenon of reciprocal phonon vector, which acts like a heat reservoir and restricts heat flow diffusion.

The resistance to phonon movement from one nanotube to another through the junction will hinder phonon movement and, hence, limit the thermal conductivity. The low thermal conductivity could be partly due to the non-uniform diameter and size, as well as, the defects in and the nano-scale dimension of UTMWNTs. However, the numerous junctions between carbon nanotubes involved in forming a conductive path and the exceptionally low thermal conductance at the interface [42] are believed to be the main reason for the low thermal conductivity.

The effect of reducing the thermal conductivity is the transfer of phonons from nanotube to nanotube. This transition occurs by direct coupling between CNTs, in the case of the improper impregnated ropes, CNT-junctions and agglomerates, or via the matrix. In all these cases, the transition occurs via an interface and, thus, the coupling losses can be attributed to an intense phonon boundary scattering. At the same time the thermal conductivity decreases with the increase in temperature (if the temperature is near the melting point of the matrix). This indicates that the thermal conductivity of the composites is dominated by the interface thermal transport between the nanotube/matrix or nanotube/nanotube interface. Thus, it is believed that the decreased effective thermal conductivity of the studied composites could be due to the high interface thermal resistance across the nanotube/matrix or nanotube/nanotube interfaces.

As shown in Figure 11 and Figure 12, the thermal conductivity of TMWCNTs reinforced TPNR matrix composites for all volume fractions studied from 30°C to 150°C is better than UTMWCNTs. The effect of temperature on the thermal conductivity is clear from 30°C to 90°C, as shown in the figures. This is because of the opposing effect of temperature on the specific heat and thermal diffusivity. Eventually, at high temperatures, as the phonon mean free path is lowered, the thermal conductivity of the matrix approaches the lowest limit and the corresponding thermal resistivity approaches the highest limit.

3.4.3. Morphological Examination

The TEM can observe the morphology of UTMWCNTs/TPNR and TPNR/TMWCNTs nanocomposites, which indicates the dispersion abilities of MWCNTs in TPNR matrix before and after treatment of MWCNTs, which summarizes the TEM images of TPNR with 1wt%, 3wt% and 7wt% UTMWCNTs as shown in figure 12-14. Figure 13 shows the good dispersion of 3wt% of UTMWCNTs inside TPNR, and exhibits the better interfacial adhesion of UTMWCNTs and TPNR, Figure 14, 7wt% of UTMWCNTs, shows the poor dispersion and the large UTMWCNTs agglomerates of UTMWCNTs. This is because of the huge surface energy of MWCNTs, as well as, the weak interfacial interaction between UTMWCNTs and TPNR, which leads to inhomogeneous dispersion in the polymer matrix and negative effects on the properties of the resulting composites that causes a decrease in the tensile strength. This supports our results for thermal behavior, which due to the kinks or twists of CNTs can affect the thermal conductivity. Therefore, so when the phonon travels along the nanotube and the phonon meets the kinks or twists, it could be blocked at those sites. The existence of those kinks or twists in CNTs would result in a decrease in the effective aspect ratio of nanotubes at 7wt% UTMWCNTs because of agglomeration compared with 3wt% of MWCNTs due to the good dispersion. The homogenous dispersion of TMWCNTs in the composites is confirmed by TEM after acid treatment. Figure 15 shows the 3% TMCWNTs, which are very well dispersed in the matrix, there by suggesting a strong polymer nanotubes interfacial. Strong interfacial adhesion is essential for efficient stress transfer from the matrix to the nanotubes; this supports our observation that the higher efficiency of carbon nanotubes assists in enhancing the properties of TPNR. Low magnification was necessary to observe the poor dispersion of 7wt% of TMWCNTs in TPNR as depicted in Figure 17. The figure clearly shows a large number of unbroken carbon nanotubes but less than Figure 14, indicating a poor polymer/nanotube adhesion which is attributed to the reduction in the properties of TPNR/MWCNTs nanocompsites.

Characterization and Morphology of Modified Multi- Walled Carbon Nanotubes Filled
Thermoplastic Natural Rubber (TPNR) Composite

135

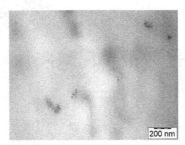

Figure 13. TPNR with 1wt% UTMWCNTs.

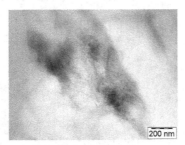

Figure 14. TPNR with 3wt% UTMWCNTs.

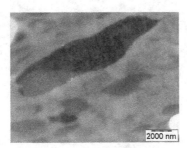

Figure 15. TPNR with 7wt% UTMWCNTs.

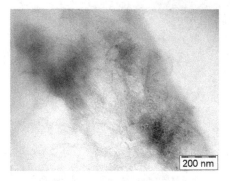

Figure 16. TPNR with 1wt% TMWCNTs.

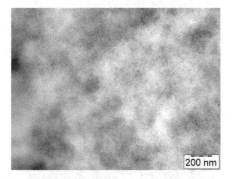

Figure 17. TPNR with 3wt% TMWCNTs.

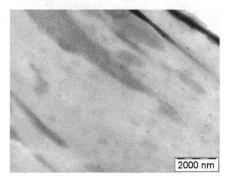

Figure 18. TPNR with 7wt% TMWCNTs.

4. Conclusion

Recently, it is believed that single-wall carbon nanotubes (SWCNTs), multi-walled carbon nanotubes (MWCNTs), coiled nanotubes and carbon nanofibers (CNFs) can be used as filler in the polymer matrix leading to composites with many enhanced properties, especially in mechanical properties. Furthermore, the inclusion of CNTs in a polymer holds the potential to improve the mechanical, electrical or thermal properties by orders of magnitude well above the performance possible with traditional fillers. In addition many researchers revealed that using functionalized MWCNTs or surface modification of MWCNTs as filler enhanced the properties of nanocomposites. This enhancement was probably suggested because of the homogenous dispersion and stronger interaction between the MWCNTs and the polymer matrix. After being treated with an acid, some functional groups were introduced onto the MWCNTs surface, which can form a physical interaction with the polymer chain. In this chapter, the effect of multi-walled carbon nanotubes with and without acid treatment on the properties of thermoplastic natural rubber (TPNR) was investigated. Two types of MWCNTs were introduced into TPNR, which are untreated UTMWCNTs (without acid treatment) and treated TMWCNTs (with acid treated MWCNTs). The acid treatment of MWCNTs removed catalytic impurities and generated functional groups such as hydroxyl, carbonyl and mainly carboxylic acid.

The results in this chapter show that the properties of MWCNTs can be improved by using this method. The TEM micrograph has shown that the effect of acid treatment has roughened the MWCNTs surface and also reduced the agglomeration. Various functional groups have been confirmed using FTIR. The TPNR nanocomposite was prepared using the melt blending method. MWCNTs are incorporated in the TPNR nanocomposite at different compositions which is 1, 3, 5 and 7 wt%. The addition of MWCNTs in the TPNR matrix improved the mechanical properties. At 3wt%, the tensile strength and Young's modulus of TPNR/UTMWCNTs increased 23% and 22%, respectively. For TPNR/TMWCNTs the optimum result of tensile strength and Young's modulus was recorded at 3% which increased 39% and 34%, respectively. In the addition the elongation of break decreased by increasing the amount of both types of MWCNTs.

The results exhibited better impact strength for UTMWCNT and TMWCNT at 3 wt% with an increase of almost 46 % and 82%, respectively. The reinforcing effect of two types of MWCNTs was also confirmed by dynamic mechanical analysis where the addition of MWCNTs have increased in the glass transition temperature (Tg) with an increase in the amount of MWCNTs (optimum at 3wt %) and it increased with the TMWCNTs more than the UTMWCNTs. Thermal conductivity improved with TMWCNTs compared to the UTMWCNTs. The homogeneous dispersion of two types of the MWNTs throughout the TPNR matrix and strong interfacial adhesion between the MWCNTs and the matrix as confirmed by the TEM images are proposed to be responsible for the significant mechanical enhancement.

Acknowledgements

The authors would like to thank the Malaysian Government and Universiti Kebangsaan Malaysia (UKM) under Science Fund Grant UKM-OUP-NBT-29-142/2011 and UKM-OUP-2012-135 for financial support.

Author details

Mou'ad A. Tarawneh* and Sahrim Hj. Ahmad

*Address all correspondence to: moaath20042002@yahoo.com

School of Applied Physics, Faculty of Science and Technology, Universiti Kebangsaan Malaysia, Malaysia

References

[1] Sakurada, I. (1985). Polyvinyl Alcohol Fibres. *International Fibre Science and Technology Series*, 6, Marcel Dekker, Inc., New York.

[2] Li, C. Y., & Chou, T. W. (2003). A structural mechanics approach for the analysis of carbon nanotubes. *Int. J. Solids Struct.*, 40, 2487-2499.

[3] Hernandez, E., Goze, C., Bernier, P., & Rubio, A. (1998). Elastic Properties of C and BxCyNz Composite Nanotubes. *Phys. Rev. Lett.*, 80, 4502-4505.

[4] Yakobson, B. I., Campbell, M. P., Brabec, C. J., & Bernholc, J. (1997). High strain rate fracture and C-chain unraveling in carbon nanotubes. *Comp. Mater. Sci*, 8, 341-348.

[5] Demcyzk, B. G., Wang, Y. M., Cumings, J., Hetman, M., Han, W., Zettl, A., & Ritchie, R. O. (2002). Direct mechanical measurement of the tensile strength and elastic modulus of multiwalled carbon nanotubes. *Mater. Sci. Eng. A*, 334, 173-178.

[6] Qian, E. C., Dickey, , Andrews, R., & Rantell, T. (2000). Load transfer and deformation mechanisms in carbon nanotube-polystyrene composites. *Appl. Phys. Lett*, 76, 2868-2870.

[7] Lopez, M. A., Valentine, L., Biagiotti, J., & Kenny, J. M. (2005). Thermal and mechanical properties of single-walled carbon nanotubes-polypropylene composites prepared by melt processing. *Carbon*, 43, 1499-1505.

[8] Liu, T., Phang, I. Y., Shen, L., Chow, S. Y., & Zhang, W. D. (2004). Morphology and mechanical properties of multi-walled carbon nanotubes reinforced nylon-6 nanocomposites. *Macromolecules*, 37, 7214-7222.

[9]   Frogley, M. D., Ravich, D., & Wagner, H. D. (2003). Mechanical properties of carbon nanoparticle-reinforced elastomers. *Compos. Sci. Technol*, 63, 1647-1654.

[10]   Koerner, H., Liu, W., Alexander, M., Mirau, P., Dowty, H., & Vaia, R. A. (2005). Deformation-morphology correlations in electrically conductive carbon nanotube- Thermoplastic polyurethane nanocomposites. *Polymer*, 46, 4405-4420.

[11]   Berber, S., Kwon, Y. K., & Tomanck, D. (2000). Unusually High Thermal Conductivity of Carbon Nanotubes. *Phys. Rev. Lett*, 84, 4613-4616.

[12]   Biercuk, M. J., Llaguno, M. C., Radosavljevic, M., Hyun, J. K., & Johnson, A. T. (2002). Carbon nanotube composites for thermal management. *Appl. Phys. Lett*, 80, 2767-2769.

[13]   Hone, J. W., Piskoti, C., & Zettl, A. (1999). Thermal conductivity of single-walled carbon nanotubes. *Phys. Rev. B*, 59, 2514-2516.

[14]   Gonnet, P., Liang, Z., Choi, E. S., Kadambala, R. S., Zhang, C., Brooks, J. S., Wang, B., & Kramer, L. (2006). Thermal conductivity of magnetically aligned carbon nanotube buckypapers and nanocomposites. *Current Applied Physics*, 6, 119-122.

[15]   Fangming, D., Csaba, G., Takashi, K., John, E. F., & Karen, I. W. (2006). An infiltration method for preparing single-wall nanotube/epoxy composites with improved thermal conductivity. *J. Poly. Sci. Part B: Poly. Phys*, 44, 1513-1519.

[16]   Sinha, S., Barjami, Iannacchione. G., Schwab, A., & Muench, G. (2005). Off-axis Thermal Properties of Carbon Nanotube Films. *J. Nanoparticle Res*, 7, 651-657.

[17]   Wen-Tai, H., & Nyan-Hwa, T. (2008). Investigations on the thermal conductivity of composites reinforced with carbon nanotubes. *Diamond & Related Materials*, 17, 1577-1581.

[18]   Moisala, Q. L., Kinloch, I. A., & Windle, A. H. (2006). Thermal and electrical conductivity of single- and multi-walled carbon nanotube-epoxy composites. *Composites Science and Technology*, 66, 1285-1288.

[19]   Shiren, W., Richard, L., Ben, W., & Chuck, Z. (2009). Dispersion and thermal conductivity of carbon nanotube composites. *Carbon*, 47, 53-57.

[20]   Assael, M. J., Antoniadis, K. D., & Tzetzis, D. (2008). The use of the transient hot-wire technique for measurement of the thermal conductivity of an epoxy-resin reinforced with glass fibres and/or carbon multi-walled nanotubes. *Composites Science and Technology*, 68, 3178-3183.

[21]   Oconnell, M. J., Boul, P., Ericson, L. M., Huffman, C., Wang, Y., Haroz, E., Kuper, C., Tour, J., Ausman, K. D., & Smalley, R. D. (2001). Reversible water-solublization of single-walled carbon nanotubes by polymer wrapping. *Chem Phys Lett*, 342, 265-271.

[22]   Niyogi, S., Hamon, M. A., Perea, D. E., Kang, C. B., Zhao, B., & Pal, S. K. (2003). Ultrasonic dispersion of single-walled carbon nanotubes. *J. Phys. Chem. B*, 107, 8799-8812.

[23] Viswanathan, G., Chakrapani, N., Yang, H., Wei, B., Chung, H., & Cho, K. (2003). Single-step in situ synthesis of polymer-grafted single-wallednanotube composites. *J Am Chem Soc*, 125, 9258-9259.

[24] Xie, X. L., Mai, Y. W., & Zhou, X. P. (2005). Dispersion and alignment of carbon nanotubes in polymer matrix: A review. *Mater Sci Eng*, 49, 89-112.

[25] Wang, Y., Wu, J., & Wei, F. (2003). A treatment method to give separated multiwalled carbon nanotubes with high purity, high crystallization and a large aspect ratio. *Carbon*, 41, 2939-2948.

[26] Abdullah, I., & Dahlan, M. (1998). Thermoplastic natural rubber blends. *Prog. Polym. Sci*, 23, 665-706.

[27] Abdullah, I., & Ahmad, S. (1992). Liquid NR as a compatibilizer in the blending of NR with PP. *Mater. Forum*, 16, 353-357.

[28] Tarawneh, Mou'ad A., Ahmad, Sahrim Hj., Rasid, Rozaidi, Yahya, S. Y., & Eh Noum, Se Yong. (2011). Thermal Behavior of a MWNT Reinforced Thermoplastic Natural Rubber Nanocomposite. *Journal of Reinforced Plastic and Composites*, 30(3), 216-221.

[29] Tarawneh, Mou'ad A., Ahmad, Sahrim Hj., Rasid, Rozaidi, Yahya, S. Y., & Eh Noum, Se Yong. (2011). Enhancement of the Mechanical Properties of Thermoplastic Natural Rubber Using Multi-walled Carbon Nanotubes. *Journal of Reinforced Plastic and Composites*, 30(4), 363-368.

[30] Tarawneh, Mou'ad A., & Ahmad, Sahrim Hj. Reinforced Thermoplastic Natural Rubber (TPNR) Composites with Different Types of Carbon Nanotubes (MWNTS), Book Name Carbon Nanotubes - Synthesis, Characterization, Applications (Chapter 21, Page 443-468). InTech. July 2011, 978-9-53307-497-9.

[31] Chang-Eui, H., Joong-Hee, L., Prashantha, K., & Suresh, G. A. (2007). *Composites Science and Technology*, 67, 1027-1034.

[32] Seung, H. L., Eunnari, C., So, H. J., & Jae, R. Y. (2007). Rheological and electrical properties of polypropylene composites containing functionalized multi-walled carbon nanotubes and compatibilizers. *Carbon*, 45, 2810-2822.

[33] Hirsch, A. (2002). Functionalization of single-walled carbon nanotubes. *Angew Chem Int Ed*, 41, 1853-1859.

[34] Sang, H. J., Young-Bin, P., & Kwan, H. Y. (2007). Rheological and mechanical properties of surface modified multi-walled carbon nanotube-filled PET composite. *Composites Science and Technology*, 67, 3434-3441.

[35] Potschke, P., Fornes, T. D., & Paul, D. R. (2002). Rheological behavior of multi-walled carbon nanotubes/polycarbonate composites. *Polymer*, 43, 3247-3255.

[36] Treacy, T. W. E., & Gibson, J. M. (1996). Exceptionally high Young's modulus observed for individual carbon nanotubes. *Nature*, 381, 678-680.

[37] Canche-Escamilla, G., Rodriguez-Laviada, J., Cauich-Cupul, J. I., Mendizabal, E., Puig, J. E., & Herrera-Franco, P. J. (2002). Flexural, impact and compressive properties of a rigid-thermoplastic matrix/cellulose fiber reinforced composites. *Compos. Part (A): Appl. Sci. & Manufact.*, 33, 539-549.

[38] Jianfeng, S., Weishi, H., Liping, W., Yizhe, H., & Mingxin, Y. (2007). The reinforcement role of different amino-functionalized multi-walled carbon nanotubes in epoxy nanocomposites. *Composites Science and Technology*, 67, 3041-3050.

[39] Kumar, S., Alam, M. A., & Murthy, J. Y. (2007). Effect of percolation on thermal transport in nanotube composites. *Appl Phys Lett*, 90, 104105-1-104105-3.

[40] Nan, C. W., Shi, Z., & Lin, Y. (2003). A simple model for thermal conductivity of carbon nanotube-based composites. *Chem Phys Lett.*, 375, 666-669.

[41] Yunsheng, X., Gunawidjaja, R., & Beckry, Abdel-Magid. (2006). Thermal behavior of single-walled carbon nanotube polymer-matrix composites. *Composites A*, 37(1), 114-121.

[42] Ramasamy, S., Shuqi, G., Toshiyuki, N., & Yutaka, K. (2007). Thermal conductivity in multi-walled carbon nanotubes/silica-based nanocomposites. *Scripta Materialia*, 56, 265-268.

Chapter 7

Mixtures Composed of Liquid Crystals and Nanoparticles

Vlad Popa-Nita, Valentin Barna, Robert Repnik and
Samo Kralj

Additional information is available at the end of the chapter

1. Introduction

The past decade has witnessed an increased interest in the study of mixtures [1–3] of various soft materials and nanoparticles. A characteristic feature of a nanoparticle is that at least one of its dimensions is limited to between 1 and 100 nm. It is of interest to find combinations where each component introduces a qualitatively different behavior into the system. Such systems are expected to play an important role in the emerging field of nanotechnology and also in composites with extraordinary material properties.

In several cases various liquid crystalline phases [4] are chosen as a soft carrier matrix. Their main advantageous properties are as follows. LCs are optically anisotropic and transparent. Their structure can be readily controlled by the confining surfaces and by applying an external electric or magnetic field. LCs exhibit a rich pallet of different structures and phases that can display almost all physical phenomena. In addition the chemistry of LCs is relatively well developed, which can mean the synthesis of LC molecules with the desired properties. As a result of these properties, even pure LC systems have found several applications, in particular in the electro-optics industry.

In our study we will confine our interest to the nematic LC phase formed by rod-like anisotropic molecules. The molecules tend to be parallel, at least locally. In bulk equilibrium nematic phase LC molecules are on the average aligned homogeneously along a single symmetry breaking direction, while translational ordering is absent. In thermotropic LCs nematic ordering is reached from the isotropic (ordinary liquid) phase by lowering the temperature via a weakly first order phase transition. Reversely, in lyotropic LCs the nematic ordering could be obtained via a first order phase transition by increasing packing density of LC molecules .

Various NPs are added to LC matrices in order to introduce additional quality into the system. It has been shown that in such mixtures one can obtain dramatically enhanced [2] or even new material properties [5] (e.g., multiferroics), which is of particular interest for composite materials with exceptional properties. Because the LC phases are reached via continuous symmetry-breaking phase transitions, the presence of NPs can stabilize the LC domain structures and consequently give rise to topological defects [3]. These can strongly interact with NPs, yielding different patterns that depend on the conditions at the LC-molecule-NP interface.

In several studies one uses as NPs carbon nanotubes (CNTs) [6–12]. Most of CNTs extraordinary properties of potential use in various applications could be realized in relatively well aligned samples. Recently it has been shown that liquid crystal alignment could trigger spontaneous ordering of CNTs with remarkably high degree of ordering [9–12]. CNTs orient parallel to average direction of liquid crystal (LC) alignment with an orientational order parameter between 0.6 to 0.9 [13–16]. Both, thermotropic [13, 14, 16] and lyotropic nematic LC phases [15] have been successfully applied as aligning solvents.

The theoretical study of the collective behavior of anisotropic nanoparticles dispersed in isotropic solvents or in liquid crystals is based on the observation that they can be consider essentially as rigid-rod polymers with a large aspect ratio [17]. The Onsager's theory for the electrostatic repulsion of long rigid rods has been used to investigate the phase behavior of SWNTs dispersed into organic and aqueous solvents [18]. In a good solvent, when the van der Waals attractive interaction between CNTs is overcome by strong repulsive interrods potential, the ordered phases of CNTs can form at room temperature. On the contrary, if the solvent is not good, the van der Waals attractive interactions between the rods are strong and as a result, only extremely dilute solutions of SWNTs are thermodynamically stable and no liquid crystal phases form at room temperature. The liquid crystallinity of CNTs with and without van der Waals interactions has been analyzed by using the density functional theory [19]. In the presence of van der Waals interaction, the nematic as well as the columnar phases occur in the temperature-packing fraction phase diagram in a wide range of very high temperatures. In the absence of van der Waals interaction, with an increase of packing fraction, the system undergoes an isotropic-nematic phase transition via a biphasic region. The isotropic-nematic packing fraction decreases with the increase of the aspect ratio of CNTs. To describe the dispersion of SWNTs in superacids, the Onsager theory for rigid rods was extended to include the length polydispersity and solvent mediated attraction and repulsion [20]. The main conclusion of these theoretical models is that to obtain liquid crystal phases of CNTs at room temperature the strong van der Waals interaction between them must be screened out. This requires a good solvent with an ability to disperse CNTs down to the level of individual tube.

In the previous papers [6–8] we have presented a phenomenological theory for predicting the alignment of length monodisperse CNTs dispersions in thermotropic nematic liquid crystals. We combined the Landau-de Gennes free energy for thermotropic ordering of the liquid crystal solvent and the Doi free energy for the lyotropic nematic ordering of CNTs caused by excluded-volume interactions between them. In the first paper [6], the interaction between CNTs and liquid crystal molecules is thought to be sufficiently weak to not cause any director field deformations in the nematic host fluid. The principal results of this first study could be summarized as follows. (i) The coupling between the CNTs and a LC seems to be dominated by an anisotropic surface tension not by any deformation of the director field because the rods are thin on the scale of the extrapolation length. This means that CNTs dispersed in NLCs are in the weak-anchoring limit. (ii) The first order nematic-isotropic phase transition of CNTs dispersed in a LC disappears for a strong enough coupling to the nematic host fluid. A tricritical point can be defined that within the Landau-de Gennes model exhibits universal characteristics if expressed in the right dimensionless variables. (iii) Although in the weak-anchoring limit, the coupling between the CNTs and the LC host is so strong that in practice one should expect CNTs always to be in the strong-coupling limit, i.e., above the tricritical point. This means CNTs in LCs are always strongly paranematic. (iv)The degree of alignment of CNTs in NLCs can be tuned by varying the CNT concentration or the temperature.

The phase and structural behavior of a mixture of CNT and LC using a mean field-type phenomenological model in the strong anchoring regime was presented in the second paper [7]. We have considered cases where the nematic director field is either nonsingular or where topological defects are present in the LC medium. The effective field experienced by CNTs yields pretransitional ordering below the critical point. Above the critical point, a gradual variation of orientational order of CNTs appears. In practice one should expect CNTs always to be in the strong-coupling limit, i.e., far above the critical point. This means CNTs in the nematic phase of LC are always strongly paranematic. The model predicts an increase of nematic-isotropic phase transition temperature of LC with the volume fraction of CNTs as well as the presence of a triple point in the phase diagram. For realistic values of the coupling constant, the degree of ordering of CNTs is enslaved by the properties of the host nematic fluid.

The comparison of the results for weak anchoring and strong anchoring regimes, respectively was presented in the third paper [8]. In both anchoring cases, the first-order nematic - isotropic phase transition of CNTs dispersed in the nematic phase transforms into a continuous transition for a strong enough coupling to the nematic host fluid. The corresponding critical value of the coupling parameter increases with increasing temperature being larger in the strong anchoring limit case. The numerical estimate of the coupling constants in the two anchoring regimes indicates that the coupling is so strong that CNTs are far above the critical point, meaning that the nematic-isotropic phase transition is a continuous one. In both cases, we have plotted the phase diagram of the homogeneous mixture for the same value of the coupling parameter. In both cases, three regions of the phase diagram could be distinguished and correspondingly the existence of triple points are shown. We mention that in both anchoring cases, the nematic-isotropic phase transition temperature of LC increases with the volume fraction of CNTs, a well-known experimental result.

Usually, after the acid treatment and ultrasonication (to enhance the dispersion and stability of the CNTs suspensions), depending on the temperature of water bath and time of ultrasonication, the obtained CNTs have different lengths and diameters [14, 21]. The influence of length bidispersity on the phase diagram and alignment of CNTs is the subject of the present paper. As the first step we extend our mesoscopic model [6–8] and consider length bidispersity of CNTs. We mention that the effect of bidispersity of the long rigid rods has been discussed in the framework of Onsager theory [22–25]. We shall refer to their results in the last section of the paper.

The plan of the paper is as follows. In the first part of the paper we focus on mixtures of nematic LCs and isotropic NPs. A simple phenomenological model is used which is sufficient to identify key mechanisms which might trigger phase separation. Conditions for efficient trapping of NPs to cores of topological defects is discussed. In the second part we confine our interest to mixtures of nematic LCs and carbon nanotubes. Using a simple model we take into account length dispersity of CNTs and analyze corresponding phase diagrams. In the last section we summarize the main results.

2. Binary nematogen - nonnematogen mixtures

We first consider a mixture of nematic LC and isotropic NPs using a relatively simple phenomenological model. We identify key phase separation triggering mechanisms. We also discuss conditions enabling efficient trapping of NPs within cores of topological defects or strongly localized elastic distortions. If these trapping sites are relatively uniformly spatially distributed they might prevent phase separation.

2.1. Free energy

We use semiphenomenological model within which the volume concentration of isotropic NPs is given by the conserved parameter ϕ. The orientational order of LC molecules is described by the symmetric and traceless tensor order parameter [4] $Q = \sum_{i=1}^{3} \lambda_i \vec{e}_i \otimes \vec{e}_i$, where λ_i and $\vec{e}_i$ stand for its eigenvalues and corresponding eigenvectors, respectively. In the case of uniaxial ordering Q is commonly expressed in terms of the nematic director field $\vec{n}$ and the uniaxial orientational order parameter S as [4]

$$Q = S \left(\vec{n} \otimes \vec{n} - \frac{1}{3}I \right). \tag{1}$$

Here I stands for the identity tensor. The unit vector $\vec{n}$ points along the local uniaxial ordering direction, where states $\pm\vec{n}$ are equivalent (the so called head-to-tail invariance). The extent of fluctuations is determined by S, where $S = 1$ and $S = 0$ reflect rigid alignment along $\vec{n}$ and isotropic liquid ordering, respectively. If strong distortions are present biaxial states could be locally entered. Degree of biaxiality is assessed via parameter [26]

$$\beta^2 = 1 - \frac{6(trQ^3)^2}{(trQ^2)^3}, \tag{2}$$

ranging in the interval $[0, 1]$. Uniaxial configurations correspond to $\beta^2 = 0$ and an ordering with the maximum degree of biaxiality is signaled by $\beta^2 = 1$.

The free energy F of a mixture is expressed as

$$F = \int \left(f_m + f_c + f_e + f_i \delta(\vec{r} - \vec{r}_i) \right) d^3 \vec{r}. \tag{3}$$

The quantity δ stands for the delta measure, $\vec{r}_i$ locates NP-LC interfaces and the integral runs over the LC volume. The role of different contributions in Eq.(3) is as follows.

The mixing term f_m describes the isotropic mixing of the two components. Within the Flory theory [27] it is expressed as

$$f_m = \frac{k_B T}{v_{lc}}(1 - \phi)\ln(1 - \phi) + \frac{k_B T}{v_{np}}\phi\ln\phi + \chi\phi(1 - \phi). \tag{4}$$

Here k_B is the Boltzmann constant, T is the absolute temperature, and χ stands for the Flory-Huggins parameter [27]. The volume of a LC molecule and of a nanoparticle is given by v_{lc} and v_{np}, respectively,

The condensation contribution f_c enforces orientational LC ordering below a critical temperature T_{NI}. Is is expressed as [4]

$$f_c = \frac{3a(T - T^*)}{2}Q_{ij}Q_{ij} - \frac{9B}{2}Q_{ij}Q_{jk}Q_{kj} + \frac{9C}{4}(Q_{ij}Q_{ij})^2, \tag{5}$$

where summation over repeated indices is assumed. The quantities a, B, C, are material constants and T^* denotes the spinodal temperature limit of the isotropic phase of the pure LC. This condensation free energy term describes a weakly first order nematic-isotropic phase transition. At $T = T_{NI} = T^* + B^2/(4aC)$, the two phases, nematic ($S^{(NI)} = B/(2C)$) and isotropic ($S = 0$) coexist in equilibrium.

The deviations from homogeneous nematic ordering are penalized by the elastic term

$$f_e = \frac{L}{2}Q_{jk,i}Q_{jk,i}, \tag{6}$$

which is expressed within a single elastic constant approximation.   Here L is the representative bare nematic elastic constant.

The conditions at the NP-LC interface are determined by the term f_i. We express it as

$$f_i = -we_k Q_{kj}e_j, \tag{7}$$

where $w > 0$ is the anchoring strength favoring the nematic ordering at an interface and $\vec{e}$ stands for the local surface normal.

2.2. Phase separation tendency

We proceed by identifying key mechanisms favoring phase separation tendency in a mixture of NPs and nematic liquid crystal. We describe global LC orientational ordering with a spatially averaged order parameter $\overline{S}$ and volume concentration of nanoparticles $\overline{\phi}$. Here the over-bar $\overline{(..)}$ denotes the spatial average. Relative presence of NPs and LC molecules in the mixture is therefore given by $\overline{\phi} = N_{np}v_{np}/V$ and $1 - \overline{\phi}$, respectively. Here N_{np} counts the number of NPs and V stands for the volume of the sample.

The resulting average free energy density is expressed as $\overline{f} = \overline{f}_m + \overline{f}_c + \overline{f}_e + \overline{f}_i$, where

$$\overline{f}_m \sim \frac{k_B T}{v_{lc}}(1 - \overline{\phi}) \ln(1 - \overline{\phi}) + \frac{k_B T}{v_{np}} \overline{\phi} \ln \overline{\phi} + \chi \overline{\phi}(1 - \overline{\phi}), \tag{8}$$

$$\overline{f}_c \sim (1 - \overline{\phi}) \left(a(T - T^*)\overline{S}^2 - B\overline{S}^3 + C\overline{S}^4 \right), \tag{9}$$

$$\overline{f}_e \sim (1 - \overline{\phi}) L\overline{S}^2 / \overline{\zeta}_d^2, \tag{10}$$

$$\overline{f}_i \sim - (1 - \overline{\phi}) \overline{\phi} w \overline{S}. \tag{11}$$

The factor $(1 - \overline{\phi})$ present in terms $\overline{f}_c$ and $\overline{f}_e$ accounts for the part of the volume not taken up by LC. Furthermore, the factor $(1 - \overline{\phi}) \overline{\phi}$ in $\overline{f}_i$ accounts for absence of this term if $\overline{\phi} = 0$ or $\overline{\phi} = 1$. Note that in general $T^* = T_*(\phi)$. Simple binary modeling [28] suggests $T^* = T_0 - \lambda \overline{\phi}$, where T_0 and λ are positive material constants. It accounts for weaker interactions among LC molecules due to presence NPs. In general NPs could introduce spatially nonhomogeneous orientational ordering of LC molecules which is taken into account by $\overline{f}_e$. On average degree of elastic distortions in $\overrightarrow{n}$ is approximated by the average domain length $\overline{\zeta}_d$.

From the expression for $\overline{f}$ one can extract the effective Flory-Huggins [27] parameter

$$\chi_{eff} = \chi + a\lambda \overline{S}^2 - w\overline{S}. \tag{12}$$

The phase transition takes place if χ_{eff} exceeds a threshold value χ_c. In typical LCs it holds [29] $\chi << a_0\lambda$ and $\chi < \chi_c$. Henceforth we limit our attention to such cases. Consequently, in the isotropic phase (where $\overline{S} = 0$) homogeneous mixtures are established. On entering orientational ordered phase different scenaria can be realized. We first consider cases where the wetting interaction between LCs and NPs is negligible weak (i.e., $w \sim 0$). In this case phase separation is very likely. It is triggered providing $\chi + a\lambda \overline{S}^2 \sim a\lambda \overline{S}^2 > \chi_c$. However, strong enough surface wetting interaction could suppress phase separation providing $\chi + a\lambda \overline{S}^2 - w\overline{S} < \chi_c$.

Next, we consider cases where there exist localized regions in LC ordering exhibiting strong local distortions. Therefore, in some parts $\overline{\zeta}_d$ entering the expression for $\overline{f}_e$ is relatively small. From Eq.(10) we infer that local free energy penalties could be reduced if they are occupied by NPs (i.e., $\overline{\phi} \sim 1$). Therefore, the structure of expression for $\overline{f}$ suggests that NPs tend to assemble at local elastic distortions in order to reduce the total free energy penalty of the system.

2.3. Interaction between NPs and topological defects

In this section we investigate in more detail interaction between NPs and localized elastic distortions. We estimate general conditions for which this interaction is attractive. For this purpose we study a specific example where we enforce a topological defect within a cell. We add NPs exhibiting different surface constraints and determine conditions for which attractive interaction is enabled.

We consider LC ordering within a cylindrical plane-parallel cell of thickness h and radius R. The cell is schematically depicted in Fig.1.

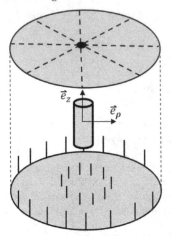

Figure 1. Schematic presentation of the hybrid plan-parallel cell hosting a nanoparticle at its symmetry axis. The diameter and the height of cylindrically shaped NP is in simulation set to be equal to the biaxial correlation length ζ_b. Furthermore, we set $h = 2R = 10\zeta_b$, where h describes the height and R the radius of the cell.

We use the cylindrical coordinate system determined by unit vectors $\{\vec{e}_\rho, \vec{e}_\varphi, \vec{e}_z\}$. Here $\vec{e}_\rho$ is the radial unit vector, $\vec{e}_z$ points along the z-coordinate, while $\vec{e}_\varphi = \vec{e}_z \times \vec{e}_\rho$. We enforce a topological defect (boojum) [30] by imposing strong hybrid anchoring conditions at the confining plates. At the top plate ($z = h$) strong uniaxial radial anchoring is set imposed, i.e., we enforce $Q(z = h) = Q_{rad} \equiv S_{eq}\left(\vec{e}_\rho \otimes \vec{e}_\rho - \frac{1}{3}I\right)$. Here S_{eq} stands for the equilibrium nematic order parameter. At the bottom plate we impose strong homeotropic anchoring, i.e. $Q(z = h) = Q_{hom} \equiv S_{eq}\left(\vec{e}_z \otimes \vec{e}_z - \frac{1}{3}I\right)$. At the lateral wall we set free boundary conditions. The resulting equilibrium equations were solved numerically, where technical details are given in [30] and [31].

The boojum structure is well characterized by its biaxial spatial profile shown in Fig. 2.

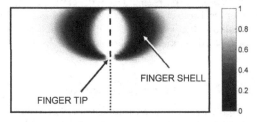

Figure 2. The cross-section through the boojum where we plot the degree of biaxiality β^2. A biaxial shell joins the isotropic finger tip and the upper surface. Along the cylindrical axis the system exhibits uniaxial ordering due to topological reasons (dashed line: negative uniaxiality, dotted line: positive uniaxiality). In the illustration the anchoring strength at the top plane is finite. The characteristic defect size is comparable to the biaxial correlation length. On the right side of the figure the grayscale bar of $1 - \beta^2$ is shown.

We plot $\beta^2(\rho, z)$ dependence (see Eq.(2)) in the plane through the defect core. Note that the system exhibits cylindrical symmetry. The boojum is characterized by a finger-like structure, where the finger tip is melted by topologically reasons. The boojum core structure is analyzed in detail in [30] and here we briefly summarize its main characteristics. The defect core is dominated by a *finger* protruding into the cell's interior along its symmetry axis. The center of the *finger*, residing at the cylinder axis, is negatively uniaxial ($S < 0$). It ends in a melted (isotropic) point ($S = 0$) to which we refer as the *finger tip*. It is placed roughly at the distance $\zeta_f \sim \zeta_b$ from the surface, where ζ_b stands for the biaxial correlation length. Below the *finger tip* the nematic configuration is positively uniaxial ($S > 0$) at the symmetry axis. The *finger* is enclosed within a biaxial shell exhibiting maximal biaxiality [26, 30] which joins the *finger* tip with the upper surface. By topological reasons the nematic order parameter melts only at the *finger tip* for realistic anchoring strengths. Note that in Fig. 2 the biaxial profile is plotted for a more realistic finite anchoring strength for which the finger-like profile is well pronounced.

At the cylinder axis we place a cylindrically shaped NP. The height and diameter of the NPs is set equal to ζ_b. Our interest is to estimate impact of NP surface treatment on interaction with its surrounding. For this purpose we impose three qualitatively different strong boundary conditions at the NP surface which is determined by the position vector $\vec{r}_i$: i) $Q(\vec{r}_i) = 0$, ii) $Q(\vec{r}_i) = Q_{rad}$, iii) $Q(\vec{r}_i) = Q_{hom}$. These conditions locally enforce melting, radial, and spatially homogeneous configuration, respectively. We calculate the LC free energy within the cell as a function of the NP position along the z axis. Note that three qualitatively different areas exist within the cell. These are: i) region surrounding the melted point at the boojum *finger tip*, ii) prevailing radial ordering at $z \sim h$, and iii) the homogeneous ordering along $\vec{e}_z$ at $z \sim 0$. We vary the position of NP along the z-axis, and for each position we calculate the free energy of the system. In Fig. 3 we plot the total free energy as a function of z-coordinate for three different surface treatments.

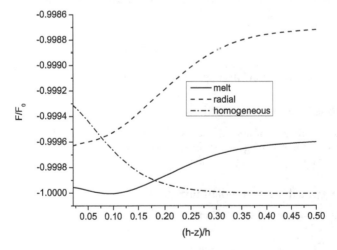

Figure 3. The free energy F of the system as a function of the nanoparticle position along the symmetry axis. The free energy is scaled with respect to the minimum of free energy F_0 calculated for the melted boundary condition. One sees that for the i) melted ii) radial and iii) homogeneous boundary condition the free energy exhibits minimum at i) the finger tip ii) $z = h$, iii) $z = 0$, respectively.

We see that NP enforcing i) melting, ii) radial, iii) homogeneous configuration tends to migrate towards the i) *finger tip*, ii) top plate, iii) bottom plate, respectively. Simulations shows that NPs tend to migrate towards regions which exhibit a similar local structure with respect to the conditions at the NP surface. Therefore, to assemble efficiently NPs at a defect core it is essential to make the surface coating such that the effective NP configuration resembles the defect core structure.

We next assume that NPs environment resembles a defect core structure and therefore tend to be trapped within the core of the defect. In the following we estimate the free energy gain if the NP is trapped within the core. For illustration we consider a line defect (disclination) of length h. Note that lattices of topological defects can be stabilized either by inherent LC property (e.g. chirality [3]) or imposed geometrically by imposing frustrating boundary conditions [32]. The corresponding condensation free energy penalty ΔF_c for introducing the defect line inside an orientational ordered medium is roughly given by

$$\Delta F_c \sim a(T_{NI} - T)\bar{S}^2 \pi \xi^2 h. \tag{13}$$

Here T_{NI} refers to I-N phase transition temperature and $T < T_{NI}$. The core average radius is roughly given by the relevant (uniaxial or biaxial) order parameter correlation length ξ. Within the core the LC order is either essentially melted (i.e. $S \sim 0$) or strongly biaxial [33].

We next assume that NPs are added to the LC medium and that they collect at the disclination line. If NPs do not apparently disrupt the defect core structure then the condensation free energy penalty is decreased due to the reduced volume occupied by the energetically costly essentially isotropic (or strongly biaxial) phase [34]. One refers to these effect as the *Defect Core Replacement mechanism* [3]. The resulting decrease in ΔF_c penalty reads

$$\Delta F_c() \sim a(T_{NI} - T)\bar{S}^2 \left(\pi \xi^2 h - N_{np}^{(def)} v_{np} \right), \tag{14}$$

where $N_{np}^{(def)}$ counts number of NPs trapped within the core.

3. Dispersions of carbon nanotubes

We next consider mixtures of nematic LCs and carbon nanotubes with length bidispersity. We furthermore assume homogeneous uniaxial orientational alignment of all components along a single symmetry breaking direction.

3.1. Free energy of three component mixtures

The mixture is characterized by the volume fractions of the three components:

$$\Phi_i = \frac{N_i v_i}{\sum\limits_{i=1}^{3} N_j v_j} \text{ with } \sum_{i=1}^{3} \Phi_i = 1, \tag{15}$$

where N_i is the number of molecules of component i ($i = 1$ defines the liquid crystal with molecules of length $L_{01} = 3$ nm and diameter $D_1 = 0.5$ nm, $i = 2$ the CNTs of length $L_{02} = 400$ nm and diameter $D = 2$ nm, and $i = 3$ the CNTs of length $L_{03} = 800$ nm and diameter $D = 2$ nm) and v_i is the volume of a particle of component i. For latter convenience we introduce scaled CNT lengths $L_i = L_{0i}/D$ ($i = 2,3$).

The degree of alignment of every component of the mixture is characterized by the scalar order parameter S_i [4]. The corresponding isotropic liquid of the component i is characterized by $S_i = 0$ while a perfectly oriented nematic phase would correspond to $S_i = 1$.

The free energy per unit volume of the mixture is expressed as

$$f = f_{CNT} + f_{LC} + f_C, \tag{16}$$

where f_{CNT} describes contribution of the two CNT components dispersed in the LC fluid, f_{LC} represents the free energy density of nematic liquid crystal order, while f_C takes into account the coupling between LC molecules and CNTs, respectively.

The free energy density of CNTs is given by [6, 7]

$$\frac{f_{CNT}}{k_B T} = \sum_{i=1}^{3} \frac{\Phi_i}{v_i} \ln \Phi_i + \sum_{i=2}^{3} \frac{L_i \Phi_i^2}{6v_i} \left[\left(\frac{3}{L_i \Phi_i} - 1 \right) S_i^2 - \frac{2}{3} S_i^3 + S_i^4 \right] - \frac{\gamma_{23}}{k_B T} \Phi_2 \Phi_3 S_2 S_3. \tag{17}$$

The first sum represents the entropic isotropic contribution due to mixing of the two CNT components and LC neglecting their orientational degree of ordering [27]. The second sum describes a first order orientational phase transition of the i-species of CNT from the isotropic phase with $S_i = 0$, to the nematic phase with $S_i = (1 + \sqrt{9 - 24/L_i \Phi_i})/4$. The first order nematic-isotropic phase transition takes place at $\Phi_i^{(NI)} = 2.7/L_i$ and $S_i^{(NI)} = 1/3$. It is obtained starting from the Onsager theory [35] and using the Smoluchovsky equation [36, 37]. The model neglects the van der Waals attractions between CNTs which are responsible for their tendency to form bundles. The last term in Eq. (17) represents the interaction energy between the different CNTs species, where the interaction parameter γ_{23} is given by

$$\gamma_{23} = 8 k_B T / \pi D^3 \approx 10^6 \text{ N/m}^2.$$

This expression for γ_{23} is obtained using the same Doi procedure [36, 37] starting with the Onsager theory for a bidisperse hard rods system.

The second term in Eq. (16) is the Landau-de Gennes free energy density [4] which describes the weakly first-order nematic-isotropic phase transition of thermotropic LC

$$f_{LC} = \Phi_1 [a(T - T^*) S_1^2 - B S_1^3 + C S_1^4]. \tag{18}$$

For representative LC material we chose pentylcyanobiphenyl (5CB), for which $T^* = 307.55$ K, $a \approx 5.2 \cdot 10^4$ J/m³K, $B \approx 5.3 \cdot 10^5$ J/m³, $C \approx 9.7 \cdot 10^5$ J/m³ [38]. This choice yields $S_1^{(NI)} = 0.275$ and $T_{NI} = 308.95$ K.

The third term in Eq.(16) represents the coupling between the liquid crystal molecules and the two CNT species. The resulting coupling term structure in both anchoring limits (weak and strong) was estimated in [6, 7]. The two limits are defined by the ratio DW/K, where W is the anchoring energy of a LC-nanotube interface, K is the average Frank nematic elastic constant. For typical values of $D = 2$ nm, $K \approx 10^{-11}$ N, $W \approx 10^{-6}$ N/m, $DW/K \ll 1$. Consequently, only the weak-anchoring limit needs to be considered, as already concluded by Lynch and Patrick [13]. The corresponding free energy density of coupling is approximately given by [6]:

$$f_C = -\gamma_{12}\Phi_2 S_1 S_2 \left(1 - \frac{1}{2}S_2\right) - \gamma_{13}\Phi_3 S_1 S_3 \left(1 - \frac{1}{2}S_3\right). \tag{19}$$

The terms in brackets ensures that $S_i \to 1$ when $\gamma_{12}(\gamma_{13}) \to \infty$ as it should. The coupling parameters γ_{12} and γ_{13} depend only on the anchoring energy of CNTs at the LC molecules surface and the diameter of CNTs [6] (no on their lengths). Therefore

$$\gamma_{12} = \gamma_{13} = \gamma_1 = 4W/3D \approx 10^3 \text{ N/m}^2$$

and the coupling free energy can be written as

$$f_C = -\gamma_1 S_1 \left[\Phi_2 S_2 \left(1 - \frac{1}{2}S_2\right) + \Phi_3 S_3 \left(1 - \frac{1}{2}S_3\right)\right]. \tag{20}$$

The free energy per unit volume of a monodisperse system (one species of CNTs of the diameter D and length L_{02} ($L_2 = L_{02}/D$, the volume v_2 and the volume fraction Φ) dispersed in a LC (with the molecular volume v_1 and the volume fraction $1 - \Phi$) is given by

$$f = k_B T \left[\frac{1-\Phi}{v_1}\ln(1-\Phi) + \frac{\Phi}{v_2}\ln\Phi\right]$$
$$+ k_B T \frac{L_2\Phi^2}{6v_2} \left[\left(\frac{3}{L_2\Phi_2} - 1\right)S_2^2 - \frac{2}{3}S_2^3 + S_2^4\right]$$
$$+ (1-\Phi)[a(T-T^*)S_1^2 - BS_1^3 + CS_1^4] - \gamma_1\Phi S_1 S_2 \left(1 - \frac{1}{2}S_2\right). \tag{21}$$

3.2. Equilibrium equations

In the bidisperse case, the equilibrium values of the order parameters are obtained by minimization of the free energy density (Eqs. (16), (17), (18), and (19))with respect to S_1, S_2, and S_3, respectively. From the corresponding equations we find the equilibrium values of the order parameters in the nematic phase (S_{in}) and paranematic phase (S_{ip}), respectively. Once the minimization procedure has been solved the volume fractions of the coexisting phases are found by solving the equilibrium conditions:

$$\mu_2(S_{1n}, S_{2n}, S_{3n}, \Phi_{2n}, \Phi_{3n}) = \mu_2(S_{1p}, S_{2p}, S_{3p}, \Phi_{2p}, \Phi_{3p}),$$
$$\mu_3(S_{1n}, S_{2n}, S_{3n}, \Phi_{2n}, \Phi_{3n}) = \mu_3(S_{1p}, S_{2p}, S_{3p}, \Phi_{2p}, \Phi_{3p}),$$
$$g(S_{1n}, S_{2n}, S_{3n}, \Phi_{2n}, \Phi_{3n}) = g(S_{1p}, S_{2p}, S_{3p}, \Phi_{2p}, \Phi_{3p}), \tag{22}$$

where the chemical potential of the two species of CNTs μ_2, μ_3, and the grand potential g are defined by the equations

$$\mu_2 = \frac{\partial f}{\partial \Phi_2}; \quad \mu_3 = \frac{\partial f}{\partial \Phi_3}; \quad g = f - \mu_2 \Phi_2 - \mu_3 \Phi_3. \tag{23}$$

There are three equilibrium equations with four variables: Φ_{2n}, Φ_{3n}, Φ_{2p}, and Φ_{2p}. We take Φ_{3p} as freely variable and calculate the other three Φ_{2n}, Φ_{3n}, Φ_{2p} from the coexistence equations (22).

In the monodisperse case, the equilibrium values of the order parameters (S_{1p}, S_{2p}, S_{1n}, and S_{2n})are obtained minimizing the free energy (21) with respect to S_1 and S_2 and the equilibrium conditions are given by

$$\mu(S_{1n}, S_{2n}, \Phi_n) = \mu(S_{1p}, S_{2p}, \Phi_p),$$
$$g(S_{1n}, S_{2n}, \Phi_n) = g(S_{1p}, S_{2p}, \Phi_p), \tag{24}$$

where the chemical potential of CNTs and the grand potential are defined as: $\mu = \partial f / \partial \Phi$, and $g = f - \mu \Phi$, respectively.

3.3. Phase behavior

In the first part of this section we present the phase behavior of monodisperse CNTs immersed in LC as a function of T, Φ and γ_1, while in the last part the analyze of the phase behavior of the bidisperse system is analyzed as a function of T, Φ_2, Φ_3 and γ_1.

3.3.1. Monodisperse CNTs

The coupling term between CNTs and the LC molecules in (21) induces two different region in the phase diagram separated by a critical line $\gamma_1^{(c)}(T)$. i) For $\gamma_1 < \gamma_1^{(c)}$, the CNTs exhibit a first order (discontinuous) phase transition between a paranematic phase (a phase with a low degree of orientational order) and a nematic phase (the *subcritical* region). ii) On the contrary, for $\gamma_1 > \gamma_1^{(c)}$, CNTs display gradual variation of S_2 with Φ (the *supercritical* regime).

The critical line $\gamma_1^{(c)}(T)$ is obtained by solving the equations $\partial f / \partial S_1 = \partial f / \partial S_2 = \partial^2 f / \partial S_2^2 = \partial^3 f / \partial S_2^3 = 0$. They yield at the critical point universal values for the order parameter $S_2^{(c)} = 1/6$ and volume fraction $\Phi^{(c)} = 18/7L_2$. The $\gamma_1^{(c)}(T)$ dependence for the two species of CNTs is presented in Figure 4.

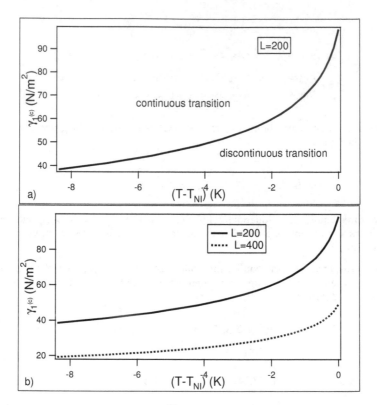

Figure 4. The critical value of the coupling parameter $\gamma_1^{(c)}$ as function of temperature calculated for both species of CNTs in the nematic LC phase.

On decreasing the temperature the external field felt by the CNTs is increasing due to increasing value of S_1 and the nematic-isotropic phase transition of CNTs becomes gradual for lower values of interaction parameter $\gamma_1^{(c)}$. Furthermore, the critical value of γ_1 decreases with increasing the length of CNTs. Therefore, the continuity of paranematic-nematic phase transition is favored by longer CNTs.

The phase diagram for the monodisperse system in the *subcritical* regime, for $L_2 = 200$ and $L_2 = 400$, respectively is shown in Figure 5.

Depending on temperature, we distinguish two regions in the phase diagram: i) for $T < T_{NI}$, the LC is in the nematic phase and the CNTs exhibit a first order phase transition with increasing Φ from a paranematic to a nematic phase. With decreasing temperature, the order parameter jump $(S_{2n} - S_{2p})$ as well as the difference in the volume fractions $(\Phi_n - \Phi_p)$ become lower and they cancel at some temperature lower for shorter CNTs. This is due to the fact that the value of $\gamma_1 = 40.33 \, \text{N/m}^2$ considered, corresponds to a critical temperature $T - T_{NI} = -7.31$ K for shorter CNTs $(L_2 = 200)$ and $T - T_{NI} = -0.45$ K for longer CNTs $(L_2 = 400)$, respectively. We emphasize that the longer CNTs become ordered at lower

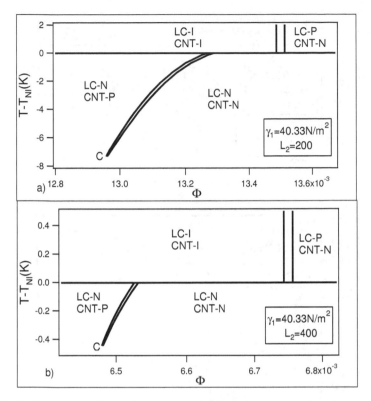

Figure 5. The (T, Φ) phase diagram of a monodisperse system in the *subcritical* regime. I means isotropic, P-paranematic, and N-nematic

volume fraction. ii) for $T > T_{NI}$, the LC is in the isotropic phase ($S_1 = 0$) and the volume fraction gap of CNTs at the transition does not depend on temperature (the Flory horn [27]). Again the longer CNTs become aligned at lower volume fractions. It is important to note also the influence of CNTs on the LC alignment in this region. Above T_{NI}, the transition of CNTs from isotropic to nematic induces the transition of LC from isotropic to a paranematic phase (with a very very small degree of ordering for this value of the coupling constant). This problem of the influence of LC properties by the CNTs is not elucidated neither theoretically, nor experimentally yet and will be a subject of a future study.

To see in more detail the orientational order developed in the system, we have plotted in Figures 6 and 7 the order parameter variation as a function of the volume fraction Φ of the CNTs at a fixed temperature.

In Figure 6a, the temperature corresponds to a *subcritical* regime ($\gamma_1 < \gamma_1^{(c)}$) (see Figure 4), and the transition of CNTs is a discontinuous one with a jump in the order parameter. On the contrary, the temperature in Figure 6b corresponds to a *supercritical* regime because for this temperature $\gamma_1 > \gamma_1^{(c)}$ (see Figure 4). As a consequence, the CNTs phase transition is

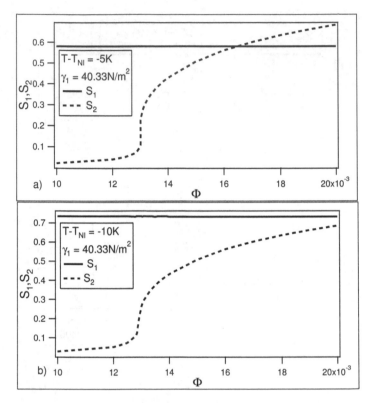

Figure 6. The order parameter variations for two temperatures lower than T_{NI} in the *subcritical* regime.

continuous, the order parameter is continuous at the transition. In both figures, the order parameter of the liquid crystal is constant. Therefore the CNTs are enslaved by LC. The degree of ordering of CNTs is present even at small volume fractions, so that the CNTs are in the paranematic phase.

In Figure 7, the temperature is greater than nematic-isotropic phase transition temperature, so that the LC is in the isotropic phase (even if the degree of ordering exists it is very small and can not be seen on the figure). For low volume fraction, the CNTs are in the isotropic phase and becomes nematic by a first order phase transition at some value of Φ.

The phase diagram for the monodisperse system in the *supercritical* regime, for $L_2 = 200$ and $L_2 = 400$, respectively is shown in Figure 8.

For a more realistic value of the coupling constant $\gamma_1 \gg \gamma_1^{(c)}$, in the nematic phase of LC, there is only a gradual variation of the order parameter of CNTs with the volume fraction and temperature. Above T_{NI}, the transition isotropic-nematic of CNTs is first order and also an induced first order isotropic-paranematic phase transition takes place in LC. With increasing temperature, the transitions takes place at a larger volume fractions. For longer CNTs, the volume fractions at the transition are lower.

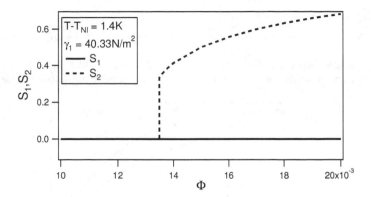

Figure 7. The order parameter variations for a temperature larger than T_{NI} in the *subcritical* regime.

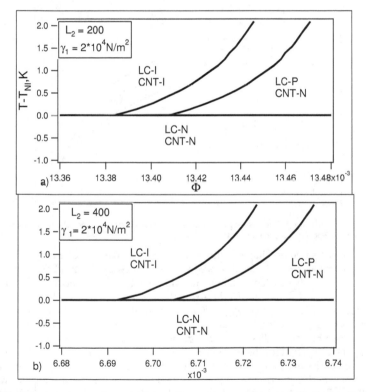

Figure 8. The (T, Φ) phase diagram of a monodisperse system in the *supercritical* regime. Signification of I, P, and N as in Figure 5.

3.3.2. Bidisperse CNTs

The (Φ_3, Φ_2) phase diagram of the bidisperse CNTs suspension in the LC in the *subcritical regime* is shown in Figure 9. In Figure 6a, the LC is in the nematic phase $(T - T_{NI} = -0.42K)$, while in Figure 9b, the LC is in the isotropic phase $(T - T_{NI} = 1.4K)$. Thick lines indicates phase boundary, while the thin lines connects the coexisting pairs (Φ_{2p}, Φ_{3p}) and (Φ_{2n}, Φ_{3n}).

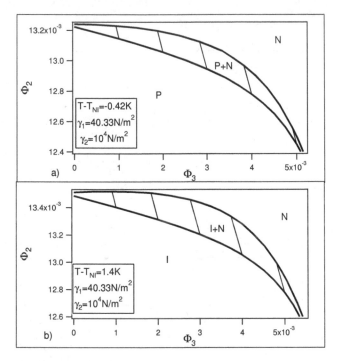

Figure 9. The (Φ_3, Φ_2) phase diagram in the *subcritical* regime for the bidisperse system, in nematic and isotropic phase of LC, respectively. I-isotropic, P-paranematic, and N-nematic.

The Figure 6 reveals two qualitatively different regimes, to which we refer as *decoupled* and *coupled* regime, respectively. In the *decoupled* regime defined by the conditions $\Phi_3 \to 0$ and $\Phi_2 \to 0$, respectively, the system exhibits monodisperse-type behavior. In the limit of very low volume fraction of the longer CNTs $(\Phi_3 \to 0)$, the system is monodisperse containing only one species of CNTs of length $L_{02} = 400$ nm. In the limit of very low volume fraction of the shorter CNTs $(\Phi_2 \to 0)$, the system is monodisperse containing only one species of CNTs of length $L_{03} = 800$ nm. In these two subregions, due to very small values of Φ_3 and Φ_2, respectively, the coupling term between CNTs species (the γ_{23} term in Eq. (17)) is relatively small and the species are independent. On the contrary, in the *coupled region* (intermediate region in Figure 6), the interaction term in Eq. (17) becomes important and the two CNTs species influence each other. In this region, the volume fraction of the longer CNTs increases in the nematic phase, while that of the shorter CNTs decreases.

In the case of a more realistic value of the coupling parameter between the LC molecules and CNTs $\gamma_1 = 2 * 10^3$ N/m^2 $>> \gamma_1^{(c)}$ (*supercritical* regime, the Φ_3, Φ_2) phase diagram for the bidisperse CNTs in the isotropic phase of LC ($T - T_{NI} = 1.4$K is plotted in Figure 10.

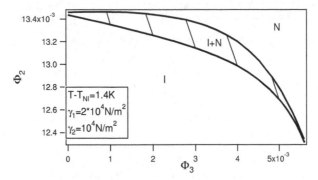

Figure 10. The (Φ_3, Φ_2) phase diagram in the *supercritical* regime for the bidisperse system, in the isotropic phase of LC, respectively. I-isotropic, and N-nematic.

The phase diagram is similar to that of Figure 9 showing again the existence of the *decoupled* and *coupled* regions that we have discussed previously.

4. Conclusions

In the paper we analyze phase behavior of mixtures consisting of LC soft carrier matrix and immersed NPs. A relatively simple phenomenological modeling is used where we focus on isotropic and nematic LC ordering.

In the first part we consider isotropic NPs. We derive the effective free energy of the mixture from which we extract the effective Flory-Huggins parameter. Its structure reveals that on entering nematic ordering phase separation is very probable. However, it could be suppressed by LC-NP interfacial contribution providing that it promotes nematic ordering. From the structure of the average elastic free energy term we also conclude that NPs have in general tendency to assemble at localized sites exhibiting relatively strong elastic distortions. We proceed by studying interaction between a nanoparticle and a topological defect, which is a typical representative of localized strong elastic distortions. It is of interest to identify conditions for which NPs could be effectively trapped to tunable localized distortions. For example, in such a way phase separation could be prevented. In addition, localized elastic distortions could be exploited controlled positional trapping of immersed NPs.

As a model system we use a cylindrical hybrid cell possessing the boojum topological defect. We consider different surface treatment of the nanoparticle and analyze where it is placed in order to minimize total free energy of the system. We find out that a nanoparticle is attracted to a region the structure of which is compatible with configuration enforced by the nanoparticle. Therefore, one could trap NPs to topological defects if its surface enforces configuration resembling the defect core structure. We further show that condensation penalty of forming defects could be in this case significantly reduced due to the *Defect Core Replacement mechanism* [3, 34].

In the second part of the paper we study interaction between nematic LC ordering and CNTs. We extend the mesoscopic model [6, 7] to include length bidispersity of CNTs dispersed in LC in the weak anchoring limit of the coupling between LC molecules and CNTs (in this limit, the coupling is dominated by the anisotropy of the surface tension not by the deformation of the director field). The main conclusions of our study can be summarized as follows:

1. Depending on the coupling between the LC molecules and CNTs (the value of the coupling constant γ_1), two different regimes can be defined: i) if $\gamma_1 < \gamma_1^{(c)}$, (the *subcritical* regime), the nematic-isotropic phase transition of CNTs dispersed in LC is first order and ii) if $\gamma_1 > \gamma_1^{(c)}$ (the *supercritical* regime), the transition is continuous. $\gamma_1^{(c)}$ is the critical value of the coupling parameter depending on the temperature (Figure 1). In both regimes, the isotropic phase of CNTs transforms into a phase with a small degree of ordering, a paranematic phase. Above the critical point this degree of orientational order is strongly increased.

2. The CNTs species are enslaved by the LC (the nematic LC order parameter depends only on the temperature not on the volume fractions of the two species).

3. The longer CNTs are driven into the nematic phase ($\Phi_{3n} - \Phi_{3p} > 0$ in Figures 6 and 7).

4. The longer CNTs induces a larger volume fractions differences $\Phi_{2n} - \Phi_{2p} > 0$ for the shorter CNTs.

5. In the nematic phase, the longer CNTs are more ordered than the shorter ones ($S_{3n} - S_{2n} > 0$).

We emphasize that the last three conclusions are similar with those obtained using the Onsager theory of nematic-isotropic phase transition of the hard rods [22–25], while the first two are specific to the dispersion of CNTs into LC.

Finally, we point out the mesoscopic model for length bidisperse CNTs dispersed in LC presented here is only a first step in considering the important influence of polydispersity on the ordering of CNTs in nematic fluids. Together with considering the attraction interaction between CNTs, this subject will be study in a future work.

Acknowledgments

V.P.-N. thanks to T. J. Sluckin for useful discussion, gratefully acknowledge the hospitality of l'Ecole Normale Supérieure de Lyon and the funding from CNRS. V.B. acknowledge support from the Romanian National Authority for Scientific Research, CNCS - UEFISCDI, project number PN-II-ID-PCE-2011-3-1007.

Author details

Vlad Popa-Nita[1,*], Valentin Barna[1], Robert Repnik[2] and Samo Kralj[2,3]

* Address all correspondence to: v.popanita@gmail.com

1 Faculty of Physics, University of Bucharest, Bucharest, Romania
2 Laboratory Physics of Complex Systems, Faculty of Natural Sciences and Mathematics, University of Maribor, Maribor, Slovenia
3 Jožef Stefan Institute, Ljubljana, Slovenia

References

[1] A.C. Balazs, T. Emrick, T.P. Russell, Science 314, 1107 (2006).

[2] F. Li, O. Buchnev, C. Cheon, A. Glushchenko, V. Reshetnyak, Y. Reznikov, T. J. Sluckin, and J. L. West, Phys. Rev. Lett. 97, 147801 (2006).

[3] B. Rozic, V. Tzitzios, E. Karatairi, U. Tkalec, G. Nounesis, Z. Kutnjak, G. Cordoyiannis, R. Rosso, E.G. Virga, I. Musevic, S. Kralj, Eur. Phys. J. E 34, 17 (2011).

[4] P.G. de Gennes and J. Prost, The Physics of Liquid Crystals, Oxford University Press, Oxford (1993).

[5] B. Rozic, M. Jagodic, S. Gyergyek, M. Drofenik, S. Kralj, G. Lahajnar, Z. Jaglicic, Z. Kutnjak, Ferroelectrics 410, 37 (2011).

[6] P. van der Schoot, V. Popa-Nita, and S. Kralj, J. Phys. Chem. B 112, 4512 (2008).

[7] V. Popa-Nita and S. Kralj, J. Chem. Phys. 132, 024902 (2010).

[8] V. Popa-Nita, M. Cvetko and S. Kralj, in Electronic Properties of Carbon Nanotubes, (Edited by Jose Mauricio Marulanda (Intech, 2011).

[9] C. Zakri, Liquid Crystals Today 16, 1 (2011).

[10] S. Zhang and S. Kumar, Small 4, 1270 (2008).

[11] J. P. F. Lagerwall and G. Scalia, J. Mater. Chem. 18, 2890 (2008).

[12] M. Rahman and W. Lee, J. Phys. D: Appl. Phys. 42, 063001 (2009).

[13] M. D. Lynch and D. L. Patrick, Nano Lett. 2, 1197 (2002).

[14] I. Dierking, G. Scalia, and P. Morales, J. of Appl. Phys. 97, 044309 (2005).

[15] J. Lagerwall, G. Scalia, M. Haluska, U. Dettlaff-Weglikowska, S. Roth, and F. Giesselmann, Adv. Mater. 19, 359 (2007).

[16] N. Lebovka, T. Dadakove, L. Lysetskiy, O. Melezhyk, G. Puchkovska, T. Gavrilko, J. baran, M. Drodz, J. Molecular Structure 887, 135 (2008).

[17] M. J. Green, N. Behabtu, M. Pasquali, W. W. Adams, Polymer 50, 4979 (2009).

[18] Y. Sabba and E. L. Thomas, Macromolecules 37, 4815 (2004).

[19] A. M. Somoza, C. Sagui, and C. Roland, Phys. Rev. B63, 081403-1 (2001).

[20] M. J. Green, A. N. G. Parra-Vasquez, N. Behabtu, M. Pasquali, J. Chem. Phys. 131, 084901 (2009).

[21] W. Song and A. H. Windle, Macromolecules 38, 6181 (2005).

[22] H. N. W. Lekkerkerker, Ph. Coulon and R. Van der Haegen, J. Chem. Phys. 80, 3427 (1984).

[23] T. Odijk, Macromolecules 19, 2313 (1986).

[24] G. J. Vroege and H. N. W. Lekkerkerker, Rep. Prog. Phys. 55, 1241 (1992).

[25] H. N. W. Lekkerkerker and G. J. Vroege, J. Philos. Trans. R. Soc. London, Ser A, 344, 419 (1993).

[26] P. Kaiser, W. Wiese, and S. Hess, J. Non-Equilib. Thermodyn. 17, 153 (1992).

[27] P. J. Flory, Proc. R. Soc. A 243, 73 (1956).

[28] S. Kralj, Z. Bradac, and V. Popa-Nita, J. Phys. Condens. Matter 20, 244112 (2008).

[29] V. J. Anderson, E. M. Terentjev, S. P. Meeker, J. Crain, and W. C. K. Poon, Eur. Phys. J E 4, 11 (2011).

[30] S. Kralj, R. Rosso, and E.G. Virga, Phys. Rev. E 78, 031701 (2008).

[31] S. Kralj, R. Rosso, and E.G. Virga, Phys. Rev. E 81, 021702 (2010).

[32] D. Coursault , J. Grand , B. Zappone , H. Ayeb , G. Levi, N. Felidj , and E. Lacaze, Adv. Mater. 24, 1461 (2012).

[33] S. Kralj, E.G. Virga, S. Zumer, Phys.Rev.E 60, 1858 (1999).

[34] H. Kikuchi, M. Yokota, Y. Hisakado, H. Yang, and T. Kajiyama, Nat. Mater. 1, 64 (2002).

[35] L. Onsager, Ann. N. Y. Acad. Sci. 51, 727 (1949).

[36] M. Doi, J. Polym. Sci., Part B: Polym. Phys. 19, 229 (1981); N. Kuzuu and M. Doi, J. Phys. Soc. Jpn. 52, 3486 (1983).

[37] M. Doi and S. F. Edwards, *Theory of Polymer Dynamics* (Clarendon, Oxford, 1989).

[38] P. Oswald and P. Pieranski, *Nematic and Cholesteric Liquid crystals; concepts and physical properties illustrated by experiments* (Taylor and Francis Group, CRC Press, in Liquid Crystals Book Series, Boca Raton, 2005).

Chapter 8

Carbon Nanotubes and Their Composites

Veena Choudhary, B.P. Singh and R.B. Mathur

Additional information is available at the end of the chapter

1. Introduction

Carbon nanotubes (CNTs), a fascinating material with outstanding properties has inspired the scientist, engineer and technologist for wide range of potential applications in many areas [1]. Since all these properties are concerned directly to the atomic structure of nanotubes, it is quite necessary to have a thorough understanding of the phenomenon to control nanotube size, the number of shells (walls), the helicities and the structure during growth. The full potential of nanotubes for applications will not be realized until the structure of nanotubes during their growth is optimized and well controlled. For utilization of CNTs properties in real world applications, like composite preparation, it is desired to obtain high quality and in bulk quantity using growth methods that are simple, efficient and inexpensive. Significant work has been carried out in this field and various methods have been studied to synthesize CNTs by several researchers.

2. Properties

Carbon nanotubes are endowed with exceptionally high material properties, very close to their theoretical limits, such as electrical and thermal conductivity, strength, stiffness, toughness and low density.

2.1. Mechanical properties of CNTs

The strength of C-C bond gives a large interest in mechanical properties of nanotubes. Theoretically, these should be stiffer than any other known substance. Young's modulus of the single walled carbon nanotubes (SWCNTs) can be as high as 2.8-3.6 TPa and 1.7-2.4 TPa for multiwalled carbon nanotubes (MWCNTs) [2]which is approximately 10 times higher than steel, the strongest metallic alloy known. Experimental values of Young's modulus for SWCNTs are reported as high as to 1470 GPa and 950 GPa [3, 4] for MWCNTs, nearly 5 times of steel. There are no direct mechanical testing experiments that can be done on individual nanotubes (nanoscopic specimens) to determine directly their axial strength. However, the indirect experiments like AFM provide a brief view of the mechanical properties as well as scanning probe techniques that can manipulate individual nanotubes, have provide some basic answers to the mechanical behavior of the nanotubes [5]. The analysis performed on several MWCNTs gave average Young's modulus values of 1.8 TPa, which is higher than the in – plane modulus for single crystal graphite. So the high stiffness and strength combined with low density implies that nanotubes could serve as ideal reinforcement in composite materials and provide them great potential in applications such as aerospace and other military applications.

2.2. Electrical properties of CNTs

The nanometer dimensions of CNTs, together with the unique electronic structure of a graphene sheet, make the electronic properties of these one-dimensional (1D) structures extraordinary. The one dimensional structure of CNTs helps them in making a good electric conductor. In a 3D conductor the possibility of scattering of electrons is large as these can scatter at any angle. Especially notable is the fact that SWCNTs can be metallic or semiconducting depending on their structure and their band gap can vary from zero to about 2 eV, whereas MWCNTs are zero-gap metals. Thus, some nanotubes have conductivities higher than that of copper, while others behave more like silicon. Theoretically, metallic nanotubes having electrical conductivity of 10^5 to 10^6 S/m can carry an electric current density of 4×10^9 A/cm2 which is more than 1000 times greater than copper metal and hence can be used as fine electron gun for low weight displays. Due to the large diameter of MWCNTs, their transport properties approaches those of turbostatic graphite. Theoretical study also shows that in case of MWCNTs the overall behavior is determined by the electronic properties of the external shell. Conductivities of individual MWCNTs have been reported to range between 20 and 2×10^7 S/m [6], depending on the helicities of the outermost shells or the presence of defects [7].The electronic properties of larger diameter MWCNTs approach those of graphite. Nanotubes have been shown to be superconducting at low temperatures. As probably CNTs are not perfect at ends and end defects like pentagons or heptagons are found to modify the electronic properties of these nanosystems drastically. There is great interest in the possibility of constructing nanoscale electronic devices from nanotubes and some progress is being made in this area. SWCNTs have been recently used to form conducting and semiconducting layers (source, drain and gate electrodes) in thin films transistors. So the high electrical conductivity of CNTs makes them an excellent additive to impart electrical conductivity in otherwise insulating polymers. Their high aspect ratio means that a very low loading is needed to form a connecting network in a polymer compared to make them conducting.

2.3. Thermal properties of CNTs

CNTs are expected to be very good thermal conductors along the tube, but good insulators laterally to the tube axis. Experiments on individual tubes are extremely difficult but measurements show that a SWCNT has a room-temperature thermal conductivity along its axis of about 3500 W m^{-1} K^{-1} and MWCNTs have a peak value of ~ 3000 W m^{-1} K^{-1} at 320 K; compare this to copper, a metal well-known for its good thermal conductivity, which transmits 385 W m^{-1} K^{-1} [8]. Although for bulk MWCNTs foils, thermal conductivity limits to 20 W m^{-1} K^{-1} suggesting that thermally opaque junctions between tubes severely limit the large scale diffusion of phonons. The thermal conductivity of CNTs across axis (in the radial direction) is about 1.52 W m^{-1} K^{-1}, which is about as thermally conductive as soil. Both SWCNT and MWCNT materials and composites are being actively studied for thermal management applications, either as "heat pipes" or as an alternative to metallic addition to low thermal conductive materials. In case of composites, the important limiting factors are quality of dispersion and interphase thermal barriers.

3. Synthesis of CNTs

A variety of synthesis methods now exist to produce carbon nanotubes. The three main production methods used for synthesis of CNTs are d.c. arc discharge, laser ablation and chemical vapor deposition (CVD).

3.1. d.c. arc discharge technique

The carbon arc discharge method, initially used for producing C60 fullerenes, is the most common and perhaps easiest way to produce carbon nanotubes as it is rather simple to undertake. In this method two carbon rods placed end to end, separated by approximately 1mm, in an enclosure that is usually filled with inert gas (helium, argon) at low pressure (between 50 and 700 mbar) as shown in Figure 1. Recent investigations have shown that it is also possible to create nanotubes with the arc method in liquid nitrogen [9]. A direct current of 50 to 100 A driven by approximately 20 V creates a high temperature (~4000K) discharge between the two electrodes. The discharge vaporizes one of the carbon rods (anode) and forms a small rod shaped deposit on the other rod (cathode). Large-scale synthesis of MWCNTs by a variant of the standard arc-discharge technique was reported by Ebbesen and Ajayan [10]. A potential of 18 V dc was applied between two thin graphite rods in helium atmosphere. At helium pressure of ~500 Torr, the yield of nanotubes was maximal of 75% relative to the starting graphitic material. The TEM analysis revealed that the samples consisted of nanotubes of two or more concentric carbon shells. The nanotubes had diameters between 2 and 20 nm, and lengths of several micrometers. The tube tips were usually capped with pentagons.

If SWCNT are preferable, the anode has to be filled with metal catalyst, such as Fe, Co, Ni, Y or Mo. Experimental results show that the width and diameter distribution depends on the composition of the catalyst, the growth temperature and the various other growth conditions. If both electrodes are graphite, the main product will be MWCNTs. Typical sizes for MWCNTs are an inner diameter of 1-3nm and an outer diameter of approximately 10nm. Because no catalyst is involved in this process, there is no need for a heavy acidic purification step. This means MWCNT can be synthesized with a low amount of defects.

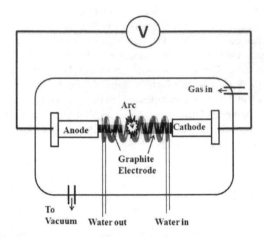

Figure 1. Schematic diagram of dc-arc discharge set-up

In most of the studies, SWCNTs are synthesized using the dc-arc discharge process by filling the catalyst powder into a hole drilled in a graphite elctrode act as an anode and arcing takes place between this anode and a pure graphite based cathode in optimized chamber conditions. In one of the study by Mathur et al. [11] SWCNTs and MWCNTs were synthesized simultaneously in a single experiment selectively. In their experiment, However, instead of filling the catalyst powder into a hole drilled in a graphite electrode; they prepared a catalyst/graphite composite electrode. Coke powder, catalyst powder, natural graphite powder and binder pitch were thoroughly mixed together in a ball mill in appropriate proportions and molded into green blocks using conventional compression molding technique. A mixture of Ni and Co powders was used as catalyst. The green blocks were heated to 1200° C in an inert atmosphere to yield carbonized blocks with varying compositions of coke, natural graphite powder, Ni and Co. These electrodes were used as the anodes in the arcing process and a high density graphite block was used as the cathode. A uniform gap of 1– 2 mm was maintained between the electrodes during the arcing process with the help of a stepper motor for a stable arc-discharge (dc voltage 20–25 V, current 100–120 A, 600 torr helium).The SWCNTs yield was found to be doubled in this case.

3.1.1. Characteristics of CNTs produced by d.c. arc discharge technique

Arc discharge is a technique that produces a mixture of components and requires separating nanotubes from the soot and the catalytic metals present in the crude product. In this technique both SWCNT and MWCNT can be produced and it has been described by several researchers.

The scanning electron microscope (SEM) and transmission electron microscope (TEM) are generally used to observe the physical appearance of any carbon based soot. Similarly, Mathur et al. [11] used SEM and TEM for the observation of SWCNT and MWCNT produced from the arc discharge technique as shown in Figure 2. In this technique, the carbon material deposits on the chamber and cathode. The arcing process resulted in the formation of web-like deposits on the inner walls of the arc chamber. A typical SEM micrograph of such deposit (Figure 2a) revealed the presence of SWNT bundles along with the amorphous carbon and catalyst particles. Rod-like microstructures aligned preferentially along the length of the cathode were also found at the tip of the cathode as shown in Figure 2b. The inset in Figure 2b shows the presence of graphitized carbon and sharp needle-like nanostructure when these rods are powdered. Upon detailed electron microscopic examination, these needles exhibited the MWCNT structure with an outer diameter of 20–25 nm (Figure 2c).

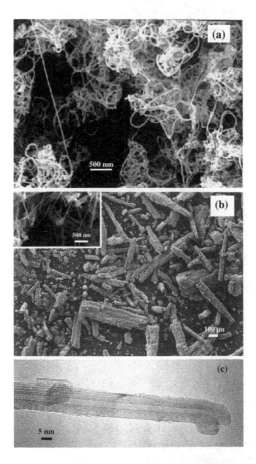

Figure 2. (a) SEM micrograph of the chamber deposit showing the presence of long and flexible carbon nanotubes. (b) SEM micrograph of the cathode deposit showing the presence of rod-like microstructures. The inset figure shows the presence of needle-like nanostructures present within each microstructure. (c) TEM micrograph of a single needle-like nanostructure (Reprinted with permission from Elsevier (11))

The nature of the soot can be identified using Raman spectrometer and generally used for confirmation of the quality of CNTs. The nature of these two deposits obtained using this arc discharge process was confirmed from their respective Raman spectrum (Figure 3). The Raman spectrum of the chamber deposit showed the characteristic radial breathing and tangential bands at 165–183 and 1591 cm^{-1}, respectively. The strong G-band at 1580 cm^{-1} in the Raman spectrum of the cathode deposit and its TEM image depicted in Figure 2c, confirmed that the cathode deposit predominantly contained MWCNTs. The prominent D-band around 1350 cm^{-1} seen in both spectrum is attributed to the presence of disordered carbonaceous material present in the as-prepared deposits. In their study, Mathur et al. [11] show that SWCNTs deposit on the arc chamber and MWCNTs on cathode deposit.

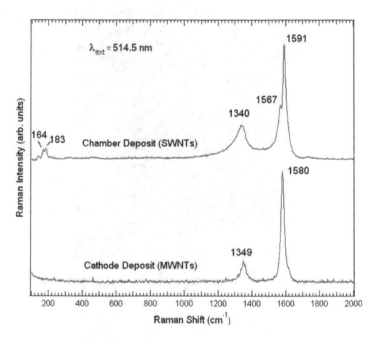

Figure 3. Room temperature Raman spectrum of the chamber and cathode deposit (Reprinted with permission from Elsevier (11))

3.2. Chemical vapor deposition

Pyrolysis of organometallic precursors such as metallocenes (e.g. ferrocene) in a furnace provides a straight forward procedure to prepare CNT by CVD technique. Different hydrocarbons, catalyst and inert gas combinations have been used by several researchers in the past for the growth of CNT by CVD technique. In one of the study by Mathur et al. [12], CNTs were grown inside the quartz reactor by thermal decomposition of hydrocarbons, e.g. toluene in presence of iron catalyst obtained by the decomposition of organometallic like ferrocene. The furnace provided a constant temperature zone of 18 cm in the centre as shown in Figure 4. The reaction zone was maintained at 750°C. Once the temperature was reached, the solution containing a mixture of ferrocene and toluene in particular proportion (0.077 g ferrocene in 1 ml toluene) was injected in the reactor at a point where the temperature was 200°C. Argon was also fed along with the charge as a carrier gas and its flow rate was adjusted so that the maximum amount of precursor is consumed inside the desired zone.

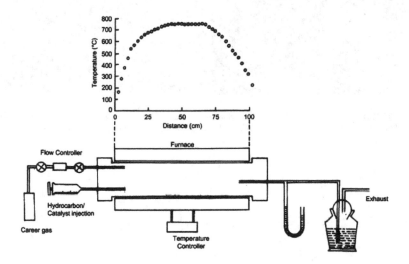

Figure 4. Schematic diagram of the CVD reactor along with the temperature profile (Reprinted with permission from Elsevier (12))

3.2.1. Characteristics of CNTs Produced by CVD Technique

CNTs are produced in the form of big bundles using CVD technique. The physical appearance of the as produced CNTs is shown in Figure 5a and Figure 5b for SEM and TEM respectively. Figure 5a shows a big CNT bundle of length >300μm and the inset image of Figure 5a shows very good quality of uniform CNTs. Figure 5b shows the TEM image of as produced CNTs confirming the presence of MWCNT with metallic catalytic impurites either on the tip of the tube or in the cavity of of CNTs (inset of Figure 5b).

Further confirmation of the quality and type of CNTs can be obtained using Raman spectrometer as shown in Figure 6. This shows the tangential band at 1580 cm^{-1} (G band) of high intensity and the disorder-induced band at 1352 cm^{-1} (D band) as a perfect MWCNT nature [13]. The ratio of intensity of G to D band gives the information regarding the quality of the CNTs. The high value of intensity ratio of G/D band confirms the better quality of CNTs.

Figure 5. (a) SEM image of aligned CNT bundle synthesized by CVD technique.The inset figure shows the very good quality of uniform CNTs (b) TEM image of as grown MWCNT and inset image shows the MWCNTs with encapsulated metallic impurities

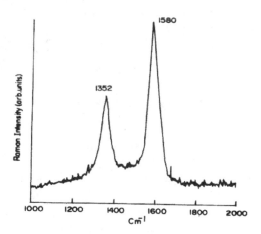

Figure 6. Raman spectrum of CVD-grown MWCNTs.

3.3. Laser ablation

In the laser ablation process, a pulsed laser vaporizes a graphite target containing small amounts of a metal catalyst [14] as shown in Figure 7. The target is placed in a furnace at roughly 1200°C in an inert atmosphere. The nanotubes develop on the cooler surface of the reactor, as the vaporized carbon condenses. The yield of nanotube synthesis by this process is roughly 70%.

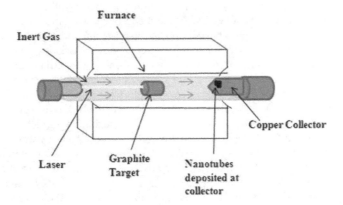

Figure 7. Schematic diagram of Laser ablation set-up for CNT synthesis

3.3.1. Characteristics of CNTs produced by laser ablation technique

The laser-ablation prepared samples usually contain >70% nearly endless, highly tangled ropes of SWCNTs along with nanoscale impurities. The SWCNTs formed in this case are bundled together by van der Waals forces. Laser vaporisation results in a higher yield for SWCNT synthesis and a narrower size distribution than SWCNTs produced by arc-discharge [15]. The nanotubes generated by the laser ablation and arc discharge technique are relatively impure, with presence of unwanted carbonaceous impurities and not operated at higher scale; therefore, the overall production costs are high.

Compared to other methods for synthesis of CNTs, more parameters, including temperature, feeding gases, flow rate, catalyst components and heating rate are accessible to control the growth process in CVD. By changing the growth conditions, we can control the properties of the produced CNTs such as length, orientation and diameter to some extent. It has been observed that the gas phase processes produces CNTs with fewer impurities and are most amenable to large scale processing. So the gas phase techniques such as CVD, for nanotube growth offer the greatest potential for scaling up nanotube production for processing of composites.

4. Purification

During CNTs synthesis, impurities in the form of catalyst particles, amorphous carbon and non tubular fullerenes are also produced. The most of the production methods involve the use of catalysts which are normally transition metals (Fe, Co, Ni or Y), these remains in the resulting nanotubes as spherical or cylindrical particles after experiments. Through careful control of process parameters one could minimize the formation of amorphous carbon particles, so that the main impurities in CNTs are the remaining catalyst particles. However as most of these catalytic particles may either hide in internal cavity or stick firmly to the walls of CNTs, it is almost impossible to get rid of these effectively without damaging the nanotubes. Several purification methods have been tried to overcome these impurities. In one of the study by Mathur et al. [11], SWCNT soot prepared by dc arc discharge process was purified by removing various forms of impurities, such as amorphous carbon, graphitic nanoshells and catalyst particles present in the chamber deposit by applying a judicious combination of wet and dry chemical methods (acid treatment and oxidation). In this process, initially SWCNT soot were oxidized at 350 °C for 6h in air which remove the amorphous carbon followed by refluxing in HCl for the removal of metallic impurities like Ni and Co and again oxidation at 550°C for 30 min for the removal of graphitic nanoshell.The final product gives 97% purified SWCNT. The MWCNTs produced by CVD technique contains mainly ~10% metallic impruites which can be removed by heating it in the inert atmosphere at 2500°C in graphitization furnace. This process gives >99% pure MWCNTs and also helps in annealing out the defects in the tubes. This graphitization process at high temperature can also be useful for removal of impurities in the arc discharge produced SWCNTs soots with the combination of other purification steps.

5. Nanocomposites

Because of the high strength and stiffness of CNTs, they are ideal candidates for structural applications. For example, they may be used as reinforcements in high strength, low weight and high performance composites. Presently there is a great interest in exploiting the exciting properties of these CNTs by incorporating them into some form of polymer matrix.

5.1. Composite fabrication techniques

A large number of techniques have been used for the fabrication of CNT-polymer nanocomposites based on the type of polymer used.

5.1.1. Solvent casting

The solution casting is most valuable technique to form CNTs/polymer nanocomposites. However, its use is restricted to polymers that are soluble. Solvent casting facilitates nanotube dispersion and involves preparing a suspension of CNTs in the desirable polymer solution via energetic agitation (magnetic stirring or sonication) and then allowing the solvent to evaporate to produce CNT-polymer nanocomposites. A lot of study is available in open literature for the formation of CNT nanocomposites by this method [16-18]. Mathur et al. [18] cast the solutions of the MWCNT/polystyrene (PS)/toluene and MWCNT/ polymethyl methacrylate (PMMA)/toluene suspensions after sonication into a petry dish to produce nanotubes composites with enhanced electrical and mechanical properties. Benoit et al. [19]

obtained electrically conductive nanocomposites by dispersing CNT and PMMA in toluene, followed by the drop casting on substrate. The choice of solvent is generally made based on the solubility of the polymer. The solvent selection for nanotube dispersion also had a significant influence on the properties of the nanocomposites and studied by Lau and co-workers [20].Their results demonstrates that, contrary to the general belief that small traces of CNTs alone will serve to strengthen the epoxy composites, the choice of the solvent used in the dispersion of CNTs also can have a significant impact. The change trend of the mechanical properties was found to be related to the boiling point of respective solvent used. In the samples observed in their study, only acetone-dispersed nanocomposites displayed improvements in flexural strength over the pure epoxy, while ethanol and DMF used in CNTs dispersion actually countered the benefits of CNTs in the resulting nanocomposites.It is reasonable that, easier the solvent can evaporate, less solvent will remain to affect the curing reaction. Their results of thermogravimetric analysis (TGA) proved the existence of residual solvent in the resulting nanocomposites. Further evidence of the solvent influence was obtained by Fourier transform infrared (FTIR) spectra, which displayed the difference in the molecular structure of the final nanocomposites depending on the solvent used. The solvent influence is attributed to the different amount of unreacted epoxide groups and the extent of cure reaction in the manufacturing process. The presence of residual solvent may alter the reaction mechanism by restricting the nucleophile–electrophile interaction between the hardener and epoxy, henceforth, affect the cross-linking density and thus degrade the transport properties [21]and mechanical properties of the cured structures. The residual solvent may absorb some heat energy from the composite systems in the pre-cured process, causing a change in local temperature. Nanocomposites with other thermoplastic materials with enhanced properties have been fabricated by solvent casting [16-18, 22]. The limitation of this method is that during slow process of solvent evaporation, nanotubes may tend to agglomerate, that leads to inhomogeneous nanotube distribution in polymer matrix. The evaporation time can be decreased by dropping the nanotube/polymer suspension on a hot substrate (drop casting) [19]or by putting suspension on a rotating substrate (spin-casting) [23]. Du et al. [24] developed a versatile coagulation method to avoid agglomeration of CNTs in PMMA-CNT nanocompositses that involves pouring a nanotube/polymer suspension into an excess of solvent. The precipitating polymer chains entrap the CNT, thereby preventing the CNT from bundling.

5.1.2. Melt mixing method

The alternative and second most commonly used method is melt mixing, which is mostly used for thermoplastics and most compatible with current industrial practices. This technique makes use of the fact that thermoplastic polymers softens when heated. Melt mixing uses elevated temperatures to make substrate less viscous and high shear forces to disrupt the nanotubes bundle. Samples of different shapes can then be fabricated by techniques such as compression molding, injection molding or extrusion. Andrews and co-workers [25] formed composites of commercial polymers such as high impact polystyrene, polypropylene and acrylonitrile–butadiene–styrene (ABS) with MWCNT by melt processing. Initially these polymers were blended in a high shear mixer with nanotubes at high loading level to form master batches that were thereafter diluted with pure polymer to form lower mass fraction samples. Compression molding was used to form composite films. A similar combination of shear mixing and compression molding is studied by many other groups discussed elsewhere [16]. Also Meincke et al. [5] mixed polyamide-6, ABS and CVD-MWCNT in a twin screw extruder at 260°C and used injection molding to make nanocomposites. Tang et al. [26] used both compression and twin-screw extrusion to form CNT/polyethylene composites. Although melt-processing technique has advantages of speed and simplicity, it is not much effective in breaking of agglomeration of CNTs and their dispersion. Bhattacharyya et al. [27] made 1 wt% CNT/polypropylene (PP) nanocomposites by melt mixing, but found that melt mixing alone did not provide uniform nanotube dispersion. Niu et el. [28] studied both methods to prepare polyvinylidene fluoride (PVDF)-CNT nanocomposites to study electrical properties and found it better in composites formed by solution casting.

5.1.3. In-situ polymerization

In addition to solvent casting and melt mixing the other method which combines nanotubes with high molecular weight polymers is in-situ polymerization starting with CNTs and monomers. In-situ polymerization has advantages over other composite fabrication methods. A stronger interface can be obtained because it is easier to get intimate interactions between the polymer and nanotube during the growth stage than afterwards [29, 30].The most common in situ polymerization methods involve epoxy in which the monomer resins and hardeners are combined with CNTs prior to polymerizing [31]. Pande and coworkers [32] performed the in-situ polymerization of MWCNT/ PMMA composites for the enhancement in flexural strength and modulus of composites. Li et al. [33] reported the fabrication and characterization of CNT/ polyaniline (PANI) composites. Xiao and Zhou [34]deposited polypyrrolre (PP) and poly(3-methylthiophene) (PMet) on the surface of MWCNTs by in situ polymerization. Saini et al. [35] reported fabrication process of highly conducting polyaniline (PANI)–(MWCNT) nanocomposites by in-situ polymerization. This material was used in polystyrene for the fabrication of MWCNT-PANI-PS blend for microwave absorption [36]. Moniruzzaman [17] reported many other studies of in-situ polymerization of CNTs with different polymers. Generally, in situ polymerization can be used for the fabrication of almost any polymer composites containing CNT that can be non-covalently or covalently bound to polymer matrix. This technique enables the grafting of large polymer molecules onto the walls of CNT. This technique is particularly important for the preparation of insoluble and thermally unstable polymers, which cannot be processed by solution or melt processing.

Some studies have been also carried out using combined methods, such as solvent casting in conjunction with sonication, followed by melt mixing. Haggenmueller et al. [37] observed considerable nanotube dispersion in CNT-polymer nanocomposites using combination of solvent casting and melt mixing. Pande et al. [32, 38] also prepared MWCNT bulk composites with PMMA and PS using a two-step method of solvent casting followed by compression molding and obtained better electrical and mechanical properties. Singh et al. [39] also prepared MWCNT-LDPE composites using solvent casting followed by compression moulding and obtained better electrical conductivity. Jindal et al. [29, 40] prepared

MWCNT-polycarbonate composite using solvent casting followed by compression mould-ing for the enhancement in the impact properties.

The other less commonly known methods for CNT- polymer nanocomposites formation are twin screw pulverization [41], latex fabrication [42], coagulation spinning [43] and electro-phoretic deposition [44].

6. Challenges in MWCNT polymer composites fabrication and possible solutions

Although these fabrication methods helped to enhance the properties of CNT reinforced composites over neat polymer but there are several key challenges that hinders the excellent CNT properties to be fruitful in polymer composite formation.

6.1. Dispersion

Disperion of nanoscale filler in a matrix is the key challenge for the formation of nanocomposite. Dispersion involves separation and then stabilization of CNTs in a medium. The methods de-scribed above for the nanocomposites fabrication require CNTs to be well dispersed either in solvent or in polymer for maximizing their contact surface area with polymer matrix. As CNTs have diameters on nanoscale the entanglement during growth and the substantial van der Waals interaction between them forces to agglomerate into bundles. The ability of bundle for-mation of CNTs with its inert chemical structure makes these high aspect ratio fibers dissolving in common solvents to form solution quite impossible. The SEM of MWCNTs synthesized by CVD technique seems to be highly entangled and the dimensions of nanotube bundles is hun-dreds of micrometer. This shows several thousands of MWCNTs in one bundle as shown in Fig-ure 5a. These bundles exhibits inferior mechanical and electrical properties as compared to individual nanotube because of slippage of nanotubes inside bundles and lower aspect ratio as compared to individual nanotube. The aggregated bundles tend to act as defect sites which ad-versely affect mechanical and electrical properties of nanocomposites. Effective separation re-quires the overcoming of the inter-tube van der Waal attraction, which is anomalously strong in CNT case. To achieve large fractions of individual CNT several methods have been employed. The most effective methods are by attaching several functional sites on the surface of CNTs through some chemical treatment or by surrounding the nanotubes with dispersing agents such as surfactant. Thereafter the difficulty of dispersion can be overcome by mechanical/physical means such as ultrasonication, high shear mixing or melt blending. Another obstacle in dispers-ing the CNTs is the presence of various impurities including amorphous carbon, spherical full-erenes and other metal catalyst particles. These impurities are responsible for the poor properties of CNTs reinforced composites [45].

6.2. Adhesion between CNTs and polymer

The second key challenge is in creating a good interface between nanotubes and the poly-mer matrix. From the research on microfiber based polymer composites over the past few decades, it is well established that the structure and properties of filler-matrix interface plays a major role in determining the structural integrity and mechanical performance of composite materials. CNTs have atomically smooth non-reactive surfaces and as such there is a lack of interfacial bonding between the CNT and the polymer chains that limits load transfer. Hence the benefits of high mechanical properties of CNTs are not utilized properly. The first experimental study focusing on interfacial interaction between MWCNT and polymer was carried out by Cooper et al. [46]. They investigated the detachment of MWCNTs from an epoxy matrix using a pullout test for individual MWCNT and observed the interfacial shear stress varied from 35-376 MPa. This variation is attributed to difference in structure and morphology of CNTs.

There are three main mechanisms for load transfer from matrix to filler. The first is weak van der Waal interaction between filler and polymer. Using small size filler and close contact at the interface can increase it. The large specific surface area of CNTs is advantageous for bonding with matrix in a composite, but is a major cause for agglomeration of CNTs. Therefore, uniformally dispersed individual nanotubes in matrix is helpful. The second mechanism of load transfer is micromechanical interlocking which is difficult in CNTs nanocomposites due to their atomically smooth surface. Although local non uniformity along length of CNTs i.e. varying diameter and bends due to non hexagonal defects contributes to this micromechanical interlocking. This interlocking can increase by using long CNTs to block the movement of polymer chains. The contribution of this mechanism may reach saturation at low CNT content. The third and best mechanism for better adhesion and hence load transfer between CNTs and polymer is covalent or ionic bonding between them. The chemical bonding between CNTs and polymer can be created and enhanced by the surface treatment such as oxidation of CNTs with acids or other chemicals. This mode of mechanism have much importance as it provides strong interaction between polymer and CNT and hence efficiently transfers the load from polymer matrix to nanotubes necessary for enhanced mechanical response in high-performance polymers.

6.3. Chemical functionalization of CNTs

The best route to achieve individual CNT to ensure better dispersion is chemical modifications of CNT surface. The chemical functionalization involves the attachment of chemical bonds to CNT surface or on end caps. Nanotube functionalization typically starts with oxidatative conditions, commonly by refluxing in nitric or sulfuric acid or combination of both to attach carboxylic acid moieties to the defect sites. The end caps of nanotubes have extra strain energy because of their high degree of curvature with pentagons and heptagonal carbon atoms are most vulnerable to reaction with acid. The side walls also containing defects like pentagon-heptagon pairs, sp^3 hybridized defects and vacancies in nanotube lattice and are easily supplemented by oxidative damage and can be stabilized by formation of functional groups mainly carboxylic acid and hydroxide group. These acid moieties and hydroxide groups can be further replaced to more reactive groups like –COCl or –CORNH$_2$. The addition of these functional groups on CNTs possesses intermolecular repulsion between functional groups on surface that overcomes the otherwise weak van der Waal attraction between CNTs. It is also vital to stabilize the dispersion to prevent reagglomeration of the CNTs. Chemical functionalization can prevent reagglomeration of CNTs also. Sen et al. [47] carried out chemical functionalization to form ester functionalized CNTs and found that it is an effective approach to exfoliate the CNTs bundles and improve their processibility with polymer matrix. Georgakilas et al. [48] observed that CNT covalently functionalized with pyrrolidone by 1,3-dipolar cycloaddition of azomethine ylides show a solubility of 50 mg/mL in chloroform, even without sonication whereas the pristine CNT is completely insoluble in this solvent. Liang et al. [49] performed reductive alkylation of CNTs using lithium and alkyl halides in liquid ammonia for sidewall functionalization of CNTs and observed their extensive debundling by inspection of HRTEM images. Kinloch et al. [50] studied the rheological behavior of oxidized CNTs and found that the composites filled with functionalized CNTs had better dispersion. It has been observed by the researchers that amine modified CNTs is very important for the enhancement in the mechanical properties with epoxy. Garg et al. [31] shows the reaction mechanism for the formation of acid functionalized and amine functionalized CNT and their interaction with the epoxy resin as Figure 8a and b respectively. Two different types of functional groups were attached on the CNT surface. In the first case MWCNTs were refluxed for 48 h in HNO_3 (400 ml, 60% concentration) to achieve reasonable surface oxidation of the tubes. The mixture was then filtered and the residue (treated material) was washed several times with distilled water till washings were neutral to pH paper. The treated MWCNT were dried in oven before use. In a second step the oxidized nanotubes were dispersed in benzene by stirring, and then refluxed with excess $SOCl_2$ along with a few drops of DMF used as catalyst for chlorination of MWCNT surfaces. After the acyl chlorination, $SOCl_2$ and DMF were removed through repeatedly washing by tetrahydrofuran (THF). 100 ml of triethylene tetra-amine (TETA) was added to react with acyl chlorinated MWCNT at 100 °C for 24 h reflux until no HCl gas evolved. After cooling to room temperature, MWCNTs were washed with deionized water 5 times to remove excess TETA. Finally, the black solid was dried at room temperature overnight in vacuum and named as amine modified CNTs. These functionalized CNTs were characterized by FTIR, TGA and HRTEM and clearly showed the presence of these types of functional group. Gojny et al. [51]achieved surface modified MWCNTs by refluxing of oxidized MWCNTs with multifunctional amines and observed from TEM images that these were completely covered by epoxy matrix that confirmed the bonding between them. Sinnott [52]has provided an in depth review of chemical functionalization of CNTs where the chemical bonds are used to tailor the interactions between nanotubes and polymers or solvent. The chemical functionalization of CNTs has also been accomplished through irradiation with electrons or ions [53]. In this manner one may hope to improve the binding of CNTs by interdigitation of active sites on its sides into polymer matrix.

Covalent bond also benefits phonon transferring between nanotubes and polymer matrix, which is a key factor for improving thermal conductivity of the nanocomposites. To ensure the adhesion between polymer and nanotubes various surfactant and chemical modification procedures have been adopted to modify the surface of otherwise inert surface of CNTs that provides bonding sites to the polymer matrix.

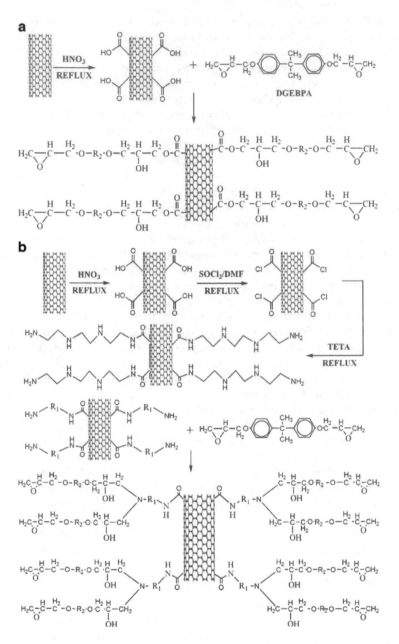

Figure 8. Reaction mechanism for (a) acid modified-MWCNT (b) amine modified-MWCNT with epoxy resin (Reprinted with permission from Springer(31))

So the surface modification of CNTs is the crucial factor that decides the effective dispersion and improves the interactions between CNTs and matrix. However there are certain drawbacks of using chemically functionalized CNTs. Chemical functionalization normally employs harsh techniques resulting in tube fragmentation and also disrupts the bonding between graphene sheets and thereby reduces the properties of CNTs. Studies revealed that different chemical treatments may decrease the maximum buckling force of nanotubes by 15% [16]. Also the chemical functionalized CNTs significantly decrease the electrical conductivity of CNTs nanocomposites due to unbalance polarization effect, shortening of length and physical structure defects during acidic treatment [54]. But it is still necessary for increased dispersion and strengthens the interfacial bonding of CNTs with polymer matrix that is more important in structural applications.

The solubility or dispersion of CNTs in certain specified solvents or polymers can also increase by non covalent association which is more fragile. The non-ionic surfactant such as sodium dodecylbenzene sulfonate (SDS) or polyoxyethylene-8-lauryl (PoEL) has two segments. The hydrophobic segment of surfactant shows strong interactions with carbon of CNTs via van der Waal force and the hydrophilic segment shows hydrogen bonding with solvent or polymer used for dispersion. Islam et al. [55] reported that ~ 65% CNT bundles exfoliated into individual nanotubes even with a very low of 20 mg (CNT)/ml of water containing SDS as surfactant. Barrau et al. [56] used palmitic acid as surfactant to disperse CNTs into epoxy resin and observed that electrical percolation threshold decreases indicating better CNT dispersion. Gong et al. [57] added PoEL as surfactant in CNT/epoxy composite to assist the dispersion of CNTs. The improvement in dispersion in chitosan with nitric acid treated CNTs was also reported by Ozarkar et al. [58] and the stability of dispersion prepared by using functionalized CNTs was observed to be better. However, CNTs treated with different surfactants are wrapped in it and hence contacts between CNTs decreases thereby the transport properties (electrical and thermal conductivities) of CNTs/polymer nanocomposites are adversely affected.

6.4. Dispersion of high loading of CNTs in polymer matrix

Dispersion of high loading of CNTs in any polymer is very difficult due to the formation of agglomerates by the conventional techniques. To maximize the improvment in properties, higher loading of CNTs is preferred [59]. However, polymer composites synthesized by using the conventional methods generally have low CNT contents. It has been observed that beyond 1 wt.-% of loading, CNTs tend to agglomerate [60] resulting in poor mechanical properties of the composites. It is therefore important to develop a technique to incorporate higher CNT loading in the polymer matrices without sacrificing their mechanical properties. Recently, several methods have been developed for fabricating CNT/polymer composites with high CNT loadings. One such technique is mechanical densification technique where vertically aligned CNTs were densified by the capillary induced wetting with epoxy resin [61]. By this technique dimensions of sample preparation are limited. In another technique, a filtration system was used to impregnate the epoxy resin into CNT bucky paper [62, 63]. However, it was very difficult to completely impregnate the bucky paper with epoxy resin.

Recently, Feng et al., [64] reported a mixed curing assisted layer by layer method to synthesize MWCNT/epoxy composite film with a high CNT loading from ~15 to ~36 wt.-%. The mixed-curing-agent consists of two types of agents, one of which is responsible for the partial initial curing at room temperature to avoid agglomeration of the CNTs, and the other for complete curing of epoxy resin at high temperature to synthesize epoxy composite films with good CNT dispersion. In another study by Feng et al. [65] upto ~39.1 wt. % SWCNT-epoxy composites were fabricated using same mixed curing layer by layer method and their mechanical properties were enhanced significantly. Bradford et al. [66] reported a method to quickly produce macroscopic CNT composites with a high volume fraction upto 27% of millimeter long, well aligned CNTs. Specifically, they used the novel method, shear pressing, to process tall, vertically aligned CNT arrays into dense aligned CNT preforms, which are subsequently processed into composites. In another study by Ogasawara et al. [67] aligned MWCNT/epoxy composites were processed using a hot-melt prepreg method. Vertically aligned ultra-long CNT arrays (forest) were converted to horizontally aligned CNT sheets by pulling them out. An aligned CNT/epoxy prepreg was fabricated using hot-melting with B-stage cured epoxy resin film. The final composites contains 21.4 vol% of MWCNTs.

6.5. Alignment of CNTs in polymer matrix

The other key challenge is to understand the effect of nanotube alignment on nanocomposites properties because the nanotubes have asymmetric structure and properties. Like other one-dimensional fiber fillers CNTs displays highest properties in the oriented reinforced direction and the mechanical, electrical, magnetic and optical performance of its composites are linked directly to their alignment in the matrix. So to take the full advantage of excellent properties of CNTs these should be aligned in a particular direction. For example, the alignment of CNT increases the elastic modulus and electrical conductivity of nanocomposites along the nanotube alignment direction.

Several methods like application of electric field during composite formation and carbon arc discharge [68], composite slicing [69], film rubbing [70], chemical vapor deposition [71, 72], mechanical stretching of CNT-polymer composites [73] and magnetic orientation [74] have been reported for aligning nanotubes in composites. Electrospinning is also an effective method for the alignment of CNTs in polymer matrix.

7. Properties of the nanocomposites

7.1. Mechanical properties of MWCNTs polymer nanocomposites

Different thermoplastic and thermoset polymer matrices have been tried to realize the superior mechanical properties of CNTs for development of light weight strong material. NASA scientists are considering CNT-polymer composite for space elevator. To date, a volume of literature is available on the improvement of mechanical performance of polymers with addition of CNTs. The first study for formation of CNT-polymer composites was carried out by Ajayan et al. [69]. CNTs were aligned within the epoxy matrix by the shear forces induced by cutting with a diamond knife, however no quantitative mechanical measurements were made. The first true study for tensile and compression properties of CNT polymer composites was carried by Schadler et al. [75] with epoxy. On addition of 5-wt% MWCNTs the tensile modulus increased from 3.1 GPa to 3.71 GPa. and compression modulus increased from 3.63 to 4.5 GPa. However, no significant increases in toughness values were observed. Bai et al. [76] observed doubling of Young's modulus from 1.2 to 2.4 GPa and significant increase in strength from 30 to 41 MPa on addition of 1 wt.% MWCNTs. Also excellent matrix–nanotube adhesion was confirmed by the observation of nanotube breakage during fracture surface studies. Zhou et al. [77] reported steady increase of flexural modulus in CNT-epoxy composite with higher CNT weight percentage and found an improvement of 11.7% in modulus with 0.4 wt% loading of CNTs and 28% enhancement in flexural strength with 0.3 wt% loading. Garg et al. [31] reported an increment of 155 % in flexural strength of epoxy with addition of merely 0.3% amine functionalized MWCNTs and an increment of 38% in flexural modulus. Mathur et al. [13] reported an increment of 158% in flexural strength of phenolic with addition of 5 vol% of MWCNT. Colemann et al. [16] reviewed the mechanical properties of a large number of CNT reinforced polymer (thermoplastic and thermosetting) composites fabricated by various methods and reported enhancement in mechanical properties. Du et al. [78] studied the experimental results for mechanical performance of CNTs nanocomposites carried out by different research groups and observed that the gains are modest and far below the simplest theoretical estimates. Haggenmueller [79] applied the Halpen Tsai composite theory to CNT nanocomposites and observed that the experimental elastic modulus is smaller than predicted by more than one order. It is attributed to the lack of perfect load transfer from nanotubes to matrix due to non uniform dispersion and small interfacial interaction. Although chemical functionalization of CNTs has sorted out those problems to an extent yet the best results have to be achieved. Also aspect ratio is other source of uncertainty in mechanical properties. Defects on the CNT surface also expected to influence the mechanical properties significantly. The methods of handling nanotubes, including acid treatments and sonication for long time are known to shorten nanotubes results in decreasing aspect ratio and are detrimental to mechanical properties. The mechanical properties of CNT based composites increased upto a certain loading of CNTs and beyond which it starts decreasing. This may be because of increase in viscosity of polymers at higher CNTs loading and also cause some surface of CNTs not to be completely covered by polymers matrix due to the large specific surface area of CNTs.Therefore, few studies have been carried out to disperse high loading of CNTs in polymer matrices with improved mechanical properties. Bradford et al. [66] reported 400 MPa (tensile strength) and 22.3 GPa (tensile modulus) for 27vol% of MWCNT–epoxy composites. Feng et al. [80] also reported 183% and 408% improvement in tensile strength and tensile modulus respectively at 39.1 wt% SWCNTs loading compared with those of the neat epoxy. Ogasawara et al. [67] found 50.6 GPa and 183 MPa modulus and ultimate tensile strength respectively of CNT (21.4 vol.%)/epoxy composite. These values were 19 and 2.9 times those of the epoxy resin respectively.

Andrews et al. [80] prepared aligned CNT/pitch composites and found the significant improvement in the electrical and mechanical properties especially due to orientation of CNTs.

Du et al. [78] compared the mechanical performance of randomly oriented and aligned CNTs polymer composites. Their study revealed that in aligned CNT polymer nanocomposites tensile strength and modulus even reached to 3600 MPa and 80 GPA respectively which is much higher than the general value of 100 MPa and 6 GPa in case of randomly oriented CNT polymer nanocomposites. They also observed that the mechanical properties are always higher for aligned CNTs composites with higher loading while the case is different for isotropic CNT polymer composites.

7.2. Electrical properties of MWCNTs polymer nanocomposites

CNTs because of their extraordinary electrical conductivity are also excellent additive to impart electrical conductivity to polymer. Many experimental results shows that the conductive CNT composites can be constructed at low loading of CNTs due to low percolation threshold originated from the high aspect ratio and conductivity of CNTs [70, 78]. Figure 9 shows the general trend of electrical conductivity of CNT- polymer nanocomposites. It can be found from almost all the experimental results and also obvious from figure that CNT nanocomposites exhibit a typical percolation behavior and CNT reinforcement to polymers can increase the conductivity of resulting composites to several order of magnitude or even some times higher than ten orders of magnitude.

According to percolation theory the conductivity follow the following power law close to threshold percolation.

$$\sigma = \sigma o (p - p_o)^t \qquad \text{for} \quad p > p_o$$

where σ is the composite conductivity, σ_o is a constant, p the weight fraction of nanotubes, p_o is the percolation threshold and t the critical exponent [81]. Theoretical and experimental results have shown that percolation laws are applicable to CNT-based composites and that the enhanced maximum conductivity and percolation can be achieved with significantly lower filler concentrations than with other carbon and other conductive fillers. Depending on the polymer matrix, the processing technology and the type of nanotubes used, recent experimental studies have achieved percolation thresholds between 0.0021 to 9.5% by weight and critical exponents varying from 0.9 to 7.6 [78].

Sandler et al. [82] observed the percolation threshold of CNTs/epoxy nanocomposites between 0.0225 and 0.04 wt %. They further observed very low percolation threshold at 0.0025 wt% for aligned CNT- epoxy composites [83]. The current voltage behavior measurements exhibited non-ohmic behavior, which is most likely due to tunneling conduction mechanism. The main mechanism of conduction between adjacent nanotubes is probably electron hopping when their separation distance is small. At concentration greater than percolation threshold, conducting paths are formed through the whole nanocomposites, because the distance between the conductive CNT filler (individual or bundles) is small enough to allow efficient electron hopping.

The electrical conductivity of CNT/polymer composites also effected by dispersion and aspect ratio of CNTs and was studied by Barrau et al. [56]. They used palmitic acid as surfactant to improve the nanotube dispersion and reduced the threshold concentration from 0.18 to 0.08 wt%. To study the effect of aspect ratio on electrical conductivity of CNT nanocomposites Bai et al. [84] pretreated MWCNTs to alter their aspect ratios before preparing epoxy/MWCNTs composites and found that the threshold concentration varied from 0.5 to > 4 wt % with decreasing aspect ratio. The effect of alignment of CNTs in polymer composites was also studied. Du et al. [24] found some contradictory results with respect to alignment of rod like fillers and observed the lowest percolation threshold and maximum conductivity with their random orientation. They found that the electrical conductivity of 2 wt% CNT/ PMMA nanocomposites decrease significantly (from ~10^{-4} to ~10^{-10} S/cm) when CNTs were highly aligned. In contrast Choi et al. [85] observed that the nanotube alignment increased the conductivity of a 3 wt% CNT/epoxy composites from ~10^{-7} to ~10^{-6}S/cm. In most of the cases the CNT nanocomposites with isotropic nanotubes orientation have greater electrical conductivity than the nanocomposites with highly aligned CNTs especially at lower CNT loadings. By alignment of CNTs in polymers, the percolation pathway is destroyed as aligned CNTs seldomly intersects each other. At higher CNTs loading the conductivity is more in case of aligned CNTs as compared to randomly oriented CNTs.

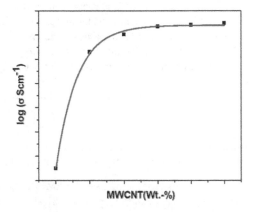

Figure 9. General trend of electrical conductivity of CNT polymer composites

The study carried out by different researchers also revealed that the composites with thermoplastic polymers have higher conductivity as compared to that of thermosetting polymers above percolation threshold. Transport properties in CNT-PMMA composites have been reported by Stephan et al. [86] and Benoit et al. [19] where low percolation threshold of 0.5 wt% and 0.33 wt% respectively were obtained. Singhai et al. [87] found that increase in number of defects lead to a decrease in conductivity. However Lau et al. [88] concluded that fuctionalization of CNTs can enhance the electrical conductivity of MWCNTs. The reason attributed to this phenomenon is electron transfer from the carbon atoms on MWCNTs to functionalized groups attached to the surface favorably promoting conductivity. The study carried out by Grimes et al. [89] revealed that the electrical response of as fabricated MWCNTs is significantly influenced by the presence of residual catalyst metal particles.

7.3. EMI shielding proerties of MWCNTs polymer nanocomposites

The electrical conductivity of CNT reinforced polymer composites makes them a very suitable candidate to be employed for electromagnetic interference (EMI) shielding. EMI is the process by which disruptive electromagnetic energy is transmitted from one electronic device to another via radiation or conduction. As we all know that the electromagnetic waves produced from some electronic instrument have an adverse effect on the performance of the other equipments present nearby causing data loss, introduction of noise, degradation of picture quality etc. The common example is the appearance of noise in television signal when a telephone or mobile rings. Also recent reports of deterious effects of electromagnetic radiations on electro medical devices have caused concern among health care providers. The overlapping of signals transmitted in air traffic system with signals from other electronic equipments became cause of several accidents in past. Also mobile phones and passing taxi radios have been known to interfere with anti-skid braking system (ABS), airbags and other electronic equipments causing drivers to lose control. In today's scenario where rapid communication is required, there is an increase in electromagnetic radiations within the spectrum in which the wireless, cordless and satellite system operates. So it a strong desire to shield electronics equipments from the undesired signals. Problems with EMI can be minimized or sometime eliminated by ensuring that all electronic equipments are operated with a good housing to keep away unwanted radio frequency from entering or leaving. The shielding effectiveness (SE) of the shielding material is its ability to attenuate the propagation of electromagnetic waves through it and measured in decibels (dB) given by

$$SE(dB) = -10\log(P_t / P_0),$$

where P_t and P_0 are, respectively, the transmitted and incident electromagnetic power. A SE of 10 dB means 90% of signal is blocked and 20 dB means 99% of signal is blocked.

One of the important criterion for a material to be used for EMI shielding material is that it should be electrically conducting. Because of their high electrical conductivity metals have been used for past several years as EMI shielding materials. But the shortcomings of metals like heavy weight, physical rigidity and corrosion restricts their use. The most notable substance that could overcome these shortcomings is the CNT-polymer composites. As discussed in previous sections these are electrically conductive, having low density, corrosion resistant and can be molded in any form. Due to easy processing and good flexibility, CNT–polymer composites have been employed for application as promising EMI shielding materials. The SE of the CNT-polymer composites depends on various factors like,type of CNTs (either SWCNT or MWCNT), aspect ratio of CNTs, quality of CNTs, thickness and electrical conductivity of the shielding material. Several studies have been reported on EMI shielding properties of randomly oriented CNT based polymer composites. Mathur and co-workers [18] have prepared MWNT-PMMA and MWNT-PS composites and observed 18dB and 17dB SE respectively with 10-wt % MWCNT loading. Singh et al. [90] reported a SE of 51 dB by using MWCNT grown carbon fibre fabric based epoxy composites with improved mechanical properties [91].The effect of the length (aspect ratio) of CNTs on EMI SE of composites was also studied by few researchers. Huang et al. [92] reported EMI SE of 18 dB with 15 wt.-% small CNTs and 23-28 dB with 15 wt% long CNTs in X band (8-12.4 GHz). Li et al. [93] also observed that SE with long length CNTs is more as compared to small length CNTs at the same 15wt % loading composites. The residual catalyst metal particle in the cavity of CNTs also effects the SE of the composites.

There are few additional advantages of using MWCNTs as EMI shielding material. The EMI SE also depends on the source of origin of electromagnetic waves. Electrically conducting material can effectively shield the electromagnetic waves generated from an electric source, whereas magnetic materials effectively shield the electromagnetic waves generated from a magnetic source. The MWCNTs exhibits electrical properties because of presence of pi electrons and magnetic properties because of the presence of catalytic iron particles in tubes. Also one common problem experienced with commonly used composite materials for EMI shielding is build up of heat in the substance being shielded. The possible solution for this is to add thermal conducting material. Composites with MWCNTs can easily overcome this problem as it has high thermal conductivity.

7.4. Thermal properties of MWCNTs polymer nanocomposites

As discussed above that the CNTs have thermal conductivity as high as 6600W/mK predicted for SWCNTs [94] at room temperature and have experimental value 3000W/mK for isolated MWCNT. So it is quite expected that the reinforcement of CNTs can significantly enhance the thermal properties of CNT-polymer nanocomposites. The improvement in thermal transport properties of CNT polymer composites leads their applications for usage as printed circuit boards, connectors, thermal interface materials, heat sinks.

8. Conclusion

Synthesis of high quality and reproducible CNTs is still remain a very importnat issue. Chemical vapor deposition has been found an efficient process for the synthseis of bulk quantity of CNTs. The CNT-polymer composites have been developed with improved mechanical properties but for actual structural applications, these have to compete with the existing carbon fibre based composites. Dispersion of high loading of CNTs and their alignment in any polymer matrix without sacrificing their mechanical properties is still a challenge for using CNTs in high performance composites for specific applications such as as automobile, defence, aerospace, sports etc. CNT- carbon fibres-polymer multiscale composites could be an alternative route for further improvement in the mechanical properties of the composites over commercially available CF-polymer composites. Till then electrical properties of CNT polymer composites provides exciting possibility as antistatic and electromagnetic interference shielding material.

Author details

Veena Choudhary[1], B.P. Singh[2] and R.B. Mathur[2]

1 Centre for Polymer Science and Engineering, Indian Institute of Technology Delhi, India

2 Physics and Engineering of Carbon, Division of Materials Physics and Engineering, CSIR-National Physical Laboratory, New Delhi, India

References

[1] Ajayan PM, Zhou OZ. Applications of carbon nanotubes. Carbon Nanotubes2001. p. 391-425.

[2] Lourie O, Wagner HD. Evaluation of Young's modulus of carbon nanotubes by micro-Raman spectroscopy. Journal of Materials Research 1998;13:2418-22.

[3] Yu MF, Files BS, Arepalli S, Ruoff RS. Tensile loading of ropes of single wall carbon nanotubes and their mechanical properties. Phys Rev Lett 2000;84:5552-5.

[4] Yu MF, Lourie O, Dyer MJ, Moloni K, Kelly TF, Ruoff RS. Strength and breaking mechanism of multiwalled carbon nanotubes under tensile load. Science 2000;287:637-40.

[5] Meincke O, Kaempfer D, Weickmann H, Friedrich C, Vathauer M, Warth H. Mechanical properties and electrical conductivity of carbon-nanotube filled polyamide-6 and its blends with acrylonitrile/butadiene/styrene. Polymer 2004;45:739-48.

[6] Ebbesen TW, Lezec HJ, Hiura H, Bennett JW, Ghaemi HF, Thio T. Electrical conductivity of individual carbon nanotubes. Nature 1996;382:54-6.

[7] Lee JO, Park C, Kim JJ, Kim J, Park JW, Yoo KH. Formation of low-resistance ohmic contacts between carbon nanotube and metal electrodes by a rapid thermal annealing method. Journal of Physics D-Applied Physics 2000;33:1953-6.

[8] Hone J. Dekker Encyclopedia of Nanoscience and Nanotechnology 2004.

[9] Jung SH, Kim MR, Jeong SH, Kim SU, Lee OJ, Lee KH, et al. High-yield synthesis of multi-walled carbon nanotubes by arc discharge in liquid nitrogen. Applied Physics a-Materials Science & Processing 2003;76:285-6.

[10] Ebbesen TW, Ajayan PM. LARGE-SCALE SYNTHESIS OF CARBON NANOTUBES. Nature 1992;358:220-2.

[11] Mathur RB, Seth S, Lal C, Rao R, Singh BP, Dhami TL, et al. Co-synthesis, purification and characterization of single- and multi-walled carbon nanotubes using the electric arc method. Carbon 2007;45:132-40.

[12] Mathur RB, Chatterjee S, Singh BP. Growth of carbon nanotubes on carbon fibre substrates to produce hybrid/phenolic composites with improved mechanical properties. Composites Science and Technology 2008;68:1608-15.

[13] Mathur RB, Singh BP, Dhami TL, Kalra Y, Lal N, Rao R, et al. Influence of Carbon Nanotube Dispersion on the Mechanical Properties of Phenolic Resin Composites. Polymer Composites 31:321-7.

[14] Guo T, Nikolaev P, Thess A, Colbert DT, Smalley RE. CATALYTIC GROWTH OF SINGLE-WALLED NANOTUBES BY LASER VAPORIZATION. Chem Phys Lett 1995;243:49-54.

[15] Thess A, Lee R, Nikolaev P, Dai HJ, Petit P, Robert J, et al. Crystalline ropes of metallic carbon nanotubes. Science 1996;273:483-7.

[16] Coleman JN, Khan U, Blau WJ, Gun'ko YK. Small but strong: A review of the mechanical properties of carbon nanotube-polymer composites. Carbon 2006;44:1624-52.

[17] Moniruzzaman M, Winey KI. Polymer nanocomposites containing carbon nanotubes. Macromolecules 2006;39:5194-205.

[18] Mathur RB, Pande S, Singh BP, Dhami TL. Electrical and mechanical properties of multi-walled carbon nanotubes reinforced PMMA and PS composites. Polymer Composites 2008;29:717-27.

[19] Benoit JM, Corraze B, Lefrant S, Blau WJ, Bernier P, Chauvet O. Transport properties of PMMA-carbon nanotubes composites. Synth Met 2001;121:1215-6.

[20] Lau KT, Lu M, Lam CK, Cheung HY, Sheng FL, Li HL. Thermal and mechanical properties of single-walled carbon nanotube bundle-reinforced epoxy nanocomposites: the role of solvent for nanotube dispersion. Composites Science and Technology 2005;65:719-25.

[21] Bryning MB, Milkie DE, Islam MF, Kikkawa JM, Yodh AG. Thermal conductivity and interfacial resistance in single-wall carbon nanotube epoxy composites. Applied Physics Letters 2005;87.

[22] Singh BP, Singh D, Mathur RB, Dhami TL. Influence of Surface Modified MWCNTs on the Mechanical, Electrical and Thermal Properties of Polyimide Nanocomposites. Nanoscale Research Letters 2008;3:444-53.

[23] de la Chapelle ML, Stephan C, Nguyen TP, Lefrant S, Journet C, Bernier P, et al. Raman characterization of singlewalled carbon nanotubes and PMMA-nanotubes composites. Synth Met 1999;103:2510-2.

[24] Du FM, Fischer JE, Winey KI. Coagulation method for preparing single-walled carbon nanotube/poly(methyl methacrylate) composites and their modulus, electrical conductivity, and thermal stability. Journal of Polymer Science Part B-Polymer Physics 2003;41:3333-8.

[25] Andrews R, Jacques D, Qian DL, Rantell T. Multiwall carbon nanotubes: Synthesis and application. Accounts of Chemical Research 2002;35:1008-17.

[26] Tang WZ, Santare MH, Advani SG. Melt processing and mechanical property characterization of multi-walled carbon nanotube/high density polyethylene (MWNT/HDPE) composite films. Carbon 2003;41:2779-85.

[27] Bhattacharyya AR, Sreekumar TV, Liu T, Kumar S, Ericson LM, Hauge RH, et al. Crystallization and orientation studies in polypropylene/single wall carbon nanotube composite. Polymer 2003;44:2373-7.

[28] Niu C, Ngaw L, Fischer A, Hoch R, Fischer AB, Ngam L. Piezoelectric high damping material for vibration suppression in vehicle, has carbon nano tube as piezoelectric electroconductive particle for packing. Hyperion Catalysis Int Inc.

[29] Jia ZJ, Wang ZY, Xu CL, Liang J, Wei BQ, Wu DH, et al. Study on poly(methyl methacrylate)/carbon nanotube composites. Materials Science and Engineering a-Structural Materials Properties Microstructure and Processing 1999;271:395-400.

[30] Velasco-Santos C, Martinez-Hernandez AL, Fisher FT, Ruoff R, Castano VM. Improvement of thermal and mechanical properties of carbon nanotube composites through chemical functionalization. Chemistry of Materials 2003;15:4470-5.

[31] Garg P, Singh BP, Kumar G, Gupta T, Pandey I, Seth RK, et al. Effect of dispersion conditions on the mechanical properties of multi-walled carbon nanotubes based epoxy resin composites. Journal of Polymer Research 2011;18:1397-407.

[32] Pande S, Mathur RB, Singh BP, Dhami TL. Synthesis and Characterization of Multi-walled Carbon Nanotubes-Polymethyl Methacrylate Composites Prepared by In Situ Polymerization Method. Polymer Composites 2009;30:1312-7.

[33] Li XH, Wu B, Huang JE, Zhang J, Liu ZF, Li HI. Fabrication and characterization of well-dispersed single-walled carbon nanotube/polyaniline composites. Carbon 2003;41:1670-3.

[34] Xiao QF, Zhou X. The study of multiwalled carbon nanotube deposited with conducting polymer for supercapacitor. Electrochimica Acta 2003;48:575-80.

[35] Saini P, Choudhary V, Singh BP, Mathur RB, Dhawan SK. Polyaniline-MWCNT nanocomposites for microwave absorption and EMI shielding. Materials Chemistry and Physics 2009;113:919-26.

[36] Saini P, Choudhary V, Singh BP, Mathur RB, Dhawan SK. Enhanced microwave absorption behavior of polyaniline-CNT/polystyrene blend in 12.4-18.0 GHz range. Synth Met;161:1522-6.

[37] Haggenmueller R, Gommans HH, Rinzler AG, Fischer JE, Winey KI. Aligned single-wall carbon nanotubes in composites by melt processing methods. Chem Phys Lett 2000;330:219-25.

[38] Pande S, Singh BP, Mathur RB, Dhami TL, Saini P, Dhawan SK. Improved Electromagnetic Interference Shielding Properties of MWCNT-PMMA Composites Using Layered Structures. Nanoscale Research Letters 2009;4:327-34.

[39] Singh BP, Prabha, Saini P, Gupta T, Garg P, Kumar G, et al. Designing of multiwalled carbon nanotubes reinforced low density polyethylene nanocomposites for suppression of electromagnetic radiation. J Nanopart Res 2011;13:7065-74.

[40] Jindal P, Pande S, Sharma P, Mangla V, Chaudhury A, Patel D, et al. High strain rate behavior of multi-walled carbon nanotubes–polycarbonate composites. Composites Part B: Engineering.

[41] Xia HS, Wang Q, Li KS, Hu GH. Preparation of polypropylene/carbon nanotube composite powder with a solid-state mechanochemical pulverization process. Journal of Applied Polymer Science 2004;93:378-86.

[42] Regev O, ElKati PNB, Loos J, Koning CE. Preparation of conductive nanotube-polymer composites using latex technology. Advanced Materials 2004;16:248-+.

[43] Vigolo B, Penicaud A, Coulon C, Sauder C, Pailler R, Journet C, et al. Macroscopic fibers and ribbons of oriented carbon nanotubes. Science 2000;290:1331-4.

[44] Dhand C, Arya SK, Singh SP, Singh BP, Datta M, Malhotra BD. Preparation of polyaniline/multiwalled carbon nanotube composite by novel electrophoretic route. Carbon 2008;46:1727-35.

[45] Montoro LA, Rosolen JM. A multi-step treatment to effective purification of single-walled carbon nanotubes. Carbon 2006;44:3293-301.

[46] Cooper CA, Cohen SR, Barber AH, Wagner HD. Detachment of nanotubes from a polymer matrix. Applied Physics Letters 2002;81:3873-5.

[47] Sen R, Zhao B, Perea D, Itkis ME, Hu H, Love J, et al. Preparation of single-walled carbon nanotube reinforced polystyrene and polyurethane nanofibers and membranes by electrospinning. Nano Letters 2004;4:459-64.

[48] Georgakilas V, Kordatos K, Prato M, Guldi DM, Holzinger M, Hirsch A. Organic functionalization of carbon nanotubes. Journal of the American Chemical Society 2002;124:760-1.

[49] Liang F, Sadana AK, Peera A, Chattopadhyay J, Gu ZN, Hauge RH, et al. A convenient route to functionalized carbon nanotubes. Nano Letters 2004;4:1257-60.

[50] Kinloch IA, Roberts SA, Windle AH. A rheological study of concentrated aqueous nanotube dispersions. Polymer 2002;43:7483-91.

[51] Gojny FH, Nastalczyk J, Roslaniec Z, Schulte K. Surface modified multi-walled carbon nanotubes in CNT/epoxy-composites. Chem Phys Lett 2003;370:820-4.

[52] Sinnott SB. Chemical functionalization of carbon nanotubes. Journal of Nanoscience and Nanotechnology 2002;2:113-23.

[53] Crespi VH, Chopra NG, Cohen ML, Zettl A, Radmilovic V. Site-selective radiation damage of collapsed carbon nanotubes. Applied Physics Letters 1998;73:2435-7.

[54] Sulong AB. MN, Sahari J., Ramli R., Deros BM., Park J. Electrical Conductivity Behaviour of Chemical Functionalized MWCNTs Epoxy Nanocomposites. European Journal of Scientific Research 2009;29:13-21.

[55] Islam MF, Rojas E, Bergey DM, Johnson AT, Yodh AG. High weight fraction surfactant solubilization of single-wall carbon nanotubes in water. Nano Letters 2003;3:269-73.

[56] Barrau S, Demont P, Perez E, Peigney A, Laurent C, Lacabanne C. Effect of palmitic acid on the electrical conductivity of carbon nanotubes-epoxy resin composites. Macromolecules 2003;36:9678-80.

[57] Gong XY, Liu J, Baskaran S, Voise RD, Young JS. Surfactant-assisted processing of carbon nanotube/polymer composites. Chemistry of Materials 2000;12:1049-52.

[58] Ozarkar S, Jassal M, Agrawal AK. Improved dispersion of carbon nanotubes in chitosan. Fibers and Polymers 2008;9:410-5.

[59] Park JG, Louis J, Cheng QF, Bao JW, Smithyman J, Liang R, et al. Electromagnetic interference shielding properties of carbon nanotube buckypaper composites. Nanotechnology 2009;20.

[60] Yang K, Gu MY, Guo YP, Pan XF, Mu GH. Effects of carbon nanotube functionalization on the mechanical and thermal properties of epoxy composites. Carbon 2009;47:1723-37.

[61] Wardle BL, Saito DS, Garcia EJ, Hart AJ, de Villoria RG, Verploegen EA. Fabrication and characterization of ultrahigh-volume-fraction aligned carbon nanotube-polymer composites. Advanced Materials 2008;20:2707-+.

[62] Gou JH. Single-walled nanotube bucky paper and nanocomposite. Polymer International 2006;55:1283-8.

[63] Wang Z, Liang ZY, Wang B, Zhang C, Kramer L. Processing and property investigation of single-walled carbon nanotube (SWNT) buckypaper/epoxy resin matrix nanocomposites. Composites Part a-Applied Science and Manufacturing 2004;35:1225-32.

[64] Feng QP, Yang JP, Fu SY, Mai YW. Synthesis of carbon nanotube/epoxy composite films with a high nanotube loading by a mixed-curing-agent assisted layer-by-layer method and their electrical conductivity. Carbon 2010;48:2057-62.

[65] Feng QP, Shen XJ, Yang JP, Fu SY, Mai YW, Friedrich K. Synthesis of epoxy composites with high carbon nanotube loading and effects of tubular and wavy morphology on composite strength and modulus. Polymer 2011;52:6037-45.

[66] Bradford PD, Wang X, Zhao HB, Maria JP, Jia QX, Zhu YT. A novel approach to fabricate high volume fraction nanocomposites with long aligned carbon nanotubes. Composites Science and Technology 2010;70:1980-5.

[67] Ogasawara T, Moon SY, Inoue Y, Shimamura Y. Mechanical properties of aligned multi-walled carbon nanotube/epoxy composites processed using a hot-melt prepreg method. Composites Science and Technology 2011;71:1826-33.

[68] Wang XK, Lin XW, Dravid VP, Ketterson JB, Chang RPH. GROWTH AND CHARACTERIZATION OF BUCKYBUNDLES. Applied Physics Letters 1993;62:1881-3.

[69] Ajayan PM, Stephan O, Colliex C, Trauth D. ALIGNED CARBON NANOTUBE ARRAYS FORMED BY CUTTING A POLYMER RESIN-NANOTUBE COMPOSITE. Science 1994;265:1212-4.

[70] Deheer WA, Bacsa WS, Chatelain A, Gerfin T, Humphreybaker R, Forro L, et al. ALIGNED CARBON NANOTUBE FILMS - PRODUCTION AND OPTICAL AND ELECTRONIC-PROPERTIES. Science 1995;268:845-7.

[71] Fan SS, Chapline MG, Franklin NR, Tombler TW, Cassell AM, Dai HJ. Self-oriented regular arrays of carbon nanotubes and their field emission properties. Science 1999;283:512-4.

[72] Li WZ, Xie SS, Qian LX, Chang BH, Zou BS, Zhou WY, et al. Large-scale synthesis of aligned carbon nanotubes. Science 1996;274:1701-3.

[73] Jin L, Bower C, Zhou O. Alignment of carbon nanotubes in a polymer matrix by mechanical stretching. Applied Physics Letters 1998;73:1197-9.

[74] Smith BW, Benes Z, Luzzi DE, Fischer JE, Walters DA, Casavant MJ, et al. Structural anisotropy of magnetically aligned single wall carbon nanotube films. Applied Physics Letters 2000;77:663-5.

[75] Schadler LS, Giannaris SC, Ajayan PM. Load transfer in carbon nanotube epoxy composites. Applied Physics Letters 1998;73:3842-4.

[76] Bai J. Evidence of the reinforcement role of chemical vapour deposition multi-walled carbon nanotubes in a polymer matrix. Carbon 2003;41:1325-8.

[77] Zhou YX, Pervin F, Lewis L, Jeelani S. Experimental study on the thermal and mechanical properties of multi-walled carbon nanotube-reinforced epoxy. Materials Science and Engineering a-Structural Materials Properties Microstructure and Processing 2007;452:657-64.

[78] Du JH, Bai J, Cheng HM. The present status and key problems of carbon nanotube based polymer composites. Express Polymer Letters 2007;1:253-73.

[79] Haggenmueller R, Zhou W, Fischer JE, Winey KI. Production and characterization of polymer nanocomposites with highly aligned single-walled carbon nanotubes. Journal of Nanoscience and Nanotechnology 2003;3:105-10.

[80] Andrews R, Jacques D, Rao AM, Rantell T, Derbyshire F, Chen Y, et al. Nanotube composite carbon fibers. Applied Physics Letters 1999;75:1329-31.

[81] D. Stauffer AA. Introduction to percolation theory. Taylor & Francis; 1992.

[82] Sandler J, Shaffer MSP, Prasse T, Bauhofer W, Schulte K, Windle AH. Development of a dispersion process for carbon nanotubes in an epoxy matrix and the resulting electrical properties. Polymer 1999;40:5967-71.

[83] Sandler JKW, Kirk JE, Kinloch IA, Shaffer MSP, Windle AH. Ultra-low electrical percolation threshold in carbon-nanotube-epoxy composites. Polymer 2003;44:5893-9.

[84] Bai JB, Allaoui A. Effect of the length and the aggregate size of MWNTs on the improvement efficiency of the mechanical and electrical properties of nanocomposites - experimental investigation. Composites Part a-Applied Science and Manufacturing 2003;34:689-94.

[85] Choi ES, Brooks JS, Eaton DL, Al-Haik MS, Hussaini MY, Garmestani H, et al. Enhancement of thermal and electrical properties of carbon nanotube polymer composites by magnetic field processing. Journal of Applied Physics 2003;94:6034-9.

[86] Stephan C, Nguyen TP, Lahr B, Blau W, Lefrant S, Chauvet O. Raman spectroscopy and conductivity measurements on polymer-multiwalled carbon nanotubes composites. Journal of Materials Research 2002;17:396-400.

[87] Singjai P, Changsarn S, Thongtem S. Electrical resistivity of bulk multi-walled carbon nanotubes synthesized by an infusion chemical vapor deposition method. Materials Science and Engineering: A 2007;443:42-6.

[88] Lau CH, Cervini R, Clarke SR, Markovic MG, Matisons JG, Hawkins SC, et al. The effect of functionalization on structure and electrical conductivity of multi-walled carbon nanotubes. J Nanopart Res 2008;10:77-88.

[89] Grimes CA, Dickey EC, Mungle C, Ong KG, Qian D. Effect of purification of the electrical conductivity and complex permittivity of multiwall carbon nanotubes. Journal of Applied Physics 2001;90:4134-7.

[90] Singh BP, Choudhary V, Saini P, Mathur RB. Designing of epoxy composites reinforced with carbon nanotubes grown carbon fiber fabric for improved electromagnetic interference shielding. AIP Adv 2012;2:6.

[91] Mathur RB, Singh, B.P,Tiwari, P.K., Gupta, T.K.Choudhary, V. Enhancement in the thermomechanical properties of carbon fibre-carbon nanotubes-epoxy hybrid composites. International Journal of Nanotechnology 2012;9:1040-1049.

[92] Huang Y, Li N, Ma Y, Feng D, Li F, He X, et al. The influence of single-walled carbon nanotube structure on the electromagnetic interference shielding efficiency of its epoxy composites. Carbon 2007;45:1614-21.

[93] Li N, Huang Y, Du F, He XB, Lin X, Gao HJ, et al. Electromagnetic interference (EMI) shielding of single-walled carbon nanotube epoxy composites. Nano Letters 2006;6:1141-5.

[94] Berber S, Kwon YK, Tomanek D. Unusually high thermal conductivity of carbon nanotubes. Phys Rev Lett 2000;84:4613-6.

Chapter 9

Kinetics of Growing Centimeter Long Carbon Nanotube Arrays

Wondong Cho, Mark Schulz and Vesselin Shanov

Additional information is available at the end of the chapter

1. Introduction

Carbon nanotubes (CNTs) are fascinating materials with outstanding mechanical, optical, thermal, and electrical properties [1-4]. CNTs also have a huge aspect ratio and a large surface area to volume ratio. Because of their unique properties, vertically aligned centimeter long CNT arrays have generated great interest for environmental sensors, biosensors, spinning CNT into yarn, super-capacitors, and super-hydrophobic materials for self-cleaning surfaces [5-11]. Yun et al. studied a needle-type biosensor based on CNTs to detect dopamine. Their results showed advantages of using CNT biosensors for detecting neurotransmitters [11]. Most of the envisioned applications require CNTs with high quality, a long length, and well aligned vertical orientation. Although many researchers have studied the synthesis of vertically aligned CNT arrays, the CNT growth mechanism still needs to be better understood. In addition, CNT lengths are typically limited to a few millimeters because the catalyst lifetime is usually less than one hour [12- 16]. Many groups have studied the kinetics of CNT growth trying to improve CNT properties. Different observation methods [17-22] were used to determine the effect of the catalyst, buffer layers, carbon precursor, and deposition conditions on nanotube growth. One of the suggested growth mechanisms postulates several steps [23]. First, the carbon source dissociates on the surface of the substrate. Next, the carbon atoms diffuse to the molten catalyst islands and dissolve. The metal-carbon solution formed reaches a supersaturated state. Finally, the carbon nanotubes start to grow from the carbon- catalyst solution. *In situ* observation of CNTs during their nucleation and growth is a useful method to understand the growth mechanism, which might help to overcome the limitation of the short length of nanotubes, and to control array growth and quality. Various remarkable approaches of *in situ* observation have been performed to affirm the growth mechanism of vertically aligned CNTs and also to obtain kinetics data such as growth rate and activation energy [12, 24-27]. Puretzky *et al.* studied the growth kinetics of CNT arrays using *in situ* time-lapse photography and laser irradiation under diffusion-limited growth conditions [28]. *In situ* transmission electron microscopy (TEM) was used by Kim *et al.* to study the dynamic growth behavior of CNT arrays [29]. Additionally, a pseudo *in situ* monitoring method was used to investigate the kinetics of CNT array growth by creating marks on the side of the CNT array during the growth. Using this method, several groups demonstrated root growth for their catalyst systems. However, their studies were limited to short lengths and also required *ex-situ* observation with SEM to obtain the growth length as a function of time [25, 27, 30, 31]. Most reported methods are designed to operate and monitor the growth length with time for relatively short (a few millimeters long) CNT arrays and also do not provide kinetics data for growing centimeter long CNT arrays.

Recently, we have developed new catalyst systems which are able to grow over 1 centimeter long vertically aligned CNT arrays [6]. In this paper we examined the growth mechanism and kinetics of centimeter long carbon nanotube arrays using a real-time photography technique and the effect of growth temperature and growth time on morphology of CNTs.

2. Experimental Method

Fig. 1 shows the schematic of the chemical vapor deposition (CVD) system used for centimeter long CNT array growth. The reactor consists of a 2 inch quartz tube placed inside a high temperature furnace (Barnstead International, type F79400 tube furnace) and four mass-flow controllers (MFC) which control the flow rate of the gas reactants such as hydrogen, ethylene, water vapor and argon. A water bubbler is also installed to provide water vapor using argon carrier gas. A window on the side of the reactor is used to acquire real-time images of CNT arrays and to record data with a digital camera (Olympus E510) controlled by a computer.

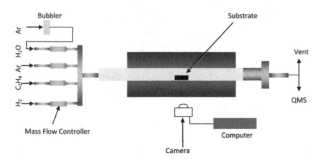

Figure 1. Schematic of the CVD system for direct observation of the centimeter long carbon nanotube arrays during their growth. The top view of the reactor is shown.

The substrates were parts of 4 inch silicon wafers (100) with SiO_2 (500 nm) on the top. The buffer and catalyst layers based on Al_2O_3 (15 nm thick)/Fe (1 nm thick) were deposited on the wafers using e-beam evaporation. After the deposition, the substrates were annealed for several hours at 400 ℃ in Air. All the experiments were performed using the following optimized recipe for centimeter long CNT arrays: 560 mmHg of argon, 60 mmHg of hydrogen, and 140 mmHg of ethylene as a carbon precursor. The water concentration in the reactor was near 900 ppm measured by a quadrupole mass spectrometer (QMS). The total pressure was kept at one atmosphere during the growth and the temperature varied from 690 ℃ to 840 ℃. Real-time images of the CNT array growth were recorded from the moment that ethylene was introduced into the reactor. The images were used to study the growth mechanism and kinetics of the CNT growth. Scanning electron microscopy (SEM, Phillips XL30 ESEM), high resolution transmission electron microscopy (HRTEM, JOEL 2000 FX) and Micro-Raman spectroscopy (Renishaw inVia Reflex Micro-Raman) were employed to characterize the CNT morphology.

3. Results and discussion

3.1. Growth evolution by real-time photography

Real-time photography was used to study the growth mechanism and kinetics of centimeter long CNT array growth. The digital camera provided clear images showing details related to the dynamic changes of the array shape during the growth. This was achieved by controlling the intervals for taking pictures from a few seconds to several hours depending on the experimental conditions.

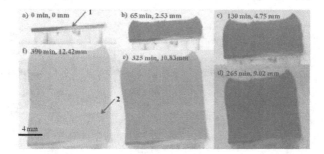

Figure 2. Real-time images of the centimeter long CNT array growth evolution with time during CVD at 780 ℃: (a) Side image of the substrate at zero growth time, (b) to (f) Images of CNT arrays grown for different times.

Fig. 2 illustrates sequential images of vertically aligned centimeter long CNT arrays grown at 780 ℃. As can be seen from Fig. 2, it is easy to distinguish the substrate from the CNT array. Arrow 1 in Fig. 2a points to a side view of the substrate. Arrow 2 in Fig. 2f shows the side view of the CNT array. The growth length can be obtained as a function of the deposition time from the images. Changes in the array shape can also be observed during the entire growth time. In Fig. 2f, the growth length was 12.47 mm and the catalyst lifetime was 450 min. This experiment was repeated several times at the same deposition conditions and the results were reproducible. Hence, Fig. 2 demonstrates the real-time images which allow measuring the growth length and observing the CNT array growth mechanism. This approach gives reproducible results for studying the kinetics of CNT array growth.

Movies composed of multiple images were created to investigate the morphological changes and growth mechanism of CNT arrays during their synthesis (Supporting material 1).

3.2. Growth mode of centimeter long CNT arrays

A carbon source interruption method combined with real-time photography was used to determine the CNT array growth mechanism (root vs. tip growth). Fig. 3 shows real-time images of CNT arrays obtained for different periods of growth employing 5 minute interruption of the ethylene supply. In this experiment, the ethylene flow was first stopped after 70 minutes growth while the rest of reactant gases such as Ar, H_2 and water vapor were continuously supplied with same partial pressure. Next, the ethylene was resupplied after the 5 min break. Fig. 3b, c and d show real-time images of CNT arrays before and after the 5 minute interruption of the carbon precursor. The growth length in Fig. 3b was 2.89 mm. The length did not change after the ethylene interruption (Fig. 3c). CNTs started to grow again after ethylene was resupplied (Fig. 3d). As can be seen in Fig. 3e and f, the first layer which grew before the 5 min ethylene interruption detached from the second layer grown after the 5 min interruption. The separation may be caused by water etching the interface between the root of the CNTs and the substrate during the interruption. Even though the first layer detached, the second layer continued to grow from the bottom. Fig. 3f shows that the growth length of the second layer was 2.85 mm. The calculated growth rate was 38.5 µm/min before and after the interruption and until the growth stopped.

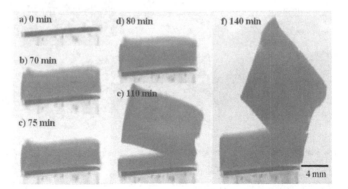

Figure 3. Real-time images of CNT array grown for different periods of time at a deposition temperature of 780 °C with 5 minutes interruption of ethylene: (a) Side image of the substrate at zero growth time, (b) to (f) Pictures of CNT arrays grown for different times.

It was obvious from the data that the interruption didn't affect the growth rate during the CNT synthesis. The images also reveal that the growth pathway of the centimeter long CNT array is "root growth" (Supporting materials 2).

The top surface of the centimeter long CNT array was studied by Energy Dispersive X-ray Spectroscopy (EDS) to determine if metal catalyst moved to the tips of the CNTs. No trace of iron catalyst was detected on the top of the CNT array. Thermogravimetric analysis (TGA) performed at a heating rate of 10 ºC/min in air showed that the change of CNT array weight was negligible at temperatures below 550 ºC. The combustion started slightly below 700 ºC and was completed at 750 ºC. The amount of residual matter was extremely small and was not measureable after the completion of combustion above 750 ºC. These results implied that the CNT array was almost "catalyst free".

3.3. Kinetics of centimeter long CNT array growth

The change in the growth length as a function of deposition time at different temperatures was investigated using real-time photography. Fig. 4 displays the dependence of growth length on the growth time in the range of 730 ºC to 840 ºC. The rest of the CVD experimental conditions were kept constant. As shown in Fig. 4a, the growth length increases linearly with the growth time within the entire temperature range of 730 ºC to 840 ºC. The data show that the growth rate remained constant until the catalyst was deactivated. Diffusion of the carbon precursor through the CNT forest apparently did not limit the growth rate even in the case of 12 mm long CNT arrays. It reveals the carbon source was able to reach the surface of catalyst particles without significant resistance. These results show that the growth rate followed a kinetic controlled mode. Zhu et al. reported that their catalyst system was controlled by a gas diffusion process and their growth rate decreased gradually with increasing the length of CNT array. They reported a diffusion controlled mode for a similar range of growth temperatures [31] used in the present paper.

In the current experiments, growth termination occurred abruptly for all experimental deposition temperatures. After growth termination, the array length remained constant with time. The reason for such abrupt catalyst deactivation is not clear and it could not be interpreted using the suggested mechanisms in the literature such as Oswald ripening, forming stable iron carbides, and depletion of the catalyst [32-34]. It is hypothesized that catalyst deactivation occurs due to several complex reasons. One reason is that amorphous carbon is built up during the CVD process and covers the catalyst active site at the surface of the substrate which results in passivation of the catalyst.

Fig. 4b shows the effect of temperature on the final growth length. The slope of the curve increases gradually as the CVD temperature increases from 730 ºC to 780 ºC, and then decreases. The longest CNT length of 12.42mm was achieved at 780 ºC. It was observed that the longest centimeter long CNT array was obtained neither at the highest growth rate nor at the longest catalyst lifetime. Analyzing Fig. 4 provides a better understanding of the catalyst lifetime at different temperatures. Fig. 4c shows that the catalyst lifetime decreases linearly with the increase of temperature in the range of 730 ºC to 840 ºC. In this experiment, the longest catalyst lifetime was 895 min at 730 ºC. The catalyst lifetime decreases dramatically down to a few minutes as the temperature approaches to 840 ºC. The results in this paper demonstrate that the catalyst lifetime and the final growth length are considerably longer than reported in the literature [12-16]. Water vapor usually plays an important role in enhancing the growth rate and also prolonging the catalyst lifetime [35] since water vapor may clean the surface of the catalyst particles by removing the amorphous carbon.

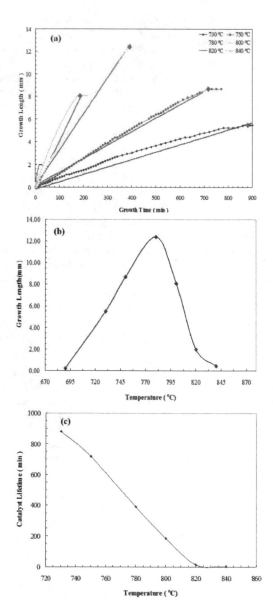

Figure 4. Experimental results from the kinetics study of centimeter long CNT array growth: (a) Growth length as a function of the growth time. The solid red lines indicate the fitting of the analytical growth equation to the experimental results. (b) Plot of the final growth length after termination vs. deposition temperature. (c) Plot of the catalyst lifetime as a function of the temperature.

Fig. 5 shows the Arrhenius plot of the CNT growth rate as a function of the temperature. The growth rate was calculated using the final growth length divided by the catalyst lifetime. The highest growth rate achieved was 193 µm/min at 840 ºC. The present results reveal that the growth rate is proportional to the concentration of ethylene, so first order reaction is assumed. The activation energy E_a was calculated based on the Arrhenius equation $k = k_0 \, exp(- E_a /RT)$ where, k_0 is the frequency factor, T is the growth temperature (K), k is the reaction rate constant (in this case k is growth rate constant), and R is the ideal gas constant. The activation energy calculated from Arrhenius equation and Fig. 5 was about 248 kJ/mol. Similarly, Zhu et al. obtained activation energy of 201.2 kJ/mol for a Fe/Al$_2$O$_3$/Si substrate using ethylene as a carbon source [27]. Li et al also reported activation energy of 158 kJ/mol for the temperature interval of 730 ºC to 780 ºC using ethylene [24]. Both groups concluded that the growth rate was not affected by the diffusion of the carbon source to the catalysts because the CNT length increased linearly with time. In the current paper, the same trend is observed although the catalyst lifetime and CNT growth length were much longer.

Based on the kinetics data, it was concluded that the centimeter long CNT array growth does not fit the diffusion controlled mode and is more reasonable to be considered as a kinetically controlled process. This conclusion is based on the specific process conditions and size of the substrate used here. Diffusion may play a role in growth for other experimental conditions.

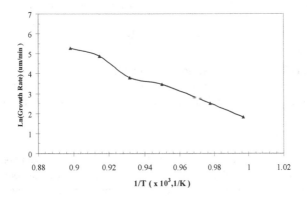

Figure 5. Arrhenius plot of centimeter long CNT array growth rate as a function of 1/temperature.

Zhu et al. adopted the silicon oxidation model for a diffusion controlled mode of CNT growth described with the equation: $h_0 = 0.5(A^2 + 4Bt)^{1/2} - 0.5A$. Futaba et al. assumed that the deactivation of catalyst was analogous to radioactive decay to model their super growth of CNT via the equation: $H(t) = \beta\tau_0 (1 - e^{-t/\tau 0})$. They demonstrated a good fit of the experimental data to the model equations. However, Futaba et al. could not extrapolate it for a longer growth time in order to obtain the final growth length and to predict the abrupt growth termination. Hence, the models described above cannot be adopted for the centimeter long CNT array growth. Based on the experimental data in this paper, it was attempted to derive a growth model using an analytical method that can be reasonably applied to the growth of centimeter long CNT arrays.

From Fig. 4, the growth length can be expressed by a linear equation:

$$GL_T(t) = GL_{Tf} - r_g(t_{Tf} - t) \tag{1}$$

where, $GL_T(t)$ is the array length at a certain growth time t and temperature T, GL_{Tf} is the final growth length of the centimeter long CNT array after termination at temperature T, r_g is the growth rate and t_{Tf} is the catalyst lifetime. The final growth length GL_{Tf} can be obtained by the following equation depending on the deposition temperature:

In case of temperatures below 780 ºC, the final CNT length

$$GL_{Tf} = 0.1367T(\text{º}C) - 94.1 \tag{2}$$

Above deposition temperatures of 780 ºC, the final CNT length is

$$GL_{Tf} = -0.216T(\text{º}C) + 181 \tag{3}$$

The catalyst lifetime t_{Tf} can also be expressed by a linear equation from Fig. 6b:

$$t_{Tf} = -10.08T(\text{º}C) + 8253.7. \tag{4}$$

The solid red lines displayed in Fig. 4a are plotted based on the equation (1). As shown in Fig. 4a, the plot fits the experimental data well. The analytical model can also predict the final growth length, growth rate, and catalyst lifetime for a certain CVD temperature.

3.4. Morphology of centimeter long CNTs

Fig. 6a, b and c display SEM images of centimeter long CNT arrays obtained at different magnifications. At low magnification of 1000x (Fig. 6a) the image shows appearance of vertically aligned CNTs. Fig. 6b and c are taken at higher magnifications and reveal individual CNTs. The images indicate that despite the long growth time and centimeter length, the tubes grew vertically without any interruption until the catalyst activity was terminated.

HRTEM was used to study changes in the structure and diameter distribution of individual CNTs at different growth temperatures. The images shown in Fig. 7 reveal well defined multi-wall CNTs without metal catalyst incorporated into the tubes. At low temperatures amorphous carbon deposited on the walls was also observed.

Fig. 7e and f show the distribution of number of walls and average tube diameter of each wall at different growth temperatures. The diameter of CNTs is in the range of 5 to 9 nm and is independent on the number of walls.

Figure 6. SEM images showing side view of a centimeter long CNT array at different magnifications. The magnification increases from (a) to (b) and (c).

This trend was observed to be similar for every growth temperature used. We found that the diameter distribution does not depend on the growth temperature. The distribution of the number of walls also revealed broad range from 1 to 10 walls for each growth temperature in Fig. 7e. Most of the tubes possess 2 to 5 walls at each growth temperature. We observed that the single wall CNTs were produced at 800 ºC and the double wall CNTs were yielded above 750 ºC. In our experiments, the distribution of the walls was not substantially affected by the growth temperature. Hence, we noted that the growth temperature is not an important factor affecting the structure of CNTs obtained in our growth process.

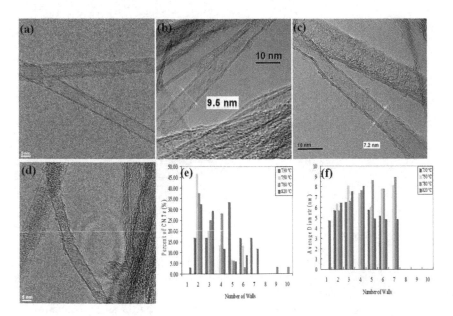

Figure 7. HRTEM images and distribution of the CNTs walls at different growth temperatures: (a) 730 ºC, (b) 750 ºC, (c) 780 ºC, (d) 820 ºC, (e) distribution of number of walls, (f) distribution of the average tube diameter of each wall for different multiwall CNTs.

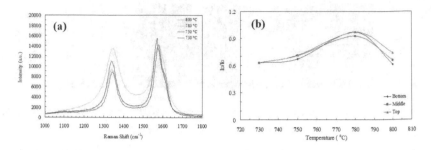

Figure 8. Raman spectra of CNTs taken from the middle position of the CNTs: (a) Spectra at various growth temperatures, (b) I_D/I_G peak intensity ratios as a function of growth temperature.

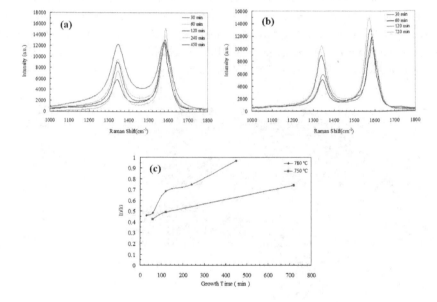

Figure 9. Raman spectra of CNTs synthesized at different temperature and growth time: (a) spectra at 780 °C, (b) spectra at 750 °C, (c) I_D/I_G peak intensity ratios vs. different growth time.

Micro-Raman spectroscopy (Renishaw inVia Reflex Micro-Raman) was used to investigate the effect of temperature and growth time on the quality of CNT arrays using a 514 nm excitation wavelength. Fig. 8 shows Raman shift and I_D/I_G ratio of the CNTs at different growth temperatures for three points along the array length. The spectra show distinguished D band peak (near 1350 cm⁻¹) which indicates presence of defects, disordered and amorphous carbon. A pronounced G band peak (at 1580 cm⁻¹) originating from graphitization of CNTs is also displayed [36-40]. As shown in Fig. 8a, the intensity of the D band increased with increasing of the growth temperature up to 780 °C. Fig. 8b displays the intensity ratio I_D/I_G as a function of the growth temperature from three different height positions of CNTs. The intensity ratio I_D/I_G was similar at bottom, middle and top positions for each growth temperature and rose from 0.62 to 1 as the temperature increases with a maximum at 780 °C. Fig. 9 shows Raman data for CNT arrays obtained at different growth time and temperatures. The intensity of D peak increases linearly with rising the growth temperature as shown on Fig. 9a and b. Fig. 9c illustrates that the I_D/I_G ratio also increases near linearly with extending the growth time, which indicates that the quality of CNTs deteriorates when the tubes reside longer in the growth zone. At 750 °C, I_D/I_G increased from 0.422 to 0.704 when prolonging the growth time. The highest value 0.968 of I_D/I_G was obtained at 780 °C. Fig. 9c indicates that I_D/I_G values at 780 °C are greater than those obtained at 750 °C.

As the growth time increases, the gap between the two plots in Fig 9c broadens. Thus, the quality of CNTs decreases faster with time at higher temperatures. The reason for this is the accumulation of the amorphous carbon at high temperatures. These results are supported by the presented HRTEM images.

4. Conclusions

Real-time photography was used to record the growth of centimeter long CNT arrays during the CVD process. The kinetics of growing vertically aligned CNTs was studied based on the photographic images. Furthermore, we found that the CNT arrays grew by the root growth mechanism which was proved by the carbon source interruption method and real-time photography. The length of the CNT arrays increased linearly with growth time for all the tested temperatures followed by an abrupt growth termination. The catalyst lifetime decreased linearly with rising the deposition temperature and varied from a few minutes up to several hundred minutes depended on the growth conditions. We found out that the formation of centimeter long CNT array could not be described by diffusion controlled or exponentially decaying growth. This study suggests that the growth in this case is governed by a kinetically controlled mode within the temperature interval from 730 °C to 840 °C. The calculated activation energy is 248 kJ/mol. An analytical model for centimeter long CNT array synthesis was proposed which can predict the growth rate, final CNT length, and the catalyst lifetime. The obtained data indicated that the wall and diameter distribution of CNTs is independent on the growth temperature. The quality of CNTs deteriorates with increasing of the growth time and temperature. We found out that the amount of amorphous carbon on the CNTs depends on the residence time of the tubes in the CVD growth zone and on the deposition temperature. Longer residence time and higher deposition temperature accumulates greater amount of amorphous carbon.

Acknowledgements

The financial support from NSF through grant CMMI-07272500 and from NCA&T through DURIP-ONR is highly acknowledged. We also would like to thank Jay Yocis who helped to set up the real-time photography system and Dr. John Robertson from Cambridge University who suggested real time photography of studying kinetics of CNT arrays.

Appendix A. Supplementary data

Movies showing the centimeter long CNT array growth inside of the CVD reactor, including the root growth are available online.

Author details

Wondong Cho[1,2], Mark Schulz[2,3] and Vesselin Shanov[1,2*]

*Address all correspondence to: vesselin.shanov@uc.edu

1 Chemical and Materials Engineering, University of Cincinnati, Cincinnati, Ohio USA

2 Nanoworld Laboratory, Rh414, University of Cincinnati, Cincinnati, Ohio USA

3 Mechanical Engineering, University of Cincinnati, Cincinnati, Ohio USA

References

[1] Puretzky, A. A., Geohegan, D. B., & Eres, G. (2008). Real-time imaging of vertically aligned carbon nanotube array growth kinetics. *Nanotechnology*, 19, 055605.

[2] Amama, P. B., Pint, C. L., Mc Jilton, L., Kim, S. M., Stach, E. A., Murray, P. T., Hauge, R. H., & Maruyama, B. (2008). Role of Water in Super Growth of Single-Walled Carbon Nanotube Carpets. *Nano Letters*, 9, 44-49.

[3] Baughman, R. H., Zakhidov, A., & de Heer, W. A. (2002). Carbon nanotubes-the route toward applications. *Science*, 297, 787-792.

[4] Bronikowski, M. (2007). Longer nanotubes at lower temperatres: the influence of effective activation energies on carbon nanotube growth by thermal chemical vapor deposition. *J. Phys. Chem. C*, 111, 17705-17712.

[5] Gommes, C., Pirard, J. P., & Blacher, S. (2004). Influence of the operating conditions on the production rate of multi-walled carbon nanotubes in a CVD reactor. *Carbon*, 42, 1473-1482.

[6] Vix-Guterl, C., Couzi, M., & Delhaes, P. (2004). Surface charactrizations of carbon mutiwall nanotubes:comparison surface active sites and raman spectroscopy. *J. Phys. Chem. B*, 108, 19361-19367.

[7] Kim, D. H., Lee, H. R., & Jang, H. S. (2003). Dynamic growth rate behavior of a carbon nanotube forest characterized by in situ optical growth monitoring. *Nano Letters*, 3, 863-865.

[8] Du, C., Yeh, J., & Pan, N. (2005). High power density supercapacitors using locally aligned carbon nanotube electrodes. *Nanotechnology*, 16, 350-353.

[9] Einarsson, E., & Maruyama, Y. M. S. (2008). Growth dynamics of vertically aligned single-walled carbon nanotubes from in situ measurements. *Carbon* , 46, 923-930.

[10] Mckee, G. S. B., , J. S. F., & Vecchio, K. S. (2008). Length and the oxidation kinetics of chemical vapor deposition generated multiwalled carbon nanotubes. *J. Phys. Chem. C*, 112, 10108-10113.

[11] Guang-yong, Xiong. , Wang, D. Z., & Ren, Z. F. (2006). Aligned millimeter-long carbon nanotube arrays grown on single crystal magnesia. *Carbon* , 44, 969-973.

[12] Nii, H., Sumiyama, Y. , & Kunishige, A. (2008). Influence of diameter on the Raman of multi-walled carbon nanotubes. *Applied Physics Express*, 1, 064005.

[13] Hata, K. (2004). Water-assisted highly efficient synthesis of impurity-free single-walled carbon nanotubes. *Science*, 306, 1362-1364.

[14] Iijima, S. (1991). Helical microtubules of graphitic carbon. *Nature*, 354, 56-58.

[15] Ishikawa, T. (2006). Overview of trends in advanced composite research and applications in Japan. *Adv. Compos. Mater.*, 15, 3-37.

[16] Iwasaki, T., Zhong, G. F., Aikawa, T., Yoshida, T., & Kawarada, H. (2005). *J. Phys. Chem. B*, 109, 19556.

[17] Benit, J. M., , J. P. B., & Lefrant, S. (2002). Low frequency raman studies of muti-wall carbon nanotubes: experiments and theory. Physical review B. Condensed matter and materials physics 073417 , 66, 1-4.

[18] Jiang, K., Li, Q., & Fan, S. (2002). Spinning continuous carbon nanotube yarns. *Nature*, 419, 801.

[19] Hasegawa, K., , S. N., & Yamaguchi, Y. (2008). Growth window and possible mechanism of millimeter-thick single-walled carbon nanotube forests. *Journal of Nanoscience and Nanotechnology*, 8, 6123.

[20] Lau, K. T., , C. G., & Hui, D. (2006). A critical review on nanotube and nanotube/ nanoclay related polymer composite materials. *Composites, Part B*, 37, 425 .

[21] Zhu, L., Xu, J. , & Wong, C. P. (2007). The growth of carbon nanotubes stacks in the kinetics-controlled regime. *Carbon*, 45, 344-348.

[22] Li, Y., Kinloch, I. A., & Windle, A. H. (2004). Direct spinning of carbon nanotube fibers from chemical vapor deposition synthesis. *Science*, 304, 276-278.

[23] Lingbo, Zhu. D. W. H., & Ching-Ping, Wong. (2006). Monitoring Carbon Nanotube growth by Formation of nanotube stacks and investigation of the diffusion-control kinetics. *J.Phys. Chem. B*, 110, 5445-5449.

[24] Liu, K., Jiang, K. L., Feng, C., Chen, Z., & Fan, S. S. (2005). *Carbon*, 43, 2850.

[25] Matthews, M. J., Pimenta, M. A., & Endo, M. (1999). Origin of disperive effects of the raman D band in carbon materials. *Physical review B*, 59, R6585.

[26] Meshot, E. R., & Hart, A. J. (2008). Abrupt self-termination of vertically aligned carbon nanotube growth. *Applied Physics Letters*, 92, 113107-113103.

[27] Noda, S., Hasegawa, K., Sugime, H., Kakehi, K., Zhang, Z., Maruyama, S., & Yamaguchu, Y. (2007). *Jpn. J. Appl. Phys*, 46, L399.

[28] Oleg, V., & Yazyev, A. P. (2008). Effect of metal elements in catalytic growth of carbon nanotubes. *Physical review letters*, 100, 156102.

[29] Li, Qingwen. X. , Xiefei, Z., & Zhu, Yuntian T. (2006). Sustain growth of ultralong carbon nanotube arrays for fiber spinning. *Advanced Materials*, 18, 3160-3163.

[30] Brukh, R., & Mitra, S. (2006). Mechanism of carbon nanotube growth by CVD. *Chemical Physics Letters*, 424, 126-132.

[31] Xiang, R., , Z. Y., & Maruyama, S. (2008). Growth deceleration of vertically aligned carbon nanotube arrays: catalyst deactivation or feedstock diffusion controlled? *J. Phys. Chem. C*, 112, 4892-4896.

[32] Maruyama, S., Einarsson, E., & Edamura, T. (2005). Growth process of vertically aligned single-walled carbon nanotubes. *Chemical Physics Letters*, 403, 320-323.

[33] Pal, S. K., , S. T., & Ajayan, P. M. (2008). Time and temperature dependence of mutiwalled carbon nanotube growth on inconel 600. Nanotechnology , 19, 045610.

[34] Shim, J. S., Yun, Y. H., Cho, W., Shanov, V., Schulz, M. J., & Ahn, C. H. (2010). Self-Aligned Nanogaps on Multilayer Electrodes for Fluidic and Magnetic Assembly of Carbon Nanotubes. *Langmuir*, 26, 11642-11647.

[35] Stadermann, M., Sherlock, S. P., In, J. B., Fornasiero, F., Park, H. G., Artyukhin, A. B., Wang, Y., De Yoreo, J. J., Grigoropoulos, C. P., Bakajin, O., Chernov, A. A., & Noy, A. (2009). Mechanism and Kinetics of Growth Termination in Controlled Chemical Vapor Deposition Growth of Multiwall Carbon Nanotube Arrays. *Nano Letters*, 9, 738-744.

[36] Shanov, V., , W. C., Schulz, M., & Malik, N. (2008). Advances in synthesis and application of carbon nanotube materials. *Materials science and technology*, 2253.

[37] Zhao, X., Saito, R., & Ando, Y. (2002). Characteristic raman spectra of mutiwalled carbon nanotubes. *Physica B*, 323, 265-266.

[38] Yun, Y. H., , A. B., Shanov, V. N., & Schulz, M. J. (2006). A nanotube composite microelectrode for monitoring dopamine levels using cyclic voltammetry and differential pulse voltammetry. 220, 53 -60.

[39] Yun, Y., Shanov, V., Tu, Y., Subramaniam, S., & Schulz, M. J. (2006). Growth Mechanism of Long Aligned Multiwall Carbon Nanotube Arrays by Water-Assisted Chemical Vapor Deposition. *The journal of physical chemistry B*, 110, 23920-23925.

[40] Zhong, G., Iwasaki, T., Robertson, J., & Kawarada, H. (2007). Growth Kinetics of 0.5 cm Vertically Aligned Single-Walled Carbon Nanotubes. *J. Phys. Chem. B*, 111, 1907 -1910.

Chapter 10

Toward Greener Chemistry Methods for Preparation of Hybrid Polymer Materials Based on Carbon Nanotubes

Carlos Alberto Ávila-Orta, Pablo González-Morones,
Carlos José Espinoza-González,
Juan Guillermo Martínez-Colunga,
María Guadalupe Neira-Velázquez,
Aidé Sáenz-Galindo and Lluvia Itzel López-López

Additional information is available at the end of the chapter

1. Introduction

Recent technological advances and the need for materials with new functionalities and better performance have generated an enormous demand for novel materials. Nanostructures such as carbon nanotubes (CNTs) possess outstanding mechanical, electrical, thermal and chemical properties which make them ideal for a wide variety of current or future applications [1], especially for the preparation of multifunctional *hybrid polymer materials*.

The incorporation of CNTs to polymer matrices have demonstrated to improve the mechanical, electrical, thermal and morphological properties of the produced nanocomposites [2]; however, the full exploitation of CNTs has been severely limited due to difficulties associated with dispersion of entangled CNTs during processing, and their poor interfacial interaction with the polymer matrix. Therefore, significant efforts have been directed toward improving the dispersion of CNTs by means of surface modification either by non-covalent functionalization or covalent functionalization [3].

Most strategies designed to functionalize CNTs involve the use of strong acids as reagents and organic solvents as reaction media, which can become environmental pollution and health hazard problems. Nowadays, the global environmental trends are seeking greener chemistry methods to prepare materials, thus, there is plenty of room for developing environmentally-friendly chemistry methods to functionalize CNTs.

"*Green*" chemistry is based on the use of a set of principles that reduces or eliminates the use of hazardous reagents and solvents in the design, preparation and application of materials [4]. In this context, functionalization of CNTs using microwaves, plasma, and ultrasound waves are strategies very promising for greener production of hybrid polymer materials, due to shorter reaction times, reduced energy consumption, and better yields.

The focus of this chapter will be on the microwaves, ultrasound and plasma assisted functionalization of CNTs as greener chemistry methods to produce hybrid polymer materials. After a brief overview on preparation of hybrid polymer materials containing CNTs, we will present the physical principles, mechanisms and processing conditions involved in the functionalization of CNTs for each of these "*Green*" chemistry methods, and then present our point of view on challenges and opportunities in both the immediate and long-term future.

2. Hybrid polymer materials

In polymer science, we can define a *hybrid polymer material* as a combination of two or more materials mixed at the nanometer level, or sometimes at the molecular level (0.1 – 100 nm) in a predetermined structural configuration, covering a specific engineering purpose. The term *hybrid material* is used to distinguish them from the conventionally known *composites* that are referred as simple mixtures of two or more materials at micro-scale level (> 1 μm).

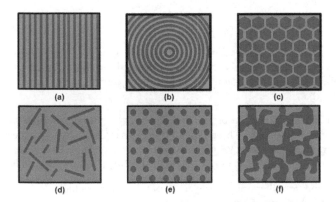

Figure 1. Some examples of structural configurations of hybrids of the composite type: (a) sandwich, (b) concentric cylindrical shells, (c) honeycomb, (d) chopped fibers, (e) particulate, and (f) amorphous blend.

An ideal hybrid polymer material requires an accurate molecular design or structural control of its components in order to obtain synergistic properties. As structural configuration of components moves away from its ideal configuration, the material properties will range from an arithmetic average value (average of the properties of each component) to below of that arithmetic value [5]. Thus, the shape and structural configuration of the components in a hybrid polymer material play a key role in determining its properties. Figure 1 shows a scheme of hybrids materials composed by two components, in which one of them is arranged so that synergistic properties can be achieved.

The hybrid polymer materials can be classified depending of the nature of interactions between their components. In particular, when structural materials in the form of particles, flakes or fibers are incorporated into polymer matrices, this type of hybrid polymer materials can be classified in (i) *class I* hybrid materials, which show weak interactions between the two components, such as van der Waals, hydrogen bonding or weak electrostatic interactions, and (ii) *class II* hybrid materials, which show covalent interactions between both components such that there is no tendency for the components to separate at their interfaces when the hybrid material is loaded [6].

Hybrid polymer materials containing CNTs have attracted considerable attention due to the unique atomic structure, high surface area-to-volume ratio and excellent electronic, mechanical and thermal properties of carbon nanotubes. Although the incorporation of CNTs to polymer matrices have significantly improved the mechanical, electrical and morphological properties of polymers, there is plenty of room for controlling the structural configuration of the hybrid polymer material, thus, different efforts have been focused in the preparation methods.

3. Polymer-CNTs hybrid materials

3.1. Structural configuration

Since the first ever materials based on polymer-CNTs were reported in 1994 by Ajayan *et al.* [7], several processing methods have been developed for fabricating polymer-CNTs hybrid materials. These methods mainly include solution mixing, *in-situ* polymerization, and melt blending [8].

Because the unique mechanical properties of CNTs, such as the high modulus, tensile strength and strain to fracture, there have been numerous efforts to obtain hybrid materials with improved mechanical properties [2]. Within the structural configurations for this specific application, the *"chopped fibers"* configuration, as seen in Figure 1(d), has been the most desired.

On the other hand, for other unique properties of CNTs such as high electrical and thermal conductivity, the obtaining of multiphase polymer amorphous blends, as seen in Figure 1(f), offers a much higher potential for the development of conductive composites containing CNTs. The selective localization of the CNTs either in one of the blend phases or at the interface of an immiscible co-continuous blend can form an ordered network of conductive phase, creating the so-called segregated systems [9]. In such systems, considerably lower value of percolation threshold compared to *"chopped fibers"* structural configuration can be achieved.

The building of polymer-CNTs hybrid materials with desired structural configurations is potentially promising to develop advanced hybrid materials; however, the full exploitation of properties of CNTs by means the manufacturing of those desired structural configurations has been severely limited, because difficulties associated with dispersion of the entangled CNTs during processing and their poor interfacial interaction with some polymer matrices.

3.2. Chemical and physical functionalization of CNTs

The efficient exploitation of the unique properties associated with CNTs depends on its uniform and stable dispersion in the host polymer matrix, as well as the nature of the interfacial interactions with the polymer. Thus, obtaining of polymer-CNTs hybrid materials with desired properties has represented a great challenge, because CNTs exhibit strong inter-tube van der Waals' forces of attraction that impede its uniform and stable dispersion in the matrix, in addition to certain properties of the polymer matrix like wetting, polarity, crystallinity, melt viscosity, among others [2, 10].

Surface modification of CNTs has been one of the most used strategies in order to improve its affinity with the polymer matrix, and therefore to achieve a better uniform dispersion. These methods have been conveniently divided into chemical functionalization and physical functionalization [3, 11].

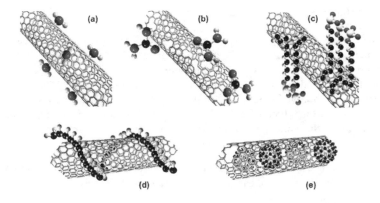

Figure 2. Strategies for chemical and physical functionalization of CNTs: a) covalent sidewall functionalization, b) covalent defect sidewall functionalization, c) non-covalent adsorption of surfactants, d) wrapping of polymers, and e) endohedral functionalization (case for C_{60}).

Chemical functionalization method is based on the covalent linkage of functional groups such as –COOH or –OH on the surface of CNTs. These methods can be also divided in sidewall functionalization and defect functionalization (see Figure 2). The reaction mechanisms that take place at their sidewall include fluorination and derivate reactions, hydrogenation, cycloaddition, and radical (R•) attachment; whilst the reaction mechanisms by amidation, esterification, thiolation, silanization, and polymer grafting (*grafting to* and *grafting from*) takes advantages of chemical transformation of defect sites on CNTs.

Physical functionalization method is based in the formation of non-covalent interactions between molecules and CNTs. These methods include the wrapping of polymer around the CNTs, the physical adsorption of surfactants and the endohedral method (see Figure 2). In the latter, molecules are stored in the inner cavity of CNTs through the capillary effect, where the insertion often takes place at defect sites localized at the ends or on the sidewalls.

In particular, the covalent functionalization of CNTs has been one of the most preferred methods since it allows an efficient interaction between polymer-CNT interface through the functional moieties of the CNTs surface and the available functional groups of the polymer. However, these methods involve rough acid treatment conditions during functionalization which damage the nanotube framework and decrease the electrical conductivity of the hybrid material. In addition, the use acids and organic solvents as the reaction media represent problems of environmental pollution and health hazard.

In this context, the global trend of seeking for "Green" chemistry methods is demanding to researchers in the field to develop environment-friendly methods to functionalize CNTs.

4. Greener production of polymer-CNTs hybrid materials

4.1. "Green" chemistry: definition and principles

Diverse definitions of "Green" chemistry can be found in the literature. According to EPA (Environment Protection Agency) "Green" chemistry philosophy speaks of chemicals and chemical processes designed to reduce or eliminate negative environmental impacts, where the use and production of these chemicals may involve reduced waste products, non-toxic components, and improved efficiency. Anastas and Warner [12], who are considered the founders of this field that born in 1990s, define "Green" chemistry as the utilization of a set of principles that reduce or eliminates the use or generation of hazardous substances in the design, manufacture and application of chemical products.

The 12 Principles of "Green" chemistry (defined by Anastas and Warner) help us think about how to prevent pollution when creating new chemicals and materials:

1. *Prevention*. It is better to prevent waste to treat or clean up waste after it has been created.

2. *Atom Economy*. Synthetic methods should be designed to maximize the incorporation of all materials used in the process into the final product.

3. *Less Hazardous Chemical Synthesis*. Synthetic methods should be designed to use and generate substances that possess little or no toxicity to people or the environment.

4. *Designing Safer Chemicals*. Chemical products should be designed to affect their desired function while minimizing their toxicity

5. *Safer Solvents and Auxiliaries*. The use of auxiliary substances should be made unnecessary whenever possible and innocuous when used.

6. *Design for Energy Efficiency*. Energy requirements of chemical processes should be recognized for their environmental and economic impacts and should be minimized. If possible, synthetic methods should be conducted at ambient temperature and pressure.

7. *Use of Renewable Feedstocks*. A raw material or feedstock should be renewable rather that depleting whenever technically and economically practicable.

8. *Reduce Derivatives*. Unnecessary derivatization like use of blocking group, protection/de-protection, and temporary modification of physical/chemical processes, should be minimized or avoided if possible, because such steps require additional reagents and can generate waste.

9. *Catalysis*. Catalytic reagents should be superior to stoichiometric reagents.

10. *Design for Degradation*. Chemical products should be designed so that at the end of their function they break down into innocuous degradation products and do not persist in the environment.

11. *Real-Time Analysis for Pollution Prevention*. Analytical methodologies need to be further developed to allow for real-time, in-process monitoring and control prior to the formations of hazardous substances.

12. *Inherently Safer Chemistry for Accident Prevention*. Substances and the form of a substance used in a chemical process should be chosen to minimize the potential for chemical accidents, including releases, explosions, and fires.

"*Green*" chemistry is a highly effective approach to pollution prevention since it applies innovative scientific solutions to real-world environmental situations. The preparation of polymer-CNTs hybrid materials can be considered as "*Green*" as more of those principles are applied to the design, production and processing of hybrid materials.

4.2. Greener processing technologies

4.2.1. Microwaves

4.2.1.1. Background and physical principles

Microwaves are electromagnetic waves with wavelengths ranging from 1 mm to 1 m and frequencies between 0.3 GHz and 300 GHz, respectively. 0.915 GHz is preferably used for industrial/commercial microwave ovens and 2.45 GHz is mostly used for household microwave ovens. Since the first ever report of a microwave-assisted organic synthesis in the 80s, it is being further developed and extended to polymer science, in particular in the field of microwave-assisted polymer synthesis and polymer nanocomposites [13].

In polymer chemistry, microwave-assisted reactions present a dramatic increasing in reaction speed and significant improvements in yield compared with conventional heating. These advantages are attributed to instantaneous and direct heating of the reactants, which lead to reduction in reaction time, energy savings and low operating costs. The principles 2, 5, 6 and 11 of '*Green*' chemistry describe these strengths.

How does microwave irradiation lead chemical reactions? When a dielectric material (i.e. molecules containing polar groups in their chemical structure) is placed under microwave irradiation, the dipolar molecules will tend to align their dipole moment along the field intensity vector. As the field intensity vector varies sinusoidally with time, the polar molecules re-align with the electro-magnetic field and generate both translational and rotational motions of the dipoles. These movements generate heat because the internal friction, so a portion of the electromagnetic field is converted in thermal energy.

The power absorbed per unit, P (V/m^3) is expressed as [14]:

$$P = 2\pi f \, \varepsilon_0 \varepsilon_r'' \mid E \mid^2 \tag{1}$$

where f is the microwave frequency (GHz), ε_0 the permittivity of free space (ε_0 = 8.86x10^{-12} F/m), ε_r'' the dielectric loss factor and E (V/m) is the magnitude of the internal field.

The dielectric loss factor is a measurement of the efficiency with which microwave energy is converted into heat, and depends on the dielectric conductivity σ and on the microwave frequency f according to

$$\varepsilon_r'' = \sigma / 2\pi f \tag{2}$$

The degree of energy coupling in the reaction system is expressed by the dissipation factor D, which is defined by the loss tangent tan δ

$$D = \tan\delta = \varepsilon_r'' / \varepsilon_r' \tag{3}$$

where ε_r' is the relative dielectric constant and describes the ability of molecules to be polarized by the electric field. Thus, the dissipation factor defines the ability of a medium at a given frequency and temperature to convert electromagnetic energy into heat.

Therefore, the absorbed microwave energy into dielectric material produces the molecular friction, which leads the rapid heating of the reaction medium and the subsequent chemical reactions. The dramatic rate enhancements of these reactions have been explained by means of very well-known Arrhenius law:

$$k = A\exp[-E_a / RT] \tag{4}$$

Some authors have suggested that, the microwave dielectric heating increases the temperature of the medium in a way that cannot be achieved by conventional heating (superheating), so the rate enhancements are considered essentially a result of thermal effects, although the exact temperature reaction has been difficult to determine experimentally [15]. Other authors, however, suggest that the microwave energy produces an increase in molecular vibrations which could affect anyway the pre-exponential factor A, and also produce an alteration in the exponential factor by affecting the activation energy [16, 17].

After 50 years of research, microwave chemistry is still a research field in expansion and also seems it as green technology; however, some questions regarding microwave heating mechanisms remain unsolved. The microwave-assisted production of polymer-CNTs hybrid materials is a recent field of research, in which additional questions have emerged. Beyond to give an overview on microwaves-assisted preparation of hybrid materials, this section is addressed under one of those questions: how could microwave energy be controlled to prepare more efficiently these hybrid materials? As discussed below, the answer to this question is still not understood.

4.2.1.2. Carbon nanotubes-microwaves interaction

Carbon nanotubes have demonstrated to act as highly efficient absorbers of microwave energy, producing heating, outgassing and light emission [18]. Over the past few years, the investigation on microwave heating mechanisms in CNTs has been a focus of interest. It has been proposed that the microwave irradiation might cause heating by two plausible mechanisms [19]: (i) Joule heating and (ii) vibrational heating.

The mechanism of *Joule heating* postulate that the electric field component of the microwave induces the motion of the electrons in electrically conductive impurities present at as-synthesised CNTs such as metallic catalysts, leading a localised superheating at the site of impurities which increase the temperature of CNTs. In addition, another suggested potential source of localized superheating has been the generation of gas plasma from absorbed gases (particularly H_2) in CNTs, introduced during the synthesis phase or via atmospheric absorption.

The sources of superheating in the *Joule heating* mechanism are focus of discussion. It has been argued that the nano-sized magnetic particles should be impacted minimally by microwave irradiation at low frequencies and therefore, plays no significant role in the microwave energy absorption. Paton *et al.* work [18], among others, demonstrated that even with the removal of iron and other catalytic particles, the CNTs still present microwave heating. On the other hand, regarding to gas plasma, it is still unclear if the plasma is directly generated by microwave irradiation or by other superheating effect. Moreover, it is doubtful that plasma be generated under presence of solvents, since their conductivity is higher than air.

Paton *et al.* [18] hypothesized that *Joule heating* mechanism in CNTs can be explained by the motion of free electrons distributed on the surface of the CNTs, induced by the electric field component of the electromagnetic field. This theory was supported by the measurements of DC conductivity of as-synthesised, heat and acid treated CNTs. The microwave energy absorption was significantly increased as the crystallinity and electrical conductivity of CNTs were improved.

Regarding to *vibrational heating* mechanism, Ye [20] described the heating of non-bounded CNTs in terms of non-linear dynamics of a vibrating nanotube. CNTs subjected to microwaves undergo superheating due to transverse vibrations attributed to parametric resonance, similarly to forced longitudinal vibrations of a stretched elastic string. Ye found that

CNTs present a resonance frequency between 2.0 – 2.5 GHz, which is in the region of the frequency of microwaves of the most operating systems used in this field (2.45 GHz). However, the intensity of vibration modes might be attenuated by the presence of impurities, a viscous environment, and highly entangled CNTs.

Both *Joule heating* and *vibration heating* mechanisms help to explain the different obtained results of microwave energy absorption, in the presence of solvents or dry conditions. However, the need of an in-depth understanding of the microwave heating mechanisms is more tangible as microwave-CNTs systems become more complex.

4.2.1.3. Microwaves-assisted functionalization of CNTs

Within the standard procedures to chemically functionalize CNTs is firstly the purification phase. The most common techniques include acid reflux, oxidation and filtration, where most of them involve long processing times or multiple stages, the use of large acid volumes and some cases the structural damage of CNTs [3]. Microwave-assisted purification of CNTs has emerged as promising technique for effective purification of CNTs with minimal damages and significant reduction of the processing times and use of harmful reactants [19].

Purification has been attributed to generation of highly localized temperatures within metallic particles which burst any amorphous carbon coating. During purification phase in conventional techniques, the use of aggressive treatments facilitates the creation of defect sites on sidewall of CNTs, in order to graft desired functional groups; however, in microwave-assisted processes the energy absorbed by CNTs leads the activation of vacancy sites on surface and the subsequent reaction with active functional groups of molecules. At the same time, the microwave irradiation can supply enough energy to reorient any "damaged" sp^3 carbon bonds into sp^2 hybridization, thus leading an increase in CNTs quality.

As described previously, the main goal of functionalization of CNTs in the preparation of hybrid materials is to improve their dispersion degree and interaction with the polymer matrix. Thus, the challenge of microwave-assisted functionalization is to achieve a desired degree of functionalization on CNTs surface, whilst avoiding damages of the structure that could compromise the properties of the final product. A review on recent works in microwave-assisted functionalization of CNTs was published by Ling and Deokar [21], and it is not our intent to duplicate that effort here. Rather, we focus on some "Green" key issues that might improve the preparation of hybrid polymer materials through control of microwave energy absorption.

In this context, microwave-assisted functionalization under solvent-free conditions is a promising approach for large-scale functionalization of CNTs and paves the way to greener chemistry, because in the absence of solvents, the CNTs and reagents absorb the microwave energy more directly and so takes full advantage of the strong microwave absorption of such components. In addition, the solvent-free conditions open the possibility to all proposed microwave heating mechanisms, and therefore increase the absorption of microwave energy.

The use of solvent-free conditions involves the use of bulk CNTs, so dealing with the entangled CNTs results more complicated. Although some works have carried out using solvent-free conditions during microwave irradiation [22-25], a pre-dispersion stage of CNTs with ultrasound in solvent systems is still used. Recently, Ávila-Orta *et al.* [26, 27] developed a method of dispersion of nanostructures in gas phase assisted by ultrasound, enhancing the dispersion of bulk CNTs under solvent-free conditions.

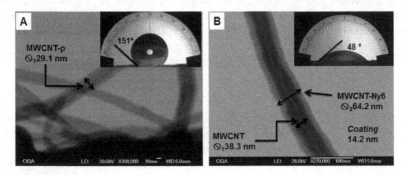

Figure 3. STEM images of MWCNTs. a) pristine MWCNTs (MWCNT-p), and b) functionalized MWCNTs with Nylon (MWCNT-Ny6). [28].

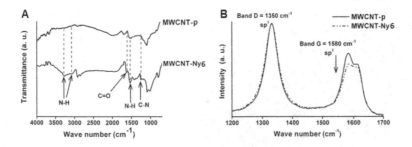

Figure 4. Evidence of the formation and grafting of Nylon-6 on surface of MWCNTs: a) FTIR spectrum, and b) RAMAN spectrum. [28].

González-Morones [28] used the dispersion method developed by Ávila-Orta *et al.* in order to functionalize multi-walled carbon nanotubes (MWCNTs) with Nylon through "*grafting from*" strategy, using ε-caprolactam and aminocaproic acid as monomers. The MWCNTs were previously dispersed into a recipient containing air and then blended with ε-caprolactam powder. The blend was treated for 30 min using a multimodal microwave oven (2.45 GHz) at 250 °C and microwave power of 600 W. Figure 3 shows a STEM image of functionalized carbon nanotubes, in which the average thickness of the polymeric coating was 14.2 nm. The contact angle measurements for pristine and functionalized MWCNTs are also showed in the Figure 3. The reduction in hydrophobic character of MWCNTs-p represented by a decreasing in their contact angle (from 151° to 48°) suggests the presence of a hydrophilic coating. Furthermore, the FTIR spectrum for functionalized MWCNTs shows the presence of the characteristic functional groups of Nylon-6 (Figure 4a), which demonstrates the formation of Nylon-6 by hydrolytic polymerization; whilst the RAMAN spectrum shows a decreasing in the G band intensity (sp^2), suggesting that Nylon-6 are grafted on the surface of MWCNTs through actives sites created during microwave irradiation.

Although the pre-dispersion stage of CNTs in gas phase assisted by ultrasound reduce the consumption of solvent ("*Green*" principle # 5), after the functionalization by microwaves, it is still necessary the use of organic solvents to eliminate the residual monomer ("*Green*" principle # 8). Thus, in order to boost the advantages of this pre-dispersion phase, the efforts should focus on pathways to increase the conversion of reagents ("*Green*" principles # 2, 6 and 9).

4.2.1.4. Preparation of polymer-CNTs hybrid materials

Some efforts have been performed in the preparation of hybrid materials under solvent-free conditions. Virtanen *et al.* developed a hybrid material with a structural configuration sandwich-like (similar to Figure 1a), composed by two polymer plates (extremes) and a film made up from functionalized CNTs (center) which were joined by microwave irradiation [29]. Lin *et al.* obtained hybrid material from microwave-assisted cured process of epoxy resin containing vertically aligned CNTs [30].

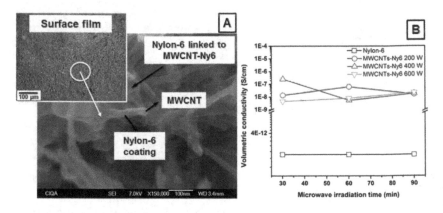

Figure 5. Nylon-6/MWCNTs hybrid material obtained by *in-situ* polymerization assisted by microwaves. a) STEM images of a film made from a hybrid material obtained at microwave power of 600 W, and b) conductivities values of hybrid materials as function of microwave power. [32].

Because a special interest is placed on one-step processes for preparation of those materials, the combined process of in-situ functionalization of CNTs by "*grafting from*" and in-situ bulk polymerization by microwave irradiation becomes a very attractive approach. In recent years, Dr. Ávila-Orta's group has focused on preparation of hybrid materials with electrical conductivity properties; so a structural configuration with an interconnection between CNTs is mostly desired. The combined process described above has demonstrated to be a good approach to enhance this structural configuration.

In this context, Yañez-Macías *et al*. [31, 32] prepared Nylon-6/MWCNTs hybrid material films with high electrical conductivities (values ranged from 10^{-9} to 10^{-7} S/cm) using this combined process. In that work, the MWCNTs were also previously pre-dispersed using the method developed by Ávila-Orta *et al*. The influence of microwave power on polymerization was studied for 200, 400 and 600 W. Figure 5 shows STEM images of a film made from Nylon-6/MWCNTs hybrid material obtained after 90 min of reaction at 600 W. The image shows as the MWCNTs are interconnected and coated by Nylon-6.

From Yañez-Macías *et al*. work, the microwave power intensity demonstrated to play a crucial role in the hydrolytic polymerization. As microwave power intensity increases the yield of Nylon-6 increases, however, at higher microwave power intensity degradation mechanism occurs. These results show that efficient production of polymer-based CNTs in solvent-free conditions can be boosted through control of microwave energy applied to bulk medium.

4.2.1.5. Future perspectives

Microwave irradiation under solvent-free conditions in combination with a pre-dispersion stage of CNTs in gas phase represents a promising approach to large-scale greener production of polymer-based CNTs hybrid materials. The pre-dispersion stage of CNTs allows increasing the efficiency in the microwave energy absorption and the available surface to their functionalization. However, although great efforts have been developed for in-situ preparation of polymer-CNTs hybrids, it is still required to improve the yield.

The control in the yield of functionalization and polymerization reactions can be performed through an in-depth understanding of the mechanisms of microwave heating and kinetic reactions studies. Since the increasing of microwave power intensity increases the temperature of medium reaction, after further research, an optimum microwave energy supply can be found as function of microwave power intensity. In addition, the use of mono-modal microwave ovens can improve the efficiency in the microwave energy absorption, because the microwaves are only concentered in a reaction volume and are not dispersed around chamber volume like multi-modal microwave ovens.

4.2.2. Ultrasound

4.2.2.1. Background and physical principles of sonochemistry

Ultrasound (US) is defined as sound that is beyond human listening range (i.e. 16 Hz to 18 kHz.). In its upper limit, ultrasound is not well defined but is generally considered to 5 MHz for gases and 500 MHz for liquids and solids, and is also subdivided according to applications of interest. The range of 20 to 100 KHz (although in certain cases up to 1 MHz) is designated as the region of high power ultrasound (*sonochemistry*), while the frequencies above 1 MHz are known as high frequency or ultrasound diagnostics (e.g. the imaging technique using echolocation, as SONAR system to detect or US in the health care).

Since the first report on the chemical effects of high power ultrasound in 1927, when Loomis and Richards [33] studied the hydrolysis of dimethyl sulfate and iodine as a catalyst; the studies on chemical effects of ultrasound have further extended to several areas such as organic and organometallic chemistry, materials science, food, and pharmaceutical, among others [34].

The use of ultrasound to accelerate chemical reactions has proven to be a particularly important tool for meeting the "*Green*" Chemistry goals of minimization of waste, reduction of energy and time requirements ("*Green*" principles # 6, 8 and 11). Thus, nowadays the applications of ultrasonic irradiation are playing an increasing role in chemical processes, especially in cases where classical methods require drastic conditions or prolonged reaction times [35].

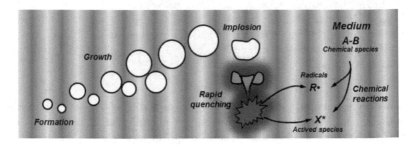

Figure 6. Chemical effects of the high power ultrasound derived from bubble collapse.

The chemical effects of ultrasound in liquids systems are derived from the formation, growth and implosion of small bubbles that appears when the liquid is irradiated by ultrasound waves, phenomenon called "acoustic cavitation" [36]. During bubble collapse, the conversion of kinetic energy of the liquid into thermal energy generates high temperatures (1000 – 10,000 K, most often in the range 4500 to 5500 K) and pressure conditions (~ 500 atm), which lead the formation of free radicals and active species as a result of the heating of the bubble content (Figure 6). On the other hand, the surrounding liquid quenches these portions of the medium in less than 10^{-6} seconds. Thus, the high local temperatures and pressures, combined with extraordinarily rapid cooling, provide a unique means for driving chemical reactions under extreme conditions [34].

A combination between the capability of ultrasonic irradiation to induce chemical reactions and also to achieve a full dispersion of nanostructures in different systems represents a synergistic approach to produce polymer-carbon nanotubes hybrid materials, because surface modification and dispersion of CNTs might take place at the same time; however, unlike microwaves and plasma technologies, there have been very few efforts for exploring it. In this section, we discuss some keys issues associated with the functionalization of CNTs, in order to foster the use of ultrasonic irradiation as greener method for preparation of hybrid materials.

4.2.2.2. Mechanisms for ultrasound activation

The studies on sonochemistry have demonstrated that the ultrasonic irradiation differs from traditional energy sources (such as heat, light or ionizing radiation), so it has been used as a source of alternating activation to assist chemical processes, such as in synthetic methods for obtaining organic molecules and macromolecules and inorganic [37, 38], extraction of natu-ral and synthetic products [39], and medicine [40]. The enormous local temperatures, pres-sures as well as the heating and cooling rates generated during bubble collapse provide an unusual mechanism for generating high-energy chemistry. However, despite chemical ef-fects of ultrasound have been studied for many years, the mechanisms underlying these ef-fects are too complex and not well-understood.

It has established that during bubble cavitation, three sites for chemical reactions can be identified [41]: i) the interior of the bubble, ii) the interface region at around the bubble sur-face, and iii) the liquid region outside the interface region (Figure 6). In the interior of a bub-ble, volatile solute is evaporated and dissociated due to extreme high temperature, where depending of the nature of the system, different free radicals and excited species are gener-ated. Those chemical species with a relatively long lifetime can diffuse out of the interface region and chemically react with solutes or the bulk medium; whilst in the interface region, in addition to high temperatures due to the thermal conduction from the heated interior of a bubble, the presence of relatively short lifetime species such as OH• and O• can lead more interesting chemical reactions.

When a solid material is present in a cavitation medium, the high speed of the liquid jet gen-erated during bubble collapse produces a violent impact on solid surface, in which some material can be removed (e.g. ultrasonic cleaning processes). On the other hand, from a chemical point of view, the shock waves emitted by the pulsating bubbles and the liquid flow around the bubble enhance a mass transfer toward the solid surface during bubble col-lapse, so the free radicals and the active species generated are available to induce different chemical reactions on solid surface.

4.2.2.3. Ultrasound-assisted functionalization of CNTs

One of the most common methods to functionalize CNTs is acid treatment at elevated tem-peratures. In this process, functional groups such as hydroxyl (-OH), carboxyl (-COOH), and carbonyl (-C=O) can be introduced into a carbon nanotube network through their physical defects and sites with imperfections. In particular, electron-spin-resonance (ESR) studies on the acid-oxidized CNTs demonstrated that sites with unpaired electrons are generated on CNTs surface, and are significantly increased when acid-treatment functionalization is as-sisted by high power ultrasound. Moreover, Cabello-Alvarado et al. [42] reported the ultra-sound-assisted functionalization oxidation of MWCNTs using H_2SO_4/HNO_3. The MWCNTs were subjected to ultrasonic radiation by 8 hours at 60 °C, obtaining similar results to those reported using high temperatures H_2SO_4/HNO_3 mixture [43].

Ultrasound-assisted acid-treatment functionalization is an ideal alternative for reducing reac-tion conditions and increase rates of reaction, but the use of strong acids as reagents does not contribute largely to "Green" chemistry. However, the high speed of the liquid jet generated during bubble collapse can be strong enough to disperse the CNTs agglomerates and also rup-ture some covalent carbon-carbon bonds of CNTs, and so, generates those sites with unpaired electrons or "active sites" and subsequently induces desired chemical reactions with the sur-rounding molecules (Figure 7). As proof of this, in 2006, Chen and Tao reported the functionalization of SWCNTs with polymethyl methacrylate by *"grafting from"* [44]. In that work, SWCNTs were irradiated by ultrasound in methyl methacrylate monomer and polymer grafted CNTs were obtained by *in-situ* sonochemically initiated radical polymerization.

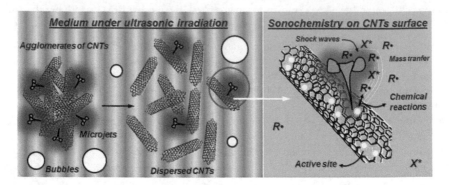

Figure 7. Scheme of the activation mechanism of the CNTs surface treated with ultrasound.

However, the damages of the sp^2-carbon network derived from rupture of carbon-carbon bonds trends to reduce both mechanical and electrical properties of CNTs. Therefore, further research on new pathways to preserve such properties is required. In this context, recently Gebhardt *et al.* [45] developed a novel covalent sidewall functionalization method of CNTs that allowing preserves the integrity of the entire σ-framework of SWCNTs in contrast to classical oxidation. The reductive carboxylation of SWCNTs under ultrasonic treatment resulted in a highly versatile reaction with respect to electronic type selectivity, since functionalization occurs preferentially on semiconducting CNTs. Also, the degree of functionalization can be controlled thought handling of external variables such as pressure.

The emerging on new pathways on ultrasound-assisted functionalization methods could displace to conventional acid treatment methods of CNTs and therefore, opening the possibility for more efficient and greener chemistry methods.

4.2.2.4. *Preparation of polymer-CNTs hybrid materials by sonochemistry*

The preparation of polymer-CNTs hybrid materials assisted by ultrasound is a feasibility approach since ultrasound has influence on the dispersion of the CNTs and activation of their surface, thereby facilitating interaction between the polymer and the CNTs. In addition, the sonochemical activation can lead polymerization reactions, so offers more attractive features such as low reaction temperatures and short reaction times compared with conventional methods.

Thus, the obtaining of polymer-CNTs hybrid materials by *in-situ* bulk polymerization assisted by ultrasound represents a viable method to exploit all these features: i) a full dispersion of CNTs can be obtained in monomer solution, at same time that ii) the effects of bubble collapse activates the surface of CNTs and lead the *in-situ* functionalization with monomer molecules, in which also iii) the polymerization is started sonochemically. Park *et al.* reported the preparation of poly(methyl methacrylate) (PMMA)-MWCNTs nanocomposites with AIBN and different content of MWCNTs [46]. The molecular weight of polymer matrix increased as MWCNTs content was increased due to generation of initiator radicals on CNTs surface. Kim *et al.* prepared polystyrene-MWCNTs nanocomposites without any added initiator, in which an electrical percolation threshold less than 1 wt% was obtained [47].

Although the role of CNTs as effective initiators and control agents for radical polymerizations have been recently demonstrated by Gilbert *et al.* [48], the ultrasound irradiation is more widely used as an alternative method of dispersion, but not as a source of sonochemical activation, leaving large areas of opportunities to explore.

Ultrasound technology is also applied in dispersion and preparation of nanocomposites in melt blending; however the concepts of sonochemistry and surface activation of CNTs that address this section might not be applied to such systems, due to cavitation phenomenon is dramatically suppressed and the chemical effects might be dominated by mechanochemical phenomenon different from cavitation [49].

4.2.2.5. Future perspectives

Ultrasound is a viable source of *"Green"* activation in the context of *"Green"* chemistry, since it has demonstrated to promote low reaction temperatures, faster reaction rates and higher yields in functionalization and polymerization processes. Notably, each of these chemical processes require a source of activation that efficiently furnishes the energy necessary to activate the carbon-carbon bonds in the CNTs network, so in order to perform the *"activation"* of the surface of the CNTs, further research on mechanism of interaction between ultrasound-CNTs has to be addressed taking into account factors such as ultrasound power, environmental solvents, temperature and being one of the most important the sonication time. In particular, the use of environmental solvents could be an interesting factor due the solvent is crucial in the process of bubble cavitation.

4.2.3. Plasma

Plasma, the fourth state of the matter, is generated when atoms and molecules are exposed to high sources of energy such as those produced by direct current (DC), alternating current (AC), microwaves and radio frequency (RF). The absorbed energy by such atoms and molecules induces particle collision processes which generate electrons, photons, and excited atoms and molecules. Because of the unique active species present in the plasma, this state of the matter is used in the synthesis [50] and surface modification of CNTs [51].

Regarding to preparation of polymer-CNTs hybrid materials, unlike microwaves and ultrasound technologies described earlier, plasma is mainly used for functionalization of CNTs in order to later use them in the preparation of hybrid materials by *in-situ* polymerization [52, 53] and melt blending [54]. As described before, the surface modification of CNTs leads to improve the dispersion and interactions between CNTs and polymer matrix and represent a critical issue to enhance properties of polymer-CNTs hybrid materials; thus, this section is focused on the role of plasma-CNTs interactions in the functionalization of CNTs by polymer grafting, in order to exploit more efficiently the unique properties of CNTs (Green principles # 2, 6).

4.2.3.1. Plasma-assisted functionalization of CNTs

The functionalization mechanisms of CNTs by plasma are carried out by both *etching* and *coating* processes [55]. A physical etching process is presented when ions bombard to CNTs leading erosion on CNTs surface and therefore vacancies or "active sites", whilst a chemical etching process is induced by surface reactions between reactive ions and CNTs. Chemical processes are limited to non-inert plasma gases, whilst both inert and reactive gases can produce the physical effects. On the other hand, the coating process results from physical or chemical deposition of the active species present at plasma environment, in which the chemical deposition is induced through the active sites generated from the etching process. A scheme of these mechanisms of functionalization of CNTs is presented in Figure 8.

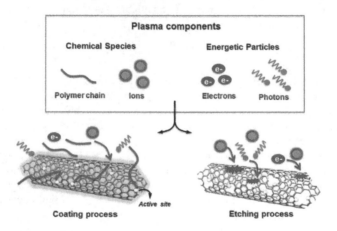

Figure 8. Etching and coating process on CNTs surface.

In order to obtain polymer-CNTs hybrid materials, the functionalization of CNTs with polymers is one of the most preferred strategies to improve the compatibility of CNTs with polymer matrices. In this context, the functionalization of CNTs by polymer grafting occurs when ionized monomers interact with active sites on CNTs surface leading the growth of polymer chains (grafting from/etching process), and also when polymer chains present into plasma environment are physically or chemically deposited on CNTs surface (grafting to/ coating process). However, both etching and coating process can occur simultaneously, thus the surface coating can retard the etching process whilst etching can remove the surface coating depending of the plasma gas chemistry.

The surface modification of CNTs based on plasma polymerization presents some advantages compared to wet chemical processes: i) surface modification without altering CNTs bulk properties ("*Green*" principles # 2, 6), ii) large amount of reagents and extreme temperatures are avoided (*"Green"* principles #3, 5 and 6), and iii) a product with no or very low amount of residual monomer can be obtained (*"Green"* principles # 2 and 8).

Despite these *"Green"* advantages, there are few reports about functionalization of CNTs based on plasma polymerization. Chen *et al.* reported the functionalization of CNTs using the monomers acetaldehyde and ethylenediamine [51]; Shi *et al.* reported the plasma deposition of polypyrrole on CNTs surface [56]; Ávila-Orta *et al.* modified MWCNTs using ethylene glycol as monomer [57]; and more recently, Chen *et al.* reported the preparation of MWCNTs grafted with polyacrylonitrile [58], and Rich *et al.* reported the surface modification of MWCNTs using methyl methacrylate and allylamine as monomers [59].

Because of the structural and chemical character of the polymer coating play an important role on interaction between CNTs and polymer matrix, the structural and chemical nature of the polymer coating obtained by plasma can be controlled through processing parameters. Recently, González-Morones [28] studied the effect of power excitation on chemical nature of the polymer deposited on CNTs surface by plasma polymerization of acrylic acid. In that work, firstly the CNTs were pre-dispersed using the method developed by Ávila-Orta *et al.* [27], then the CNTs were exposed to acrylic acid plasma. It was observed that at low power excitation (20 W) the CNTs surface is partially coated by polyacrylic acid and –COOH groups. At power excitation of 40 W, the polyacrylic acid and -COOH groups are mostly removed since the etching process is favored. Whilst at power excitation of 100 W, the CNTs are partially coated by the polymer and functional groups, thus suggesting a competition between etching and coating process. Figure 9a shows FTIR spectra of the functionalized CNTs at different power excitation. For each case, the solubility in water of the functionalized CNTs is showed in a photograph. A STEM image of the polymer coating on the CNT surface obtained at 20 W is also showed in Figure 9b.

Within the actions that can be executed in order to increase the efficiency of the plasma polymerization process and achieving some of the *"Green"* principles are the following:

1. *Frequency and power excitation:* An increase in value of these factors will result in an increase in both the level of ionization of the species and polymer deposition rates. (*"Green"* principles # 2, 6 and 8).

2. *Monomer flow rate/pressure:* Low flow rates and pressure lead an increase in the level of ionization of the species. (*"Green"* principles # 2 and 8).

3. *Geometry of the plasma reactor:* Reactors with cylindrical geometry (without corners) distribute homogeneously the generated plasma. (*"Green"* principles # 6, 8 and 12).

4. *Temperature of substrate:* Very low temperatures promote a higher insertion of monomer molecules on CNTs surface with minimal molecular modifications. (*"Green"* principles # 2, 6, 8).

5. *Treatment time:* This factor will depend on desired final structure and morphology. Long treatment times could produce thick polymer coatings or a rugous surface. The later, due to competition between the etching and coating process. (*"Green"* principle # 2).

Another interesting modification to plasma polymerization process includes the pre-treatment of CNTs with an inert of non-inert gas in order to induce the etching process and the saturation of the CNTs surface with active sites. Subsequently, the active CNTs are subjected to monomer plasma so a better polymer coating can be achieved [58].

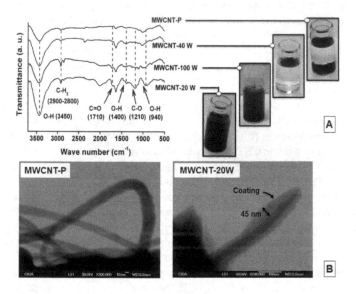

Figure 9. Functionalized CNTs with polyacrylic acid by plasma polymerization. a) FTIR spectra and tests of solubility in water. b) STEM images of pristine MWCNTs (left) and MWCNTs functionalized at 20W (right). [28].

Plasma polymerization is a complex process which makes difficult to achieve an efficient functionalization of CNTs. Due to high number of involved factors, one or two factors have been only studied.

4.2.3.2. Future perspectives

Plasma-assisted functionalization has demonstrated to be a successful method for creating environmentally friendly polymer coatings on CNTs surface. The functionalization of CNTs with desired structural and chemical characteristics can be performed by means of control of the involved processing conditions; however, there is a need for a complete understanding of the interactions between plasma-CNTs which allow controlling successfully the etching and coating process.

A better understanding of the plasma-CNTs interactions can be enhanced if the efficiency of the plasma-CNTs interactions is improved. The stages of pre-dispersion and pre-activation of CNTs should be added previous to plasma polymerization, which allow increasing the surface area exposed to plasma as well as the interactions with the active species, respectively. Further research on such issues could launch to plasma polymerization process to large-scale applications.

5. Summary and final remarks

The properties of CNTs and its application in the preparation of polymer hybrid materials have been very active research fields over the last decade. The preparation of polymer-CNTs hybrid materials faces considerable research challenges because their performance depends on the dispersion of CNTs in the polymer matrix and interfacial interactions between the CNTs and the polymer. In order to achieve the maximum performance, the surface modification of CNTs through several functionalization methods is one of the most used strategies to achieve this goal. Nowadays, the global environmental trends provide convincing reasons for exploring greener chemistry methods for functionalization of CNTs, which represent additional research challenges.

Microwave, ultrasound and plasma technology have emerged as promising green approaches since they have demonstrated to reduce the energy consumption, shorter reaction times and increase yields. These greener methods offer a range of energies that are not available from other conventional sources such as thermal, so leave open the possibility to explore more efficient functionalization routes, taking into account the 12 principles of *"Green"* chemistry as a framework.

Despite the progress described within this chapter, there are still considerable research challenges within this field that remain to be addressed. The successful functionalization of CNTs depends on availability of active sites on CNTs surface; however, the mechanisms that generate these sites remain unclear. Therefore, an in-depth understanding of the mechanisms of interaction between microwave/ultrasound/plasma-CNTs is required. For satisfying this need, further studies on correlation between the irradiation conditions and the level of the functionalization are suggested, in order to establish the most efficient and greener conditions for each particular system.

In addition, the incorporation of a dispersion stage of CNTs in gas phase previous to irradiation is highly recommended, because the surface area exposed to irradiation is increased and therefore the interactions with the matter are highly promoted.

As a general statement, the *"Green"* chemistry methods based on microwave, ultrasound and plasma energy can be easily incorporated to emerging fields of nanotechnology, in particular into strategies for preparation polymer-CNTs hybrid materials. However, there are still a number of important research challenges to study.

Acknowledgements

The authors acknowledge the financial support from project 132699 funded by CONACyT, México.

Author details

Carlos Alberto Ávila-Orta[1*], Pablo González-Morones[1], Carlos José Espinoza-González[1], Juan Guillermo Martínez-Colunga[2], María Guadalupe Neira-Velázquez[3], Aidé Sáenz-Galindo[4] and Lluvia Itzel López-López[4]

*Address all correspondence to: cavila@ciqa.mx

1 Department of Advanced Materials, Research Center for Applied Chemistry

2 Department of Plastics Transformation Processing, Research Center for Applied Chemistry

3 Department of Polymer Synthesis, Research Center for Applied Chemistry

4 Department of Organic Chemistry, Autonomous University of Coahuila, Saltillo, México

References

[1] Jorio, A., Dresselhaus, G., & Dresselhaus, M. S. (2008). *Carbon Nanotubes: Advanced Topics in the Synthesis, Structure, Properties and Applications*, Berlin, Springer.

[2] Bose, S., Khare, R. A., & Moldenaers, P. (2010). Assessing the Strengths and Weaknesses of Various Types of Pre-treatments of Carbon Nanotubes on the Properties of Polymer/Carbon Nanotubes Composites: A Critical Review. *Polymer*, 51(5), 975-993.

[3] Jeon, I. Y., Chang, D. W., Kumar, N. A., & Baek, J. B. (2011). Functionalization of Carbon Nanotubes. In: Yellampalli S, (ed). Carbon Nanotubes- Polymer Nanocomposites Rijeka InTech Available from http://www.intechopen.com/books/carbon-nanotubes-polymer-nanocomposites/functionalization-of-carbon-nanotubes(accessed 25 May 2012), 91-110.

[4] Dahl, J. A., Maddux, B. L. S., & Hutchison, J. E. (2007). Toward Greener Nanosynthesis. *Chemical Reviews*, 107(6), 2228-2269.

[5] Ashby, M. F., & Bréchet, Y. J. M. (2003). Designing Hybrid Materials. *Acta Materialia*, 51(19), 5801-5821.

[6] Kickelbick, G. (2007). *Hybrid Materials*, Weinheim, Wiley-VCH.

[7] Ajayan, P. M., Stephan, O., Colliex, C., & Trauth, D. (1994). Aligned Carbon Nano-tube Arrays Formed by Cutting a Polymer Resin-Nanotube Composite. *Science*, 265(5176), 1212-1214.

[8] Spitalsky, Z., Tasis, D., Papagelis, K., & Galiotis, C. (2010). Carbon Nanotube-Poly-mer Composites: Chemistry, Processing, Mechanical and Electrical Properties. *Progress in Polymer Science*, 35(3), 357-401.

[9] Mamunya, Y. Carbon Nanotubes as Conductive Filler in Segregated Polymer Com-posites- Electrical Properties. In: Yellampalli S, (ed). Carbon Nanotubes- Polymer Nanocomposites Rijeka InTech (2011). Available from http://www.intechopen.com/ books/carbon-nanotubes-polymer-nanocomposites/carbon-nanotubes-as-conductive-filler-in-segregated-polymer-composites-electrical-properties(accessed 25 May 2012), 173-196.

[10] Rahmat, M., & Hubert, P. (2011). Carbon Nanotube-Polymer Interactions in Nano-composites: A Review. *Composites Science and Technology*, 72(1), 72-84.

[11] Chen, L., Xie, H., & Yu, W. Functionalization Methods of Carbon Nanotubes and Its Applications. In: Marulanda JM, (ed). Carbon Nanotubes Applications on Electron Devices Rijeka InTech (2011). Available from http://www.intechopen.com/books/ carbon-nanotubes-applications-on-electron-devices/functionalization-methods-of-carbon-nanotubes-and-its-applications(accessed 25 May 2012)

[12] Anastas, P. T., & Warner, J. C. (1998). *Green Chemistry. Theory and Practice*, New York, Oxford University Press.

[13] Kempe, K., Becer, C. R., & Schubert, U. S. (2011). Microwave-Assisted Polymeriza-tions: Recent Status and Future Perspectives. *Macromolecules*, 44(15), 5825-5842.

[14] Sutton, W. H. (1989). Microwave Processing of Ceramic Materials. *American Ceramic Society Bulletin*, 68(2), 376-386.

[15] Kappe, C. O. (2004). *Controlled Microwave Heating in Moderm Organic Synthesis* (Ange-wandte Chemie International Edition), 43(46), 6250-6284.

[16] Jacob, J., Chia, L. H. L., & Boey, F. Y. C. (1995). Thermal and Non-Thermal Interaction of Microwave Radiation with Materials. *Journal of Materials Science*, 30(21), 5321-5327.

[17] Binner, J. G. P., Hassine, N. A., & Cross, T. E. (1995). The Possible Role of the Pre-exponential Factor in Explaining the Increased Reaction Rates Observed During the Microwave Synthesis of Titanium Carbide. *Journal of Materials Science*, 30(21), 5389-5393.

[18] Paton, K. R., & Windle, A. H. (2008). Efficient Microwave Energy Absorption by Car-bon Nanotubes. *Carbon*, 46(14), 1935-1941.

[19] Mac, Kenzie. K., Dunens, O., & Harris, A. T. (2009). A Review of Carbon Nanotube Purification by Microwave Assisted Acid Digestion. *Separation and Purification Tech-nology*, 66(2), 209-222.

[20] Ye, Z. (2005). Mechanism and the effect of microwave-carbon nanotube interaction. *PhD thesis*, University of North Texas.

[21] Ling, Y. C., & Deokar, A. (2011). Microwave-Assisted Preparation of Carbon Nanotubes with Versatile Functionality. In: Marulanda JM, (ed). Carbon Nanotubes Applications on Electron Devices Rijeka InTech Available from http://www.intechopen.com/books/carbon-nanotubes-applications-on-electron-devices/microwave-assisted-preparation-of-carbon-nanotubes-with-versatile-functionality(accessed 15 May 2012), 127-142.

[22] Brunetti, F. G., Herrero, M. A., Muñoz, JdM., Giordani, S., Díaz-Ortiz, A., Filippone, S., et al. (2007). Reversible Microwave-Assisted Cycloaddition of Aziridines to Carbon Nanotubes. *Journal of American Chemical Society*, 129(47), 14580-14581.

[23] Brunetti, F. G., Herrero, M. A., Muñoz, JdM., Díaz-Ortiz, A., Alfonsi, J., Meneghetti, M., et al. (2008). Microwave-Induced Multiple Functionalization of Carbon Nanotubes. *Journal of American Chemical Society*, 130(25), 8094-8100.

[24] Economopoulos, S. P., Pagona, G., Yudasaka, M., Iijima, S., & Tagmatarchis, N. (2009). Solvent-Free Microwave-Assisted Bingel Reaction in Carbon Nanohorns. *Journal of Materials Chemistry*, 19, 7326-7331.

[25] Rubio, N., Herrero, M. A., Meneghetti, M., Díaz-Ortiz, Á., Schiavon, M., Prato, M., et al. (2009). Efficient Functionalization of Carbon Nanohorns via Microwave Irradiation. *Journal of Materials Chemistry*, 19, 4407-4413.

[26] González-Morones, P., Yañez-Macias, R., Navarro-Rodríguez, D., Ávila-Orta, C., & Quintanilla, M. L. (2011). Study of Dispersion of Carbon Nanotubes in Gas Phase by Ultrasound and its Effect on Plasma Surface Modification. *Ide@s CONCYTEG*, 6, 727-738.

[27] Ávila-Orta, C., Neira-Velázquez, M., Borgas-Ramos, J., Valdéz-Garza, J., González-Morones, P., & Espinoza-González, C. (2010). CIQA, assignee Process of desagglomeration, fragmentation and size reduction of agglomerated nanoparticles in gas phase assisted by ultrasound. México patent MX/a/2010/014326 (20 December 2010)

[28] González-Morones, P. (2011). Nylon-6/MWCNTs nanocomposites: Functionalization and hybridization. *PhD thesis*, Research Center for Applied Chemistry.

[29] Virtanen, J., Tilli, M., & Keinanen, . (2006). Novel hyibride materials and related methods and devices. U.S.A. patent WO 2006040398 (2 July 2009)

[30] Lin, W., Moon, K. S., & Wong, C. P. (2009). A Combined Process of In-Situ Functionalization and Microwave Treatment to Achieve Ultrasmall Thermal Expansion of Aligned Carbon Nanotube-Polymer Nanocomposites: Toward Applications as Thermal Interface Materials. *Advanced Materials*, 21(23), 2421-2424.

[31] Yañez-Macías, R., González-Morones, P., Ávila-Orta, C., Torres-Rincón, S., Valdéz-Garza, J., Rosales-Jasso, A., et al. (2011, August). Polymer nanohybrids with high

electrical conductivities. Cancún, México. *IMRC meeting, MRS proceedings*, 14-19, Cambridge University Press, 2012.

[32] Yañez-Macías, R. (2011). Polymerization of Nylon-6 on surface of MWCNTs by microwave irradiation. *Master thesis*, Research Center for Applied Chemistry.

[33] Richards, W. T., & Loomis, A. L. (1927). The Chemical Effects of High Frequency Sound Waves. I. A Preliminary Study. *Journal of the American Chemical Society*, 49(12), 3086-3089.

[34] Suslick, K. S. (1989). The Chemical Effects of Ultrasound. *Journal of the American Chemical Society*, 111(6), 2342-2344.

[35] Pizzuti, L., Franco, M. S. F., Flores, A. F. C., Quina, F. H., & Pereira, C. M. P. Recent Advances in the Ultrasound-Assisted Synthesis of Azoles. In: Kidwai M, (ed). Green Chemistry- Environmentally Benign Approaches Rijeka InTech (2012). Available from http://www.intechopen.com/books/green-chemistry-environmentally-benign-approaches/recent-advances-in-the-ultrasound-assisted-synthesis-of-azoles(accessed 25 May 2012), 81-102.

[36] Peters, D. (1996). Ultrasound in Materials Chemistry. *Journal of Materials Chemistry*, 6, 1605-1618.

[37] Ovalle, R. J. (2009). Synthesis and study of aromatic and aliphatic amides obtained through green chemistry. *Master thesis*, Autonomous University of Coahuila.

[38] Sánchez, E. B. N. (2010). Obtaining such anilia aromatic amides using green methods with potential pharmacological application. *Master thesis*, Autonomous University of Coahuila.

[39] Keith, L. H., Gron, L. U., & Young, J. L. (2007). Green Analytical Methodologies. *Chemical Reviews*, 107(6), 2695-2708.

[40] Qu, W., Ma, H., Jia, J., He, R., Luo, L., & Pan, Z. (2012). Enzymolysis Kinetics and Activities of ACE Inhibitory Peptides from Wheat Germ Protein Prepared with SFP Ultrasound-Assisted Processing. *Ultrasonics Sonochemistry*, 19(5), 1021-1026.

[41] Mason, T. J. (1999). *Sonochemistry*, Oxford University Press.

[42] Cabello, C., Sáenz, A., López, L., & Ávila, C. (2011). Surface Modification of (MWCNTs) with Ultrasound H_2SO_4/HNO_3 . *Afinidad*, 68(555), 370-374.

[43] Aravind, S. S. J., Baskar, P., Baby, T. T., Sabareesh, R. K., Das, S., & Ramaprabhu, S. (2011). Investigation of Structural Stability, Dispersion, Viscosity, and Conductive Heat Transfer Properties of Functionalized Carbon Nanotube Based Nanofluids. *Journal of Physical Chemistry C*, 115(34), 16737-16744.

[44] Chen, W., & Tao, X. (2006, 27 Nov- 1 Dec). Ultrasound-induced functionalization and solubilization of carbon nanotubes for potential nanotextiles applications. Boston, MA. *MRS Fall Meeting: MRS Proceedings*, Cambridge University Press, 0920-S0902-0902.

[45] Gebhardt, B., Hof, F., Backes, C., Muller, M., Plocke, T., Maultzsch, J., et al. (2011). Selective Polycarboxylation of Semiconducting Single-Walled Carbon Nanotubes by Reductive Sidewall Functionalization. *Journal of American Chemical Society*, 133(48), 19459-19473.

[46] Park, C., Ounaies, Z., Watson, K. A., Crooks, R. E., Smith, J., Lowther, S. E., et al. (2002). Dispersion of Single Wall Carbon Nanotubes by In Situ Polymerization under Sonication. *Chemical Physics Letters*.

[47] Kim, S. T., Choi, H. J., & Hong, S. M. (2007). Bulk Polymerized Polystyrene in the Presence of Multiwalled Carbon Nanotubes. *Colloid and Poymer Science*, 285(5), 593-598.

[48] Gilbert, A. C. C., Derail, C., Bounia, N. E. E., & Billon, L. (2012). Unexpected Behaviour of Multi-Walled Carbon Nanotubes During "In-Situ" Polymerization Process: When Carbon Nanotubes Act as Initiators and Control Agents for Radical Polymerization. *Polymer Chemistry*, 3, 415-420.

[49] Ávila-Orta, C., Espinoza-González, C., Martínez-Colunga, J. G., Bueno-Baqués, D., Maffezzoli, A., & Lionetto, F. (2012). An Overview of Progress and Current Challenges in Ultrasonic Treatments of Polymer Melts. *Advances in Polymer Technology*, Accepted for publication (13 July 2012).

[50] Lee, S. J., Baik, H. K., Yoo, J. E., & Han, J. H. (2002). Large Scale Synthesis of Carbon Nanotubes by Plasma Rotating Arc Discharge Technique. *Diamond and Related Materials*, 11(3-6), 914-917.

[51] Chen, Q., Dai, L., Gao, M., Huang, S., & Mau, A. (2001). Plasma Activation of Carbon Nanotubes for Chemical Modification. *Journal of Physical Chemistry B*, 105(3), 618-622.

[52] Cruz-Delgado, V. J., España-Sánchez, B. L., Ávila-Orta, C. A., & Medellín-Rodríguez, F. J. (2012). Nanocomposites Based on Plasma-Polymerized Carbon Nanotubes and Nylon-6. *Polymer Journal*, doi:10.1038/pj.2012.49.

[53] Lee, J. I., Yang, S. B., & Jung, H. T. (2009). Carbon Nanotubes–Polypropylene Nanocomposites for Electrostatic Discharge Applications. *Macromolecules*, 42(21), 8328-8334.

[54] Ruelle, B., Peeterbroeck, S., Bittencourt, C., Gorrasi, G., Patimo, G., Hecq, M., et al. (2012). Semi-Crystalline Polymer/Carbon Nanotube Nanocomposites: Effect of Nanotube Surface-Functionalization and Polymer Coating on Electrical and Thermal Properties. *Reactive & Functional Polymers*, 72(6), 383-392.

[55] Hou, Z., Cai, B., Liu, H., & Xu, D. (2008). Ar, O2, CHF3, and SF6 Plasma Treatments of Screen-Printed Carbon Nanotube Films for Electrode Applications. *Carbon*, 46(3), 405-413.

[56] Shi, D., Lian, J., He, P., Wang, L. M., Ooij, W., Jv, Schulz. M., et al. (2002). Plasma Deosition of Ultrathin Polymer Films on Carbon Nanotubes. *Applied Physics Letters*, 81(27), 5216-5218.

[57] Ávila-Orta, C. A., Cruz-Delgado, V. J., Neira-Velázquez, M. G., Hernández-Hernán-

dez, E., Méndez-Padilla, M. G., & Medellín-Rodríguez, F. J. (2009). Surface Modificat-

ion of Carbon Nanotubes with Ethylene Glycol Plasma. *Carbon*, 47(8), 1916-1921.

[58] Chen, I. H., Wang, C. C., & Chen, C. Y. (2010). Preparation of Carbon Nanotube

(CNT) Composites by Polymer Functionalized CNT under Plasma Treatment. *Plasma*

Processes and Polymers, 7(1), 59-63.

[59] AbouRich, S., Yedji, M., Amadou, J., Terwagne, G., Felten, A., Avril, L., et al. (2012).

Polymer Coatings to Functionalize Carbon Nanotubes. *Physica E*, 44(6), 1012-1020.

Chapter 11

Carbon Nanotubes from Unconventional Resources: Part A: Entangled Multi-Walled Carbon Nanotubes and Part B: Vertically-Aligned Carbon Nanotubes

S. Karthikeyan and P. Mahalingam

Additional information is available at the end of the chapter

1. Part A: Entangled Multi-walled carbon nanotubes

1.1. Introduction

Nanotechnology is a topic attracting scientists, industrialists, journalists, governments and even a common people alike. Carbon nanotubes (CNTs) and other carbon nanostructures are supposed to be a key component of this nanotechnology. Having realized its tremendous application potential in nanotechnology, a huge amount of efforts and energy is invested in CNT projects worldwide. Till date, the art of CNT synthesis lies in the optimization of parameters for selected group materials on a particular experimental set-up. As viewed from the perspective of green chemistry, sustaining the environment implies sustaining the human civilization. The long-term key of a sustainable society lies in stable economy that uses energy and resources efficiently. Therefore, it is high time to evaluate the existing CNT techniques on these parameters.

Let us examine three popular methods of CNT synthesis viz Arc discharge, Laser-vaporization and CVD method. Arc-discharge method, in which the first CNT was discovered [1], employs evaporation of graphite electrodes in electric arcs that involve very high temperatures around 4000º C. Although arc-grown CNTs are well crystallized, they are highly impure. Laser-vaporization technique employs evaporation of high-purity graphite target by high-power lasers in conjunction with high-temperature furnaces [2]. Although laser-grown CNTs are of high purity, their production yield is very low. Thus it is obvious that these two methods score too low on account of efficient use of energy and resources. Chemical vapor deposition (CVD), incorporating catalyst-assisted thermal decomposition of hydrocarbons, is the most popular method of producing CNTs, and it is truly a low-cost and scalable technique for mass production of CNTs [3]. Unfortunately, however, till date only purified petroleum products such as methane, ethylene, acetylene, benzene, xylene are in practice, as precursor, for synthesizing CNTs. Apart from petroleum based hydrocarbons, carbon nanotubes have been synthesized from polymers, metallocenes and domestic fuels such as kerosene and liquefied petroleum gas [4-7].

According to the principle of green chemistry, the feed stock of any industrial process must be renewable, rather than depleting a natural resource. Moreover, the process must be designed to achieve maximum incorporation of the constituent atoms (of the feed stock) into the final product. Hence, it is the time's prime demand to explore regenerative materials for CNT synthesis with high efficiency. This well-valued material of biotechnology research was successfully brought to nanotechnology research with the first report of CNTs from natural precursors in 2001. Since then, investigators involved with this environment-friendly source of CNTs and established the conditions for growing multiwalled nanotubes (MWNTs), single-wall nanotubes (SWNTs) and vertically aligned MWNTs on the suitable catalytic support by a simple and inexpensive CVD technique. Researchers prepared good quality of Multi-walled carbon nanotubes (MWNTs) and vertically aligned ones by thermal decomposition of turpentine oil and Camphor [8,9]. Andrews et al. synthesized pure Single-walled carbon nanotubes (SWNTs) by catalytic decomposition of camphor and its anologes [10]. Ghosh et al. prepared single-walled carbon nanotubes from turpentine oil and eucatlyptus oil [11]. We have succeeded in growing of CNTs from pine oil [12] and methyl ester of *Jatropha curcas* oil [13], a botanical hydrocarbon extracted from plant source, having boiling point around 200 - 280º C. Being a green-plant product, these oils are eco-friendly source and can be easily cultivated in as much quantity as required. Unlike any fossil/petroleum product, there is no fear of its ultimate shortage as it is a regenerative source. These sources are readily available, cheap and also user-friendly for chemical vapor deposition due to its volatile and non-toxic nature.

United states Environmental Protection Agency has developed 12 principles of green chemistry [14] that explain what the green chemistry means in practice. Using those definitions as a protocol, we can evaluate the CNT precursor materials such as Pine oil, Methyl ester of *Jatropha curcas* oil and Methyl ester of *Pongamia pinnata* oil, as follows.

Prevention : With the highest CNT-production efficiency, plant based precursor complies with waste-prevention rule significantly.

Atom Economy : Botanical hydrocarbons gets maximum incorporation of its constituent atoms into the final product, CNTs.

Less Hazardous Chemical Synthesis : All the substances involved in this technique (carbon source, catalyst and support material) possess little or no toxicity to human health and the environment.

Designing Safer Chemicals: Our final product is common-grade CNTs that are apparently safe.Safer Solvents and Auxiliaries : The only auxiliary used in our method is the metal catalyst that are apparently safe.

Design for Energy Efficiency : By virtue of a low-temperature atmospheric pressure CVD process, the energy requirements of our technique are significantly low.

Use of Renewable Feedstock : The raw material is purely a regenerative material; so there is no danger of depleting a natural resource.

Reduce Derivatives: No derivative is formed in this technique; solely catalyst-assisted in-situ decomposition of plant based precursor leads to CNTs.

Catalysis: It is a purely catalytic process, no stoichiometric reagents are involved.

Design for Degradation: CNTs as such are non-biodegradable; however, their intentional degradation can be achieved by introducing certain functional groups.

Real-time analysis for pollution Prevention: This technique as such should be fully compatible with real-time analysis for pollution prevention, if executed at an industrial laboratory.

Inherently Safer Chemistry for Accident Prevention : Substances and their form used in our process are chosen so as to minimize the potential for chemical accidents, including releases, explosions and fires.

In this chapter the authors discuss the morphology of the carbon nanostructures produced using Pine oil, Methyl ester of *Jatropha curcas* oil and Methyl ester of *Pongamia pinnata* oil as natural carbon precursors by spray pyrolysis method, a modified chemical vapor deposition method. It is similar to chemical vapor deposition method, the only difference is the vaporization and pyrolysis of carbon source occurs simultaneously in spray pyrolysis whereas in CVD it is a two step process. Since these precursors evaporate at relatively higher temperatures, spray pyrolysis method was adopted for synthesis of CNTs. The results show that the morphology closely correlates with precursor concentration and the authors propose that the effect is related to the competition between rate of decomposition of precursor and the diffusion of carbon species through the catalyst particle.

1.2. Experimental Methods

A catalytic supporting material Fe, Co and Mo with silica (Fe:Co:Mo:SiO$_2$ = 1:0.5:0.1:4) was prepared by wet impregnation method [15]. Appropriate quantities of metal salts (Merk) i.e. Fe(NO$_3$)$_3$ 6H$_2$O, Co(NO$_3$)$_3$ 3H$_2$O and (NH$_4$)$_6$Mo$_7$O$_{24}$ 4H$_2$O were dissolved in methanol and mixed thoroughly with methanol suspension of silica (Merk). The solvent was then evaporated and the resultant cake heated to 90-100 °C for 3 h, removed from the furnace and ground to fine power. The fine powders were then calcined for 1 h at 450 °C and then re-ground before loading into the reactor. The prepared catalyst was directly placed in a quartz boat and kept at the centre of a quartz tube which was placed inside a tubular furnace. The carrier gas nitrogen was introduced at the rate of 100 mL per minute into the quartz tube to remove the presence of any oxygen inside the quartz tube. The temperature was raised from room temperature to the desired growing temperature. Subsequently, the carbon precursor was introduced into the quartz tube through spray nozzle and the flow was maintained at the rate of 0.1 - 0.5 ml/min. Spray pyrolysis was carried out for 45 minutes and thereafter furnace was cooled to room temperature. Nitrogen atmosphere was maintained throughout the experiment. The morphology

and degree of graphitization of the as-grown nanostructures were characterized by high resolution transmission electron microscopy (JEOL-3010), Raman spectroscopy (JASCO
NRS-1500W) green laser with excitation wavelength 532 nm) and thermo gravimetric analysis
(TGA). The as-grown products were subjected to purification process as follows [16]. The sample material was added to 20 ml 1N HCl to form an acidic slurry. This slurry was heated to 60
°C and stirred at 600 rpm. To this heated acidic slurry 20 ml H_2O_2 was added to form oxidative
slurry and stirred for 30 minutes. The sample was filtered and washed with 1N HCl and distilled water. The collected sample was calcined at 400 °C in nitrogen atmosphere for 2 hours. The
experiment at optimum reaction conditions were repeated several times to ensure the reproducibility of formation of carbon nanostructures.

1.3. Result and Discussion

Pine oil, Methyl ester of *Jatropha curcas* oil and Methyl ester of *Pongamia pinnata* oil were
used for synthesis of carbon nano structures by spray pyrolysis method. These oils were individually sprayed at different rate (10 mL, 20 mL and 30 mL per hour) over silica supported Fe, Co and Mo catalyst at different reaction temperatures (550 °C, 650 °C and 750 °C) to
synthesis carbon nanostructures. The as-grown nanostructures were characterized by SEM,
TEM, TGA and Raman spectroscopy techniques.

The result shows low yield of carbon nano structures with three types of precursors, Pine
oil, Methyl ester of *Jatropha curcas* oil and Methyl ester of *Pongamia pinnata* oil, when sprayed
individually at the precursor flow rate of 10mL per hour over silica supported Fe, Co and
Mo catalyst at 550 °C. An improved carbon nanostructure formation at 650 °C for the precursor flow rate of 20 mL per hour was observed invariably for the three precursor materials. When the experimental condition was either 750 °C or precursor flow rate of 30 mL per
hour, decrease in carbon nanotube yield observed.

Figure 1 illustrates the SEM image of the CNTs samples grown at different temperatures using pine oil as precursor at a flow rate of 20 mL per hour. At 550 °C few CNTs were found to
grow because this temperature is not sufficient to pyrolyse the carbon source (Fig. 1a). On
the other hand, at 650 °C (Fig. 1b) the formation of CNTs is high because at this temperature
the carbon source decomposes effectively. At 750 °C, the quantity of CNTs has decreased
and thick nanotubes have been formed (Fig. 1c). Experiments with methyl ester of *Jatropha
Curcas* oil as a carbon precursor at different temperatures were conducted and the SEM images of the sample were recorded. The SEM images of sample synthesized at 650 °C revealed the formation of homogeneous and dense distribution of CNTs in a web-like network
(Fig. 1e), while very few nanotubes formation at 550 °C (Fig. 1g).

At higher temperature (750 °C) an increase in diameter of the CNTs was observed (Fig. 1f). The
morphology of carbon nanotube synthesized at different temperatures using methyl ester of
Pongamia pinnata oil are recorded using SEM. Few carbon nanostructures were observed at a
lower deposition temperature of 550 °C (Fig. 1g). When the temperature was increased to 650
°C chain like carbon nanostructure was dominant (Fig. 1h). At 750 °C, a more uniform distribution of MWNTs was observed with a diameter of around 150 nm (Fig. 1i).

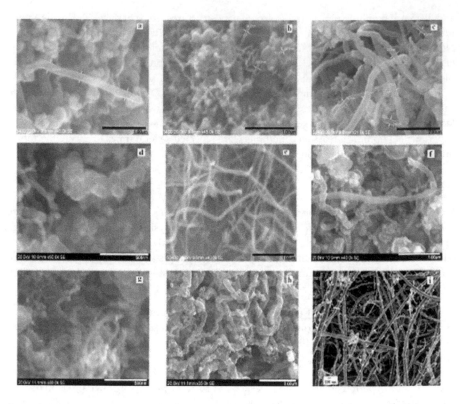

Figure 1. Representative SEM images of as-grown CNTs at different temperatures using Pine oil (a-c), Methyl ester of *Jatropha carcus* oil (d-f) and Methyl ester of *Pongamia pinnata* oil (g-i).

The TEM result shows varied morphologies of carbon nanostructures synthesized at different experimental conditions. All the three precursors used in this study produced mostly amorphous carbon at 550 °C for different flow rate of precursor materials. When the reaction temperature was 650 °C, Pine oil at a flow rate of 20 mL per hour, produced MWNTs of good quality (Fig.2a). The HRTEM result of sample obtained with pine oil as a precursor at a flow rate of 30 mL per hour and reaction temperature of 650 °C, shows (Fig.2b) that the inner and outer diameter of the synthesized MWNTs was about 6.6 and 14 nm and consist of 11 graphene layers with inter layer distance of 0.342 nm. The outer layer of MWNTs were covered with amorphous carbon. Methyl ester of *Jatropha curcas* oil produced well crystalline MWNTs of inner and outer diameter of 3 nm and 9 nm respectively, when the precursor sprayed at a flow rate of 20 mL per hour over silica supported Fe, Co and Mo catalyst at reaction temperature of 650°C (Fig. 2d). However, an increase of precursor flow rate to 30 mL per hour at the same temperature produced largely amorphous carbon and small quantity of metal encapsulated MWNTs of size around 40 nm (Fig. 2c). A metal filled MWNTs of inner and outer diameter of 4 nm and 24 nm respectively was observed when methyl ester

Carbon Nanotubes from Unconventional Resources: Part A: Entangled Multi-Walled Carbon Nanotubes and Part B: Vertically-Aligned Carbon Nanotubes

239

of *Pongamia Pinnata* oil, at a flow rate of 20 mL per hour, was spray pyrolysed at 650 °C over silica supported Fe, Co and Mo catalyst. When flow rate was increased to 30 mL per hour, a short growth of metal encapsulated carbon nanostructures was observed (Fig. 2e). The TEM and TGA studies revel that the carbon nanostructures are not well graphitized.

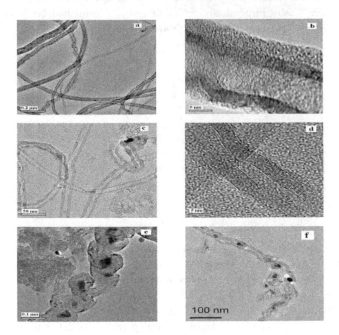

Figure 2. Representative TEM images of as-grown CNTs under constant reaction temperature of 650 °C for flow rate of precursor: Pine oil (a) 20 mL per hour (b) 30 mL per hour; Methyl ester of *Jatropha carcus* oil (c) 30 mL per hour (d) 20 mL per hour and Methyl ester of *Pongamia pinnata* oil (e) 30 mL per hour (f) 20 mL per hour.

The Fig. 2a shows TEM images of sample prepared using Pine oil as carbon source at a flow rate of 20 mL per hour over silica supported Fe, Co and Mo catalyst at a reaction temperature of 650°C. The TEM images indicate that the inner and outer diameter was uniform over entire length of the tube. It is also evident that carbon nanotubes do not contain encapsulated catalyst but amorphous carbon at the outer wall of the tube. The well crystallization of the graphene layer was confirmed from the I_G/I_D ratio of 2.5 (Fig.3a). The I_G/I_D ratio increases with increase of precursor flow rate from 10mL per hour to 20 mL per hour, but further increase of precursor flow rate to 30mL per hour results in decrease of I_G/I_D ratio to 0.5 (Fig.3b). The improved quality can be attributed to increase of precursor concentration in the tube. The higher flow rate of precursor (20 mL per hour) increases precursor concentration in the tube which leads to increase in decomposition of precursor over catalyst. Thus quality carbon nanotubes were formed at a reaction temperature of 650 °C. However, too high a concentration of precursor (30 mL per hour) or a reaction temperature of 750 °C leads to the formation of amorphous product as vapor phase decomposition of precursor is promoted than the decomposition over catalyst. The results are in good accordance with the reports by Afre et al. [17]. They have synthesized MWNTs using turpentine oil as a precursor by spray pyrolysis method in the temperature range of 600 to 800 °C. While Ghosh and co workers [18] reported synthesis of single walled carbon nanotube at a reaction temperature of 850 °C and abundant amount of MWCNTs at lower temperatures using turpentine oil as precursor by CVD method. No lower frequency RBM peaks in Raman spectra of our samples shows the absence of SWNTs.

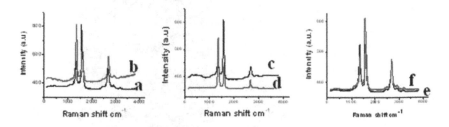

Figure 3. Typical Raman spectrum (green laser with excitation wavelength 532 nm) of as-grown MWNTs at different flow rate of precursors: Pine oil (a) 20 mL per hour (b) 30 mL per hour; Methyl ester of *Jatropha carcus* oil (c) 20 mL per hour (d) 30 mL per hour and Methyl ester of *Pongamia pinnata* oil (e) 20 mL per hour (f) 30 mL per hour.

Thermogravimetric analysis was performed to characterize the thermal behavior of the carbon nanotube synthesized using Pine oil as precursor. The Thermogravimetric graph shows the wt% vs. temperature. Upto a certain period of time there is no weight loss. The oxidation of carbon deposit starts after this point which is indicated by a dip in curve. With increase in temperature the weight loss increases till all the carbon get burnt. The residue is the catalyst and support. The TGA curves gives the temperature corresponds to the maximum mass decrease, which is considered to be a measure of the level of crystallinity of carbon nanotube. The TGA results (Fig.4a,b,c) shows the variation in decomposition temperature for the products synthesized at 650 °C for different flow rate of Pine oil precursor. It was found that the decomposition temperature varied between 568 and 591 °C. The relatively high decomposition temperature of 591 °C for the product synthesized at a reaction temperature of 650 °C for precursor flow rate of 20 mL per hour indicates well crystalline structure formation.

A HRTEM images (Fig. 2d) of sample prepared using Methyl ester of *Jatropha curcas* oil as carbon precursor at a flow rate of 20 mL per hour over silica supported Fe, Co and Mo catalyst at a reaction temperature of 650°C clearly shows well graphitized layers of a typical MWNTs with uniform inner and outer diameter. The TEM image also revels that encapsulated catalyst or amorphous carbon is rarely seen in the sample. The image indicates that the MWNTs are composed of around 26 walls and layers grow perpendicular to the growth axis of the tube. An increase of precursor flow rate to 30 mL per hour at the reaction temperature of 650 °C produced largely amorphous carbon and small quantity of metal incorporated MWNTs of size around 40 nm is evident from the TEM images (Fig. 2c). An additional confirmation for high degree graphitization and formation of metal filled MWNTs for sample prepared using the precursor flow rate of 20 mL per hour and 30 mL per hour respectively is shown by Raman spectra (Fig. 3c,d). The G band at 1571 cm^{-1} was attributed to well crystallized carbon structure, while the D band at 1359 cm^{-1} was attributed to defects in the structure [19]. The decrease in relative intensity of the G band and D band (I_G/I_D ratio) for sample prepared with precursor flow rate of 30 mL per hour indicates more defects in as-grown sample (Fig. 3d). The defects in MWNTs can be attributed to increase of precursor concentration in the tube and encapsulation of catalyst particles. An increase of precursor concentration in the tube leads to increase in decomposition of precursor over catalyst. Above the critical concentration of precursor, rate of decomposition of precursor exceeds rate of diffusion of carbon into the catalyst particle and thus encapsulation of metal particle occurs. The TGA results of MWNTs sample grown using methyl ester of *Jatropha curcas* oil are shown in (Fig.4d,e,f). Higher decomposition temperature and 37% residue observed in the TGA studies for the product synthesized at a reaction temperature of 650 °C for precursor flow rate of 20 mL per hour shows the sample contain around 60% MWNTs with well crystallized structure. A decrease in decomposition temperature and high residue shows defects in structure and metal encapsulation for the sample prepared at 650 °C for the precursor flow rate of 30 mL per hour.

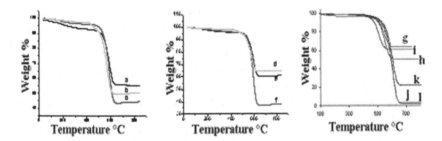

Figure 4. TGA curves of as-grown CNTs samples at different flow rate of precursors: Pine oil (a) 10 mL per hour (b) 20 mL per hour (c) 30 mL per hour, Methyl ester of *Jatropha carcus* oil (d) 10 mL per hour (e) 20 mL per hour (f) 30 mL per hour, Methyl ester of *Pongamia pinnata* oil, as-grown, (g) 10 mL per hour (h) 20 mL per hour (i) 30 mL per hour and purified (j) 10 mL per hour (k) 20 mL per hour (l) 30 mL per hour.

The TEM images of sample synthesized for 20 mL per hour flow rate of methyl ester of *pongamia pinnata* oil over silica supported Fe, Co and Mo catalyst at a reaction temperature of 650 °C was shown in Fig. 2f. The average outer diameter of the tube synthesized varied randomly as the reaction temperature was changed. At 550 °C, the diameter of the arborization-like nanostructures were around 65 nm, whilst at 650 °C, the formed MWNTs has inner and outer diameter in the range of 7nm & 33nm. The HRTEM image of the sample synthesized at 650 °C (Fig.2f) indicates the metal particles, seen as a dark spot on the image, were tightly covered by carbon layers with a thickness of a few nanometers. When precursor flow rate was reduced from 30 to 20 mL per hour, due to effective decomposition of precursor and the fluid nature of catalyst particle, the morphology of the product changed from arborization-like nanostructure to magnetic nanopartical encapsulated in multi-walled carbon nanotubes.

A similar observation was reported by Kang et al. [20]. Also, morphology change from magnetic nanopartical encapsulated in multi-walled carbon nanotubes to multi-walled carbon nanotubes structure was attributed to the high fluid nature of catalyst at 750 °C. The inner diameter of the carbon structure formed is same as that observed for products synthesized at 650 °C indicates the prevention of agglomeration of catalyst particles by Mo [21].

Amorphous carbon formation in large quantity, at 750 °C or at precursor flow rate of 30mL per hour, may be due to increased thermal decomposition of precursor material. The TGA results are shown in Fig.4g,h,i. It is evident that the weight loss continues to increase rapidly with temperature until reaches a constant value. Residue of the as-grown sample for precursor flow rate of 10 mL, 20 mL and 30 mL per hour at 650 °C was found to be 63.2, 60.9 and 50.5% respectively by weight (Fig.4g,h,i). The TGA results of same samples after purification shows (Fig. 4j,k,l) weight of residue as 3.5, 22.5 and 2.5% mass fraction. The more decline in mass fraction was caused by the acid leaching of catalyst particles that was not encapsulated by carbon. The low residue observed for products synthesized at 550 °C and 750 °C are attributed to low catalytic activity and high thermal decomposition of precursors respectively, which leads to formation of high amorphous carbon and low encapsulated products. The products synthesized at 650 °C shows minimum mass loss in TGA studies, even after purification, is due to better encapsulation of metal particles by carbon layers. The I_G/I_D value of 0.9 (Fig.3e) for samples prepared at 650 °C with flow rate of 20 mL per hour indicates that magnetic nanoparticals encapsulated in carbon nanotubes structure had defects and moderate crystallization of graphene planes [22,23]. This supports the HRTEM results. The increased I_G/I_D value of 1.87 (Fig. 3f) for the same sample, after purification indicates the removal of amorphous carbon and defective structures during purification. The removal of amorphous carbon and defective structures were further supported by higher ignition temperature 610 °C in TGA studies. According to TGA curves, weight of residue for the purified sample decreased to about 13 to 31% mass fraction comparing to unpurified sample due to leaching of metal particle and amorphous carbon removal during purification process. This shows that carbon layers covering the metal particles prevent their dissolution during purification process.

The mechanism of CNT nucleation and growth is one of the challenging and complex topics in current scientific research. Presently, various growth models based on experimental and quantitative studies have been proposed. It is well established, that during CNT nucleation and growth the following consecutive steps were taken place [24].

1. Carbon species formation by decomposition of precursor over the catalyst

2. Diffusion of carbon species through the catalyst particle

3. Precipitation of the carbon in the form of CNTs

The first step involves formation of carbon species by catalytic vapor decomposition of vapors of the precursor material over the catalyst. In the second step the diffusion of carbon species through the catalyst particle takes place. The catalyst surface may exert a diffusion barrier. It is still unclear whether carbon species diffuse on the particle surface [25], on the particle bulk [24] or whether surface and diffusion compete. The most accepted growth

Carbon Nanotubes from Unconventional Resources: Part A: Entangled Multi-Walled Carbon
Nanotubes and Part B: Vertically-Aligned Carbon Nanotubes

243

model suggests bulk diffusion of carbon species into the metal particles [26]. The third step is the precipitation of the carbon in the form of CNTs from the saturated catalyst particle.

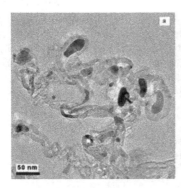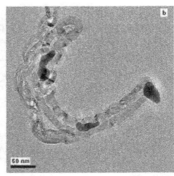

Figure 5. TEM images of as-grown CNTs sample synthesized under constant reaction temperature of 650 °C usingr Methyl ester of *Pongamia pinnata* oil at a flow rate of 20 mL per hour. (a) indicates the reshaping of catalyst particle (b) metal particle at the tip of tube.

Based on experimental results, a possible growth mechanism of MWNTs was proposed. It is known from the fact that Catalytic centers on catalyst particle act as nucleation site for the growth of MWNTs [22]. The precursor vapor decomposed on surface of the catalyst particle produces carbon. As the reactivity between the catalyst and the carbon exceeds the threshold value, carbon atoms loose their mobility in the solid solution, forming metal carbides [23]. These meta stable Fe and Co carbides decomposed and produce carbon which dissolve in these metal particles. The dissolved carbon diffuses through the metal particle and gets precipitated in the form of crystalline graphene layer. This carburized surface acts as a barrier for further carbon transfer from the gas phase to the bulk of the catalyst since carbon diffusion is slower through metal carbides [27]. The saturated metal carbide have lower melting point and they are fluid like during the growth process [28].

If the rate of precursor decomposition and the rate of diffusion of carbon are equal, then the metal raise through a capillary action and tube growth occurs. The fact that long carbon nanotubes observed have their catalyst particles partially exposed indicates that the direct contact of catalyst surface with carbon precursor is essential for continuous CNT growth (Fig. 5a). This is consistent with the growth mechanism proposed by Rodriguez [29]. In case the decomposition rate exceeds the diffusion rate, more of carbon produced forms a thick carbide layer over the surface of metal which acts as a barrier for further carbon transfer from the gas phase to the bulk of the catalyst. However, the thick carbide layer crystallizes out as graphene layer which encapsulate the metal particle. When a catalyst particle is fully encapsulated by layers of graphene sheets, the carbon supply route is cut and CNT growth stops resulting in short MWCNTs. The catalyst particle undergoes several mechanical reshaping during the tip growth of multi-walled nanotubes [30, 31]. This gives the impression that the catalyst is in liquid state during reaction. The catalyst particle seen inside and at the tip of tube could be the solidified form of the liquid phase metal particle. Thus the growth process is by the vapor–liquid–solid (VLS) mechanism [32].

The CNTs grow with either a tip growth mode or a base growth mode. Base growth mode is suggested when the catalyst particle remain attached to the support, while tip growth happens when the catalyst particle lifts off the support material. These growth modes depends on the contact forces or adhesion forces between the catalyst particle and support [33], while a weak contact favors tip-growth mechanism, a strong interaction promotes base growth [34]. Catalyst particle seen at tip of CNTs (Fig. 5b) indicate tip growth mode. These catalyst particles have lifted off the support and elongated due to the flow nature and stress induced by the carbon surrounding the catalyst.

2. Part B: Vertically-aligned carbon nanotubes

2.1. Introduction

Aligned carbon nanotubes were first reported by Thess et al.[2]. In the same year the Chinese academy of science reported that a 50 μm thick film of highly aligned nanotubes had successfully grown by chemical vapor deposition (CVD) [35]. Vertically aligned CNTs are quasi-dimensional carbon cylinders that align perpendicular to a substrate [36]. Vertically aligned with high aspect ratios [37] and uniform tube length made it easy spinning into macroscopic fibres [38] Aligned CNTs are widely used in nano electronics, composite materials as reinforcing agents and self-cleaning applications [39-41]. Aligned CNTs are ideal electrode material for biosensors over entangled CNTs, may be due to its high electrical conductive property [42]. Large CNT arrays have successfully been grown on different substrates, such as mesoporous silica [43] planar silicon substrate [44] and quartz glass plate [45]. Substrate provides a solid foundation for growing aligned CNTs. The substrate must able to inhibit the mobility of the catalyst particles in order to prevent agglomeration. The most commonly used active catalyst for growing CNTs are magnetic elements such as Fe, Co or Ni. Gunjishima et al. [46] used Fe-V bimetallic catalyst for synthesize of aligned DWCNTs. Recently, there have been appreciable attempt of using ferrocence as a catalyst for synthesis of aligned carbon nanotubes[47]. Here we report fabrication of aligned CNTs by spray pyrolysis on silicon wafer using mixture of Pine oil, Methyl ester of *Jatropha curcas* oil and Methyl ester of *Pongamia pinnata* oil with ferrocence.

2.2. Experimental Methods

The syntheses of aligned CNTs were carried out using the spray pyrolysis method. In this spray pyrolysis method, pyrolysis of the carbon precursor with a catalyst take place followed by deposition of aligned CNTs occur on silicon substrate. Pine oil, Methyl ester of *Jatropha curcas* oil and Methyl ester of *Pongamia pinnata* oil were used as carbon source and ferrocene [Fe$(C_5H_5)_2$] (Sigma Aldrich, high purity 98 %) was used as a source of Fe which acts as a catalyst for the growth of CNTs. n type silicon wafer (100) of size (1x1cm ^{2}) was used as a substrate and kept inside the quartz tube. In a typical experiment, the quartz tube was first flushed with argon (Ar) gas in order to eliminate air from the quartz tube and then heated to a reaction temperature. The precursor mixture was sprayed into the quartz tube, using Ar gas. The concentration of ferrocene in carbon precursor was ~25 mg/ml. The flow rate of Ar was 200 sccm/min. The experiments were conducted at 650 ºC with reaction time of 45 min was maintained for each deposition. After deposition, the furnace was switched off and allowed to cool down to room temperature under Ar gas flow. A uniform black deposition on the silicon substrate was observed. Finely, the substrate containing aligned CNTs was removed from the quartz tube for characterization. The experiments were repeated several times to ensure the reproducibility of the formation of vertically aligned carbon nanotubes.

2.3. Result and Discussion

The morphology of carbon sample grown on silicon substrate using a mixture of Pine oil and ferrocene at 650 °C can be observed in Figure 6a. The image revel the formation of high abundance of carbon nanotubes which are forest like and vertically-aligned to the substrate surface. The growth of carbon nanotubes seems to be uniform and reaches up to a length of 10μm. Figure 6b shows the SEM image of carbon sample grown on silicon substrate using a mixture of Methyl ester of *Jatropha curcas* oil and ferrocene at 650 °C. The dense, aligned but non-uniform growth of carbon nanotubes was observed. The length of carbon nanotubes grown was found to be varied from 12.5 to 7.5μm. Figure 6c illustrates the SEM image of the carbon naotubes grown at 650 °C using Methyl ester of *Pongamia pinnata* oil. A thick carbon nanotube with poor structure and alignment was observed.

Figure 6. Representative SEM images of as-grown vertically-aligned carbon nanotubes at 650 °C using Pine oil (a), Methyl ester of *Jatropha carcus* oil (b) and Methyl ester of *Pongamia pinnata* oil (c).

From the experimental results we suggest that the synthesis of aligned CNTs is very sensitive to the carbon precursors used. Ferrocene on thermal decomposition at high temperature forms Fe nano particles on the silicon substrate surface. During the chemical vapor deposition process, the carbon precursor is catalytically decomposed and the carbon fragments formed diffuse into the Fe catalyst. The Fe particles may thus easily become saturated or supersaturated with carbon atoms, and the precipitation of the carbon from the surface of the Fe particle leads to the formation of dense carbon nanotubes [48]. The high surface density of the growing nanotubes serves as an additional advantage for the constituent nanotubes to be "uncoiled". The Vander waals forces between the tube keep them aligned. Thus, the Fe catalysts can effectively catalyze the growth of highly dense vertically aligned carbon nanotubes on silicon substrate. Further investigation is going on in our laboratory for a better understanding of the actual growth mechanism of vertically aligned carbon nanotubes.

3. Conclusions, challenges and future prospects

In view of the perspective of green chemistry, we attempt to explore regenerative materials for CNT synthesis with high efficiency. In this research work a well graphitized MWNTs were synthesized from Pine oil and Methyl ester of *Jatropha curcas* oil using silica supported Fe, Co and Mo catalyst by spray pyrolysis method. The optimum reaction conditions for synthesis of MWNTs were 650 °C and precursor flow rate of 20 mL per hour. Spray pyrolysis of Methyl ester of *Pongamia pinnata* oil over silica supported Fe, Co and Mo catalyst results in formation of MWNTs filled with magnetic nanoparticles, which find potential applications in magnetic recording, biomedical and environmental protection. Vertically aligned carbon nanotubes were obtained by spray pyrolysis of Pine oil and Methyl ester of *Jatropha curcas* oil and ferrocene mixture, at 650 °C on silicon substrate under Ar atmosphere. The use of natural precursors gives sensible yield and makes the process natural world friendly as well. A thick carbon nanotube with poor structure and alignment was observed with mixture of Methyl ester of *Pongamia pinnata* oil and ferrocene.

The studies in this work demonstrate that the carbon materials are potential precursor for CNTs production under suitable experimental conditions and comply with green chemistry principles. It is clear that specific carbon nanostructures can be synthesized by suitably altering the experimental parameters. However, it is a challenge to consistently reproduce CNT of same quality and quantity form the precursor of inconsistent composition. Designing of catalyst material and optimization of reaction parameters which is suitable for synthesis of specific morphological CNTs from a renewable natural precursor of inconsistent chemical composition is one of the future prospects in this area of research.

Acknowledgements

The authors acknowledge the UGC New Delhi for financial support, the Institute for Environmental and Nanotechnology for technical support and IITM for access to Electron microscopes.

Author details

S. Karthikeyan[1*] and P. Mahalingam[2]

*Address all correspondence to: skmush@rediffmail.com

1 Chikkanna Government Arts College, TN, India

2 Arignar Anna Government Arts College, TN, India

Carbon Nanotubes from Unconventional Resources: Part A: Entangled Multi-Walled Carbon
Nanotubes and Part B: Vertically-Aligned Carbon Nanotubes

247

References

[1] Iijima, S. (1991). Helical microtubules of graphitic carbon. *Nature*, 354, 56-58.

[2] Thess, A., Lee, R., Nikolaev, P., Dai, H., Petit, P., Robert, J., Xu, C., Smalley, R., et al. (1996). Crystalline ropes of metallic carbon nanotubes. *Science*, 273, 483-487.

[3] Karthikeyan, S., Mahalingam, P., & Karthik, M. (2008). Large scale synthesis of carbon nanotubes. *E-Journal of Chemistry*, 6, 1-12.

[4] Parathasarathy, R. V., Phani, K. L. N., & Martin, C. R. (1995). Template synthesis of graphitic nanotubules. *Advanced Materials*, 7, 896-897.

[5] Sen, R., Govindaraj, A., & Rao, C. N. R. (1997). Carbon nanotubes by the Metallocene Route. *Chemical Physics Letters*, 267, 276-280.

[6] Pradhan, D., & Sharon, M. (2002). Carbon nanotubes, Nanofilaments and Nanobeads by thermal Chemical Vapor Deposition Process. *Material Science and Engineering: B*, 96, 24-28.

[7] Qian, W., Yu, H., Wei, F., Zhang, Q., & Wang, Z. (2002). Synthesis of carbon nanotubes from liquefied petroleum gas containing sulfur. *Carbon*, 40, 2968-2970.

[8] Aswasthi, K., Kumar, R., Tiwari, R. S., & Srivastava, O. N. (2010). Large scale synthesis of bundles of aligned carbon nanotubes using a natural precursor: Turpentine oil. *Journal of Experimental Nanoscience*, 5(6), 498-508.

[9] Ghosh, P., Soga, T., Tanemura, M., Zamri, M., Jimbo, T., Katoh, R., & Sumiyama, K. (2009). Vertically aligned carbon nanotubes from natural precursors by spray pyrolysis method and their field electron emission properties. *Applied Physics A: Materials Science & Processing*, 94, 151-156.

[10] Andrews, R. J., Smith, C. F., & Alexander, A. J. (2006). Mechanism of carbon nanotube growth from camphor and camphor analogs by chemical vapor deposition. *Carbon*, 44(2), 341-347.

[11] Ghosh, P., Afre, R. A., Soga, T., & Jimbo, T. (2007). A simple method of producing single-walled carbon nanotubes from a natural precursor: Eucalyptus oil. *Materials Letters*, 61(17), 3768-3770.

[12] Karthikeyan, S., & Mahalingam, P. (2010). Studies of yield and nature of multi-walled carbon nanotubes synthesized by spray pyrolysis of pine Oil at different temperatures. *International Journal of Nanotechnology and Applications*, 4, 189-197.

[13] Karthikeyan, S., & Mahalingam, P. (2010). Synthesis and characterization of multi-walled carbon nanotubes from biodiesel oil: green nanotechnology route. *International Journal of Green Nanotechnology: Physics and Chemistry*, 2(2), 39-46.

[14] Anastas, P., & Warner, J. C. (1998). *Green Chemistry: Theory and Practice*, Oxford University Press, Oxford, 30.

[15] Sabelo, D. M., Kartick, C. M., Robin, C., Michael, J. W., & Neil, J. C. (2009). The effect of synthesis parameters on the catalytic synthesis of multiwalled carbon nanotubes using Fe-Co / CaCO3 catalysts. *South African Journal of Chemistry*, 62, 67-76.

[16] Smalley, R. E., Marek, I. M., Wang, Y., & Hange, R. H. (2007). Purification of carbon nanotubes based on the chemistry of fenton's reagent. *USPTO Patent Application No. 20070065975*.

[17] Afre, R. A., Soga, T., Jimbp, T., Mukul, Kumar., Anto, Y., Sharon, M., Prakash, R., Somani, , & Umeno, M. (2006). Carbon nanotubes by spray pyrolysis of turpentine oil at different temperatures and their studies. *Microporous and Mesoporous Materials*, 96, 184-190.

[18] Ghosh, P., Soga, T., Afre, R. A., & Jimbo, T. (2008). Simplified synthesis of single-walled carbon nanotubes from a botanical hydrocarbon: Turpentine Oil. *Journal of alloys and Compounds*, 462-92008.

[19] Dresselhausa, M. S., Dresselhausc, G., Joriob, A., Souza Filhob, A. G., & Saito, R. (2002). Raman spectroscopy on isolated single wall carbon nanotubes. *Carbon*, 40, 2043-2061.

[20] Kang, J. L., Li, J. J., Du, X. W., Shi, C. S., Zhao, N. Q., Cui, L., & Nash, P. (2008). Synthesis and growth mechanism of metal filled carbon nanostructures by CVD using Ni/Y catalyst supported on copper. *Journal of Alloys and Compounds*, 456(1-2), 290-296.

[21] Abello, M. C., Gomez, M. F., & Ferretti, O. (2001). Mo/γ-Al2O3 catalysts for the oxidative dehydrogenation of propane: Effect of Mo loading. *Applied Catalysis A*, 207(1-2), 421-431.

[22] Lee, M. H., & Park, D. G. (2003). Preparation of MgO with High Surface Area, and Modification of Its Pore Characteristics. *Korean Chemical Society*, 24(10), 1437-1443.

[23] Sinclair, R., Itoh, T., & Chin, R. (2002). In situ TEM studies of metal-carbon reactions. *Microscopy and Microanalysis*, 8(4), 288-304.

[24] Brukh, R., & Mitra, S. (2006). Mechanism of carbon nanotube growth by CVD. *Chemical Physics Letters*, 424, 126-32.

[25] Hofmann, S., Csanji, G., Ferrari, A. C., Payne, M. C., & Robertson, J. (2005). Surface diffusion: The low activation energy path for nanotube growth. *Physical Review letters*, 95, 036101-036104.

[26] Ducati, C., Alexandrou, I., Chhowalla, M., Robertson, J., & Amaratunga, G. A. J. (2004). The role of the catalytic particle in the growth of carbon nanotubes by plasma enhanced chemical vapor deposition. *Journal of Applied Physics*, 95(11), 6387-6391.

[27] Ozturk, B., Fearing, Vl., Ruth, J. A., & Simkovich, G. (1982). Self-Diffusion Coefficients of Carbon in Fe3C at 723 K via the Kinetics of Formation of This Compound. *Metallurgical and Materials Transactions A*, 13(10), 1871-1873.

[28] Chakraborty, A. K., Jacobs, J., Anderson, C., Roberts, C. J., & Hunt, M. R. C. (2005). Chemical vapor deposition growth of carbon nanotubes on Si substrates using Fe catalyst: What happens at the nanotube/Fe/Si interface. *Journal of Applied Physics*, 100(8), 084321.

[29] Rodriguez, N. M. (1993). A review of catalytically grown carbon nanofibers. *Journal of Materials Research*, 8, 3233-3250.

[30] Hofmann, S., Sharma, K., Ducati, C., Du, G., Mattevi, Cepek C., Cantoro, M., Pisana, S., Parvez, A., Cervantes-sodi, F., Ferrari, A. C., & Dunin-Borkowski, R. E. (2007). In situ Observations of Catalyst Dynamics during Surface-Bound Carbon Nanotube Nucleation. *Nano Letters*, 7(3), 602-608.

[31] Helveg, S., Lopez-Cartes, C., Sehested, J., Hensen, P. L., Clausen-Nielsen, Rostrup., Abild-Pederson, J. R., , F., & Norskov, J. K. (2004). Atomic-scale imaging of carbon nanofibre growth. *Nature*, 427, 426-429.

[32] Kukovitsky, E. F., L'vov, S. G., & Sainov, N. A. (2000). VLS-growth of carbon nanotubes from the vapor. *Chemical Physics Letters*, 317, 65-70.

[33] Leonhardt, A., Hampel, S., Büchner, B., et al. (2006). Synthesis, properties and applications of ferromagnetic filled carbon. *Chemical Vapor Deposition*, 12(6), 380-387.

[34] Bower, C., Otto, Z., Wei, Z., Werder, D. J., & Jin, S. (2000). Nucleation and growth of carbon nanotubes by microwave plasma chemical vapor deposition. *Applied Physics Letters*, 77(17), 2767-2768.

[35] Li, W. Z., Xie, S. S., Qian, L. X., Chang, B. H., Zou, B. S., Zhou, W. Y., Zhao, R. A., & Wang, G. (1996). Large-scale synthesis of aligned carbon nanotubes. *Science*, 274(5293), 1701-1703.

[36] Feng, W., Bai, X. D., Lian, Y. Q., Liang, J., Wang, X. G., & Yoshino, K. (2003). Well aligned polyaniline/carbon-nanotube composite films grown by in-situ aniline polymerization. *Carbon*, 41(8), 1551-1557.

[37] Zhang, Q., Zhao, M. Q., Huang, J. Q., Liu, Y., Wang, Y., Qian, W. Z., & Wei, F. (2009). Vertically aligned carbon nanotube arrays grown on a lamellar catalyst by fluidized bed catalytic chemical vapor deposition. *Carbon*, 47(11), 2600-2610.

[38] Zhang, Q., Zhou, W. P., Qian, W. Z., Xiang, R., Huang, J. Q., Wang, D. Z., & Wei, F. (2007). Synchronous growth of vertically aligned carbon nanotubes with pristine stress in the heterogeneous catalysis process. *Journal of Physical Chemistry*, 111(40), 14638-14643.

[39] Afre, R. A., Soga, T., Jimbo, T., Kumar, M., Ando, A., & Sharon, M. (2005). Growth of vertically aligned carbon nanotubes on silicon and quartz substrate by spray pyrolysis of a natural precursor: Turpentine oil. *Chemical Physics Letters*, 414(1-3), 6-10.

[40] Teo, K. B. K., Singh, C., Chhowalla, M., & Milne, W. I. (2003). Catalytic synthesis of carbon nanotubes and nanofiber. *In: Nalwa H S, eds. Encyclopedia of nanoscience and nanotechnology,* American Scientific Publishers, 1-22.

[41] Bu, I. Y. Y., & Oei, S. P. (2010). Hydrophobic vertically aligned carbon nanotubes on Corning glass for self cleaning applications. *Applied surface science,* 256(22), 6699-6704.

[42] Yang, L., Xu, Y., Wang, X., Zhu, J., Zhang, R., He, P., & Fang, Y. (2011). The application of b-cyclodextrin derivative functionalized aligned carbon nanotubes for electrochemically DNA sensing via host-guest recognition. *Analytica Chimica Acta,* 689(1), 39-46.

[43] Pan, Z. W., Zhu, H. G., Zhang, Z. T., Im, H. J., Dai, S., Beach, D. B., & Lowndes, D. H. (2003). Patterned growth of vertically aligned carbon nanotubes on pre-patterned iron/silica substrates prepared by sol-gel and shadow masking. *Journal of physical chemistry B,* 107(6), 1338-1344.

[44] Jung, Y. J., Wei, B. Q., Vajtai, R., & Ajayan, P. M. (2003). Mechanism of selective growth of carbon nanotubes on SiO2/Si patterns. *Nano Letters,* 3(4), 561-564.

[45] Gong, Q. M., Li, Z., Li, D., Bai, X. D., & Liang, J. (2004). Fabrication and structure: a study of aligned carbon nanotube/carbon nanocomposites. *Solid State Communications,* 131(6), 399-404.

[46] Gunjishima, I., Inoue, T., Yamamuro, S., Sumiyama, K., & Okamoto, A. (2007). Synthesis of vertically aligned, double-walled carbon nanotubes from highly active Fe-V-O nanoparticles. Carbon ., 45(6), 1193-1199.

[47] Kumar, M., & Ando, Y. (2003). A simple method of producing aligned carbon nanotubes from an unconventional precursor- Camphor. *Chemical Physics Letters,* 374(5-6), 521-526.

[48] Liu, Y., Yang, X. C., Pu, Y., & Yi, B. (2010). Synthesis of aligned carbon nanotube with straight-chained alkanes by nebulization method. *Transaction of Nonferrous Metals Society of China,* 20(6), 1012-1016.

Chapter 12

Toroidal and Coiled Carbon Nanotubes

Lizhao Liu and Jijun Zhao

Additional information is available at the end of the chapter

1. Introduction

The perfect graphite and carbon nanotube (CNT) are composed of hexagonal rings of carbon atoms. However, non-hexagonal rings like pentagons and heptagons usual exist in the realistic CNT. Due to the change of topology, different arrangements of the pentagons and heptagons would lead to various structures, such as CNTs with Stone-Wales defects [1], CNT junctions [2], toroidal CNTs [3], and coiled CNTs [4, 5]. Each type of these CNT-based structures has its unique physical and chemical properties; as a consequence, the diversity in morphology extends the applications of CNTs. In this chapter, we will review the current progress on two important members of the CNT family, i.e., the toroidal CNTs at the first and coiled CNTs in the second.

The toroidal CNT (also known as carbon nanotorus or carbon nanoring) is a kind of zero-dimensional CNT-based nanostructure. In other words, a carbon nanotorus can be considered as a giant molecule and directly used as a nanoscale device. As for the synthesis of the toroidal CNTs, numerous methods have been proposed, including laser-growth method, ultrasonic treatments, organic reactions, and chemical vapour deposition (CVD), which will be illustrated in the following. In addition to experimental synthesis, various theoretical efforts have been devoted to construct the structural models of the toroidal CNTs. In general, there are two kinds of toroidal CNTs: one is formed by pristine nanotube with pure hexagon networks, and the other contains certain amount of pentagon and heptagon defects. Due to the circular geometry of the carbon nanotorus and incorporation of pentagon/heptagon defects, it may exhibit novel mechanical, electronic and magnetic properties different from the straight CNTs.

Another kind of curved CNT-based nanostructure is the coiled CNT, which is also known as carbon nanocoil or carbon nanospring. Different from the zero-dimensional toroidal CNT, the coiled CNT is a kind of quasi one-dimensional CNT-based nanostructures with a certain spiral angle. Intuitionally, a carbon nanocoil is like a spring in geometry. Therefore, mechanic properties of the coiled CNTs attract lots of attentions. Among various methods to produce the coiled CNTs, CVD approach is predominant due to the high quality and good controllability. Besides, several methods have been proposed to build the structural models of the coiled CNTs. An important feature of the carbon nanocoil models is the periodic incorporation of pentagons and heptagons in the hexagonal network. In addition, due to the excellent properties of the coiled CNTs, they have promising applications in many fields, such as sensors, electromagnetic nano-transformers or nano-switches, and energy storage devices.

2. Toroidal CNTs

In this section, we summarize experimental fabrication and theoretical modelling of the toroidal CNTs, as well as their physical and chemical properties. The toroidal CNTs are predicted to be both thermodynamic and kinetically stable. Due to the circular geometry, the toroidal CNTs possess excellent properties, especially the electronic and magnetic properties.

2.1. Fabrication

Synthesis and characterization of the toroidal CNTs are of key importance in the carbon nanotorus related fields. Early in 1997, Liu et al. reported synthesis of the toroidal CNTs with typical diameters between 300 and 500 nm by using the laser growth method [3]. From the measurement of scanning force microscopy (SEM) and transmission electron microscopy (TEM), it was shown that the toroidal CNTs were form by single-walled carbon nanotube (SWNT) ropes consisting of 10 to 100 individual nanotubes. Soon after, the toroidal CNTs were also found in the CNT samples prepared by catalytically thermal decomposition of hydrocarbon gas [6] and an ultrasound-aided acid treatment [7, 8]. Later, a variety of experimental approaches were developed to fabricate carbon nanotori, such as organic reactions [9, 10], chemical vapor deposition (CVD) [11, 12], and depositing hydrocarbon films in Tokamak T-10, the facility for magnetic confinement of high-temperature plasma [13]. In addition, incomplete toroidal CNTs [13, 14], large toroidal CNTs with diameters of~200–300 nm, sealed tubular diameters of 50–100 nm [15], and patterning of toroidal CNTs [16, 17] were also achieved in laboratory. In particular, the tubular diameter of a carbon nanotorus is now controllable. Toroidal CNTs from single-walled [7-10, 18, 19], double-walled [20], triple-walled [21], and multi-walled CNTs [6] have been achieved. Combining the experimental measurements and a simple continuum elastic model, formation of the toroidal CNTs was supposed to involve a balance between the tube-tube van der Waals adhesion, the strain energy resulting from the coiling-induced curvature and the strong interaction with the substrate [8, 14]. Various kinds of the toroidal CNTs are presented in Figure 1.

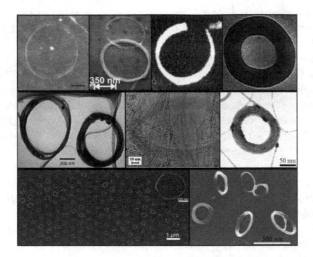

Figure 1. Experimental fabrications of various kinds of toroidal CNTs.

2.2. Structural models and thermodynamic stabilities

Prior to the experimental synthesis, Dunlap proposed to construct the structural model of a carbon nanotorus by connecting two CNTs with different diameters [22]. Almost at the same time, researchers in Japan built a C_{360} nanotorus from C_{60} fullerene [23] and then generated a series of toroidal CNTs with 120 to 1920 carbon atoms using the prescription of Goldberg [24, 25]. So far, there have been six major approaches to construct the structural models of toroidal CNTs: (1) bending a finite CNT and connecting its ends together [26-29]; (2) connecting CNTs with different diameters by introducing pentagons and heptagons [22, 30-32]; (3) constructing from fullerenes [23-25] by employing the prescription of Goldberg [33]; (4) built through the connection of one zigzag-edged chain of hexagons and another armchair-edged chain of hexagons [34]; (5) sewing the walls of a double-walled CNT at both ends [35]; (6) constructing from only pentagons and heptagons [36]. To summarize, there are two kinds of toroidal CNTs: one is formed by pure hexagonal networks and the other is a hexagonal structure with pentagon-heptagon defects. In a more detailed way, Itoh et al. classified the toroidal CNTs into five types using the parameters of the inner (r_i) and outer (r_o) diameters, and the height (h) [37]. As depicted in Figure 2, type (A) indicates a nanotorus with $r_i \approx r_o$, h $\ll r_i$, and h $\approx (r_o - r_i)$, type (B) is the case of $r_i \sim r_o$ ~h and h $\approx (r_o \odot r_i)$, type (C) denotes h $\ll (r_o \odot r_i)$, type (D) is the case of $r_i < r_o$, $r_o \sim$ h, and h ~ $(r_o \odot r_i)$, and type (E) means $(r_o \odot r_i) \ll$ h, respectively.

After establishing the structural models, one important issue is to examine the thermodynamic stabilities of the toroidal CNTs. Many groups demonstrated that toroidal CNTs are more stable than C_{60} fullerene through comparing their binding/cohesive energies calculated by means of empirical potential methods [22-25]. Besides, molecular dynamics (MD) simulations also demonstrated that toroidal CNTs can survive under high temperature [23, 29, 38,

39]. Generally, the thermodynamic stability of a carbon nanotorus depends on its geometric parameters, such as ring and tubular diameter, symmetry, curvature, and position of the pentagons and heptagons. Ihara et al. showed that the cohesive energy of a carbon nanotorus derived from C_{60} fullerene decreased with increasing number of carbon atoms in the carbon nanotorus [24]. The ring and tubular diameter can also affect the thermodynamic stability of a carbon nanotorus [40-42]. At a fixed tubular diameter, there was a preferable ring diameter where the nanotorus possesses the lowest formation energy [40]. Besides, dependence of the stability on the rotational symmetry was also reported for the toroidal CNTs [32, 37]. Among the toroidal CNTs constructed from (5, 5), (6, 6), and (7, 7) armchair CNTs, the one with D_{6h} symmetry is energetically favourable [32]. Despite the dependence on the geometric details, it was believed [43, 44] that for the toroidal CNTs with large ring diameters, the pure hexagonal structure is energetically more stable, but for the ones with small ring diameters, the mixture of hexagonal networks and pentagon-heptagon defects is energetically more favourable. In [44], this critical ring diameter is given by the equation $R_c = \pi r^2 Y/(4\sigma)$, where r is the tubular diameter of the initial CNT, Y is the Young's modulus of the initial CNT, and σ is the surface tension of graphite perpendicular to the basal planes. For example, taking the Y = 1.0 TPa, a R_c of 90 nm can be obtained for a carbon nanotorus made of a (10,10) nanotube (r = 0.68 nm).

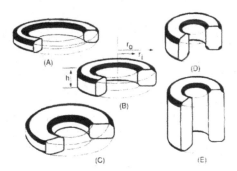

Figure 2. Schematic diagram for five types of toroidal CNTs classified by the parameters of the inner diameter r_i, the outer diameter r_o, and the height h, respectively. Reprinted with permission from [37]. Copyright (1995) Elsevier.

2.3. Mechanical properties

Mechanical property is of fundamental importance for the applications of a material. Employing MD simulation with a reactive force field, Chen et al. investigated the mechanical properties of zero-dimensional nanotorus, one-dimensional nanochain and two-dimensional nanomaile constructed from toroidal CNTs [45]. For a nanotorus constructed from bending a (5, 5) CNT, its Young's modulus increases monotonically with tensile strain from 19.43 to 121.94 GPa without any side constraints and from 124.98 GPa to 1560 GPa with side constraints, respectively, where the side constraint means fixing the position of small regions of carbon atoms at left and right sides. Besides, the tensile strength of the unconstrained and constrained nanotorus was estimated to be 5.72 and 8.52 GPa, respectively. In addition, the maximum elastic strain is approximate 39% for the nanochain and 25.2% for thenanomaile. For a nanotorus obtained from bending a (10, 10) CNT, its Young's modulus along the tube axis was 913 GPa by taking [46]. Later, buckling behavior of toroidal CNTs under tension was investigated using the molecular mechanics (MM) computations, including the toroidal CNTs formed from (5, 5), (8, 8) and (9, 0) CNTs [47, 48]. It was found that the buckling shapes of the toroidal CNTs constructed from both armchair and zigzag CNTs with an odd number of units are unsymmetrical, whereas those with an even number of units are symmetrical. Recently, reversible elastic transformation between the circular and compressed nanotorus in a colloid has been observed under TEM [17]. This geometric reversibility was also predicted theoretically by using a nonlinear continuum elastic model [49, 50], suggesting the potential application of toroidal CNTs as ultrasensitive force sensors and flexible and stretchable nanodevices.

2.4. Electronic properties

It is well-known that a CNT can be expressed by a chiral vector C_h (n, m) and a translation vector T (p, q) and can be either metallic or semiconducting, depending on its chirality [51]. Since a carbon nanotorus can be considered as a bended CNT or a CNT incorporated with pentagons and heptagons, it would be interesting to explore will the bending behavior or inclusion of pentagons and heptagons affect the electronic properties of the pristine CNT. For a carbon nanotorus formed by bending a (n, m) CNT, it can be divided into three types: (1) if m − n = 3i, and p − q = 3i (i is an integral), the carbon nanotorus is metallic; (2) if m − n = 3i, and p − q ≠ 3i, the carbon nanotorus is semiconducting; (3) if m − n ≠ 3i, and p − q = 3i, the carbon nanotorus is insulating [52]. This classification was partly confirmed by the tight-binding (TB) calculation that a metallic carbon nanotorus can be constructed by bending a metallic CNT and also follows the rule of divisibility by three on the indices of chiral and twisting vectors [53]. Moreover, delocalized and localized deformations play different roles on the electronic properties of a carbon nanotorus built bending a CNT [27]. The delocalized deformations only slightly reduce the electrical conductance, while the localized deformations will dramatically lower the conductance even at relatively small bending angles. Here the delocalized deformation means the deformation induced by the mechanical bending of the CNT, and the localized deformation indicates the deformation induced by the pushing action of the tip of AFM. In addition, Liu et al. reported the oscillation behavior of the energy gap during increasing size of the nanotorus and the gap was eventually converged to that of the infinite CNT [54].

Meanwhile, in the case of incorporation of pentagons and heptagons, a HOMO-LUMO gap can be expected for the carbon nanotorus. For a carbon nanotorus C_{1960} constructed by connecting (6, 6) and (10, 0) CNTs, a gap of 0.05 eV was calculated by a TB approach [44]. Using both the TB and semiempirical quantum chemical approaches, a series of toroidal CNTs with total number of atoms ranging from 120 to 768 were investigated and most of them have HOMO-LUMO gaps [55]. Besides, employing the extended-Hückel method, energy gaps of 0.4-0.32µB eV were predicted for the toroidal CNTs of C_{170}, C_{250}, C_{360}, C_{520}, and C_{750} [56]. Further accurate DFT examination also showed that the nanotorus C_{444} has a gap of 0.079 eV and the nanotorus C_{672} has a gap of 0.063 eV, respectively [57].

2.5. Magnetic properties

The unique circular geometry endows its advantage to study the magnetic response when ring current flows in a carbon nanotorus. Early in 1997, Haddon predicted that the nanotorus C_{576} has an extremely large and anisotropic ring-current diamagnetic susceptibility, which can be 130 times larger than that of the benzene molecule [58]. Afterwards, colossal paramagnetic moment was also reported in the metallic toroidal CNTs, which was generated by the interplay between the toroidal geometry and the ballistic motion of the π-electrons [28], as shown in Figure 3. For example, the nanotorus C_{1500} built from a (5, 5) CNT possesses a large paramagnetic moment of 88.4 µB at 0 K. Similarly, the nanotorus C_{1860} built from a (7, 4) CNT has a giant zore-temperature magnetic moment of 98.5 µB. In addition to the paramagnetic moments, existence of ferromagnetic moments at low temperatures in the toroidal CNTs without heteroatoms was also predicted by using a π-orbital nearest-neighbor TB Hamiltonian with the London approximation, which is attributed to the presence of pentagons and heptagons [59]. Another important phenomenon, i.e., the Aharonov–Bohm effect can be also observed in the toroidal CNTs [60-64]. Indeed, the magnetic properties of the toroidal CNTs are affected by many factors. Liu et al. pointed out that the paramagnetic moments of the toroidal CNTs decrease distinctly as temperature increases [28]. Such temperature dependence was also confirmed by several successive studies [65-68]. Moreover, the magnetic properties of a toroidal CNT also rely on its geometric parameters, such as ring diameter, curvature, chirality, and the arrangement of pentagons and heptagons [65-67].

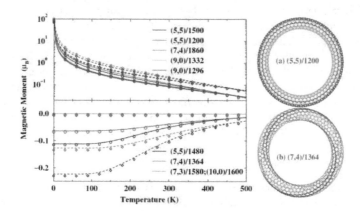

Figure 3. Induced magnetic moment as a function of temperature for various toroidal CNTs in a perpendicular magnetic field of 0.1 T (solid line) and 0.2 T (dashed line), respectively. Reprinted figures with permission from [Liu L, Guo GY, Jayanthi CS, Wu SY. Colossal Paramagnetic Moments in Metallic Carbon Nanotori. 88, 217206 (2002)]. Copyright (2002) by the American Physical Society. http://prl.aps.org/abstract/PRL/v88/i21/e217206.

2.6. Modification of the toroidal CNTs

Chemical modification is an important approach to tailor the properties of materials. A common approach of chemical modification is doping. It was found that doping electrons or holes into a carbon nanotorus could vary its magnetic properties through altering the band-filling configuration [69]. Our previous work also demonstrated that substitutional doping of boron or nitrogen atoms could modify the electronic properties of the toroidal CNTs due to change of the six π-electron orbitals [32]. Moreover, compared with the hexagonal rings, existence of pentagons favours the doping of nitrogen atoms and existence of heptagons prefers the doping of boron atoms. Besides, the toroidal CNTs coated with beryllium can be used as candidates for hydrogen storage. Each beryllium atom can adsorb three H_2 molecules with moderate adsorption energy of 0.2-03 eV/H_2 [70].

Since the toroidal CNTs also have the hollow tubular structures similar to the CNTs, atoms or molecules can be encapsulated into the toroidal CNTs. Early in 2007, Hilder et al. examined the motion of a single offset atom and a C_{60} fullerene inside a carbon nanotorus to explore its application as high frequency nanoscale oscillator [71]. They demonstrated that the C_{60} fullerene encapsulated carbon nanotorus can create high frequency up to 150 GHz, which may be controlled by changing the orbiting position. By inserting the chains of Fe, Au, and Cu atoms into a carbon nanotorus, Lusk et al. investigated the geometry, stability and electronic magnetic properties of this nano-composite structure [72]. Reduced HOMO-LUMO gap and ferromagnetism of the nanotorus were predicted by encapsulating chains of metal atoms. In addition, diffusion behavior of water molecules forming two oppositely polarized chains in a carbon nanotorus was studied by MD simulations. It was demonstrated that Fickian diffusion is in the case of a single chain and the diffusion for two or more chains is consistent with single-file diffusion [73].

3. Coiled carbon nanotube

Similar to the case of the toroidal CNTs, we first introduce the experimental synthesis and theoretical methods to construct the structural models, as well as their formation mechanism and stabilities. Then the mechanic properties and electronic properties of the coiled CNTs are summarized. Finally, the promising applications of coiled CNTs in various fields compared with their straight counterparts owing to their spiral geometry and excellent properties will be discussed in the end of this section.

3.1. Fabrication and formation mechanism

The coiled CNTs were first experimentally produced through catalytic decomposition of acetylene over silica-supported Co catalyst at 700 °C in 1994 [4, 5]. Afterwards, numerous methods have been proposed to synthesize the coiled CNTs, including the laser evaporation of the fullerene/Ni particle mixture in vacuum [74], opposed flow flame combustion of the fuel and the oxidizer streams [75], electrolysis of graphite in fused NaCl at 810 °C [76], self-assembly from π-conjugated building blocks [77, 78], and CVD method [79-83]. Among these various methods, the CVD approach is predominant due to its high quality, which has been reviewed by several literatures [84-86]. To fabricate the coiled CNTs, CVD process involves the pyrolysis of a hydrocarbon (e.g. methane, acetylene, benzene, propane) over transition-metal catalysts (e.g. Fe, Co, Ni) at high temperatures. Compared to the high growth temperature (> 2000 °C) of CNT through arc discharge and laser evaporation process, the relatively low growth temperature of CVD method (500–1000 °C) allows carbon atoms move slowly and form non-hexagonal carbon rings [84]. In 2006, Lau et al. reviewed the three major CVD-based methods to fabricate the coiled CNTs, including the catalyst supported CVD growth, on substrate CVD growth and template-based CVD growth [84]. Later, synthetic parameters of CVD growth of the coiled CNTs, such as catalyst, gas atmosphere and temperature, were introduced and catalogued by Fejes et al. [85] and Shaikjee et al. [86], respectively. Moreover, Shaikjee et al. [86] presented different types of the coiled CNTs with non-linear morphology, which are shown in Figure 4.

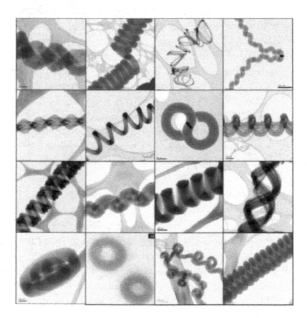

Figure 4. Experimental fabrications of various kinds of the coiled CNTs. Reprinted with permission from [86]. Copyright (2011) Elsevier.

As for the formation mechanism of the coiled CNTs, Fonseca et al. presented a formation of (chiral and achiral) knees on a catalyst particle to further form toroidal and coiled CNTs, which can be described by a simple formalism using the heptagon-pentagon construction [87]. In addition, formation of the coiled CNTs is closely related to the catalyst. Pan et al. suggested that the catalyst grain is crucial to the geometry of a carbon nanocoil and the non-uniformity of carbon extrusion speed in the different parts of the catalyst grain leads to the helical growth of the coiled CNTs [88]. Chen et al. pointed out that the driving force of coiling straight CNTs was the strong catalytic anisotropy of carbon deposition between different crystal faces [89]. For growth of carbon microcoils, the catalyst grain rotates around the coil axis which is on the symmetric face of the deposition faces; while for the twisted carbon nanocoils, the catalyst grain rotates around the axis which is perpendicular to the symmetric face of the deposition faces. Taking both the energy and entropy into account, Bandaru et al. proposed a mechanism that for a given volume of material, the helical form occupies the least amount of space and the entropy of the ambient conditions should increase to compensate for the close packing of the helices, which in turn is facilitated by nonwetting catalyst particles or induced by catalyst/ambient agitation in the CVD growth [90].

3.2. Structural models and thermodynamic stabilities

An important feature of a carbon nanocoil is incorporation of pentagons and heptagons in the hexagonal network. Dunlap [22, 91] showed that connecting two CNTs with pentagons and heptagons could result in a curved structure or knee structure. Based on the knee structure, Fonseca et al. was able to construct the toroidal and coiled CNTs using the knee segments as building blocks, where the former is an in-plane structure and the latter is out of plane [92]. In addition, researchers in Japan proposed two kinds of methods to construct structures of the coiled CNTs. One approach is to cut the toroidal CNTs into small pieces and recombine them to form the coiled CNTs with one pitch containing one nanotorus [37, 93]. For the coiled CNTs built from toroidal segments, Setton et al. suggested that the toroidal segments were only feasible for single-shell or at best two-shell nanocoils [94]. The other way is to insert pentagons and heptagons into a perfect graphene network and then roll up this structure to form a carbon nanocoil [95, 96]. Similarly, Biró et al. proposed to build the coiled CNTs from rolling up the Haeckelite structure, a graphite sheet composed of polygonal rings [97]. Recently, we were able to construct the carbon nanocoils from segment of CNTs in which the tube chirality is maintained [98]. Through introducing a pair of pentagons in the outer side and another pair of heptagons in the inner side into the segment of an armchair CNT, a curved structure can be obtained. Using this curved structure as a building block, a carbon nanocoil can be formed by connecting the building blocks with a rotate angle. This method was also employed to construct the structural models of the toroidal CNTs, as mentioned above [32]. A simple schematic diagram of this method is presented in Figure 5. Usually, a carbon nanocoil can be expressed by the parameters of inner coil diameter (D_i), outer coil diameter (D_o), tubular diameter (D_t) and coil pitch (λ) [84, 86], as illustrated in Figure 6.

In addition to the structural models of the coiled CNTs, several works have been devoted to investigating their thermodynamic stabilities. Employing MD simulation, Ihara et al. [93] obtained the cohesive energies of 7.41, 7.39 and 7.43 eV/atom for C_{360}, C_{540} and C_{1080} nanocoils, respectively, which are close to that of graphite sheet (7.44 eV/atom) and lower than that of the C_{60} fullerene (7.29 eV/atom). Therefore, these carbon nanocoils are more stable than C_{60} fullerene. Moreover, these carbon nanocoils can maintain the coiled geometry without collapse at a temperature up to 1500 K, which further confirms their thermodynamical stability. By taking into account the volume free energy, the surface energy, and the curva-ture elastic energy, it was found that there is a threshold condition for the formation of straight multiwall CNTs [99]. Below that the straight multiwall CNTs become unstable and would undergo a shape deformation to form the coiled CNTs.

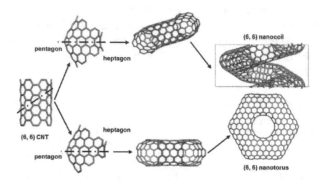

Figure 5. Schematic diagram for constructing the structural models of (6, 6) carbon nanotorus and nanocoil by intro-ducing pairs of pentagons (highlighted in blue) and heptagons (highlighted in red).

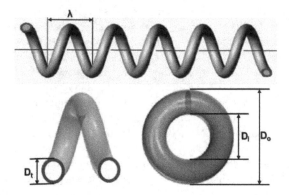

Figure 6. Parameters of inner coil diameter D_i, outer coil diameter D_o, tubular diameter D_t and coil pitch λ to describe a carbon nanocoil.

3.3. Mechanical properties

Intuitively, a carbon nanocoil is similar to a spring in geometry. It is well-known that spring exhibits excellent mechanic properties and is very useful in the mechanics-based devices. Therefore, mechanical properties of the coiled CNTs as "nanospring" have attracted lots of attentions. Early in 2000, Volodin et al. measured the elastic properties of the coiled CNTs with atomic force microscopy (AFM) and showed that the coiled CNTs with coil diameters

(> 170 nm) possess high Young's modulus of 0.4–0.9 TPa [100]. Using a manipulator-equipped SEM, Hayashida reported the Young's modulus of 0.04–0.13 TPa and the elastic spring constants of 0.01–0.6 N/m for the coiled CNTs with coil diameters ranging from 144 to 830 nm [101]. Remarkable spring-like behavior of an individual carbon nanocoil has been demonstrated by Chen et al. [102], as presented in Figure 7. A spring constant of 0.12 N/m in the low-strain regime and a maximum elastic elongation of 33% were obtained from AFM measurement. In contrast to the high measured Young's modulus, the shear modulus of the coiled CNTs is extremely low. Chen et al. [102] considered the coiled CNTs with a D_o of ~126±4 nm but different D_i. For the case of D_i = 3/4 D_o, a shear modulus of ~2.5±0.4 GPa was estimated; if D_i = 1/2 D_o, the corresponding shear modulus was ~2.3±0.4 GPa; and if D_i = 0, a shear modulus of ~2.1±0.3 GPa can be obtained. Afterwards, Huang [103] studied the coiled CNTs under uniaxial tension in simple explicit expressions and obtained a maximum elastic elongation of ~30%, a shear modulus of 2.8–3.4 GPa and a spring constant of ~0.1–0.4 N/m for the double nanocoils formed by twisting two single nanocoils, which is comparable to the experimental result [102]. Later, Chang et al. reported a shear modulus of 3±0.2 GPa for the double coiled CNTs [104]. In addition, Poggi et al. demonstrated the compression behavior of the coiled CNTs and presented that repeated compression/buckling/decompression of the nanocoil was very reproducible with a limiting compression of 400 nm [105].

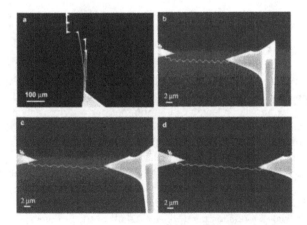

Figure 7. Measurement of the mechanical properties of a carbon nanocoil using the AFM cantilevers: (b) the initial state, (c) at a tensile strain of 20%, and (d) at a tensile strain of 33%. Reprinted with permission from [102]. Copyright (2003) American Chemical Society.

In addition to the experimental measurements, numerous theoretical simulations were carried out to investigate the mechanical properties of the coiled CNTs. Using the Kirchhoff rod model, Fonseca et al. derived a series of expressions to obtain the Young's modulus and Poisson's ratios for the coiled CNTs. Taking the parameters for the carbon nanocoil reported by Chen et al. [102], Fonseca et al. estimated the Young's modulus of 6.88 GPa for a nanocoil with a coil diameter of 120 nm, a Poisson's ratio of 0.27 and a shear modulus of 2.5 GPa [106,

107]. Besides, equations were derived to calculate the elastic constants of the forests of the coiled CNTs, which shows that the entanglement among neighboring nanocoils will contribute to the mechanical properties of the nanocoil forests [108]. Employing the DFT and TB calculations, we computed the Young's modulus and elastic constant of a series of single-walled carbon nanocoils built from the armchair CNTs [98]. The Young's modulus ranges from 3.43 to 5.40 GPa, in good agreement with the Fonseca's reports [106, 107] and the elastic constant lies between 15.37 to 44.36 N/m, higher than the experimental values [100, 102]. Furthermore, superelastic behavior of the coiled CNTs was also predicted from our computations where the coiled CNTs can undertake an elastically tensile strain up to ~60% and compressive strain up to ~20–35%. Such superelasticity is due to the invariance of bond length under strain associated with the strong covalent C-C bonding. In a recent computation on the mechanic properties of the single-walled carbon nanocoils using the finite element ANSYS code, spring constants ranging from 15–30 N/m were obtained for the armchair carbon nanocoils with different tubular diameters [109]. As the tubular diameter increases, the spring constant increases accordingly. Generally speaking, the calculated Young's modulus and elastic constants for the coiled CNTs are more or less different from that of the experimental measurements. This difference may be attributed to the structural details of the synthesized carbon nanocoils, especially the larger sizes of experimental nanocoil samples.

3.4. Electronic and transport properties

Similar to the toroidal CNTs, pentagons and heptagons exist in the coiled CNTs, which may lead to different electric properties with regard to that of the pristine CNTs. Using the two and four probes methods, Kaneto et al. measured the electric conductivity of the micro carbon nanocoils, which lies in 30–100 S/cm [110]. Later, it was found that for a carbon nanocoil with a coil diameter of 196 nm and a length of 1.5 mm, the conductivity is about 180 S/cm [101], which is much smaller than a straight CNT (~10^4 S/cm) [111]. Recently, Chiu et al. reported a very high conductivity of 2500 S/cm and an electron hopping length of ~5 nm for the single carbon nanocoils measured at low temperature [112]. An even higher electron hopping length of 5–50 nm was predicted by Tang et al. [113]. Moreover, the temperature dependence of the electric resistance was also observed where resistivity of the carbon nanocoil decreases as the annealing temperature increases [114]. Therefore, the measured electric properties of the coiled CNTs are closely related to the temperature and details of the samples.

Theoretically, employing a simple TB model, Akagi et al. [95, 96] calculated the band structures and electron density of states of the carbon nanocoils and suggested that the coiled CNTs could be metallic, semiconducting and semimetallic, depending on the arrangement of the pentagons and heptagons. Compared with the pristine CNTs, the semimetal property is unique for the carbon nanocoil [96]. Recently, we investigated the electric conductance of a series of armchair carbon nanocoils through using a π-orbital TB model combined with the Green's function approach [115]. Using the metallic armchair CNTs as the electrodes, we calculated the quantum conductance of the (5, 5), (6, 6) and (7, 7) carbon nanocoils, as presented in Figure 8. Clearly, there is a transport gap in the conductance spectrum. Further

analysis of the electronic states indicates that only incorporation of pentagons and heptagons (such as Stone-Wales defects) can not lead to gap opening, and thus creation of the
band gap should be attributed to the existence of sp³ C-C bonds caused by coiling the CNTs.
In addition, change of quantum conductance for the armchair carbon nanocoils under uniaxial elongation or compression is not distinct due to the nearly invariant bond length under
strain, i.e. superelasticity [98].

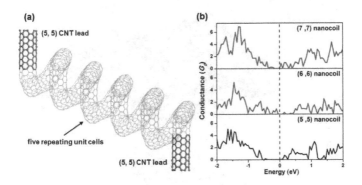

Figure 8. Structural model to calculate conductance of the (5, 5) carbon nanocoil (a) and conductance of the (5, 5), (6, 6) and (7, 7) carbon nanocoils (b). Reprinted with permission from [115]. Copyright (2011) Science China Press and Springer-Verlag Berlin Heidelberg.

3.5. Applications

Owing to the spiral geometry and unusual properties, the coiled CNTs have promising applications in various fields compared with their straight counterparts [84-86, 116]. One important application of the carbon nanocoils is to act as the sensors. In 2004, Volodin et al.
[117] reported the use of coiled nanotubes as self-sensing mechanical resonators, which is
able to detect fundamental resonances ranging from 100 to 400 MHz, as illustrated in Figure
9. The self-sensing carbon nanocoil sensors are sensitive to mass change and well suited for
measuring small forces and masses in the femtogram range. After measuring the mechanical
response of the coiled CNTs under compression using AFM, Poggi et al. pointed out that a
nonlinear response of the carbon nanocoil can be observed, which is associated with compression and buckling of the nanocoil [105]. Bell et al. demonstrated that the coiled CNTs
can be used as high-resolution force sensors in conjunction with visual displacement measurement as well as electromechanical sensors due to the piezoresistive behavior without an
additional metal layer [118]. Besides, the applications of carbon nanocoils as magnetic sensors [119], tactile sensors [120], and gas sensors [121] were also exploited.

Another kind of major applications of the carbon nanocoils is to form composites with other materials. It was found that incorporation of carbon nanocoils in epoxy nanocomposites
can enhance the mechanic properties of the epoxy nanocomposites [122-125]. Besides, the
coiled CNT/silicone–rubber composites show high resistive sensitivity, relying on the densi-

ty of the carbon nanocoil [126, 127]. In addition, metal-coated carbon nanocoils can also display some properties different from the pristine coiled CNTs. Tungsten-containing carbon nanocoils can expand and contract as flexibly as macro-scale springs and the elastic constants of the tungsten-containing carbon nanocoils rises along with increasing content of tungsten [128]. Bi et al. [129] found that the coiled CNTs coated with Ni have enhanced microwave absorption than the uncoated ones, which is results from stronger dielectric and magnetic losses.

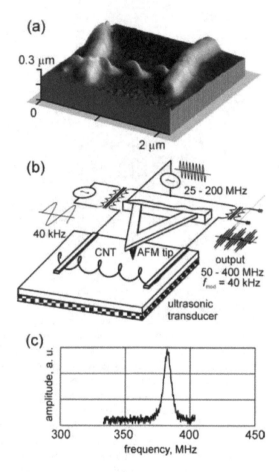

Figure 9. The carbon nanocoil to act as a mechanical resonator: (a) AFM image of the carbon nanocoil, (b) circuit contains two broad-band radio frequency transformers and the carbon nanocoil, and (c) resonant response of the carbon nanocoil device to electromechanical excitation. Reprinted with permission from [117]. Copyright (2004) American Chemical Society.

In addition, field emission [79, 130], energy storage [131, 132] and biological applications [133] of the coiled CNTs were also reported. Nowadays, the coiled CNT have been used as sensors [117-121], flat panel field emission display [79], microwave absorbers [134] and additives in the cosmetic industry [86].

4. Conclusion

Experimental fabrication and theoretical modelling of the toroidal and coiled CNTs were reviewed in this chapter. Compared with the pristine CNTs, the zero-dimensional toroidal CNTs exhibit excellent electromagnetic properties, such as persistent current and Aharonov–Bohm effect. Moreover, electronic properties of the toroidal CNTs can be tuned by chemical modification. In contrast to the toroidal CNTs, the coiled CNTs are quasi one-dimensional CNT-based nanostructures. Due to the spring-like geometry, the coiled CNTs possess fascinating mechanical properties, which are known as superelastic properties. This superelasticity allows the carbon nanocoils to act as electromechanical, electromagnetic, and chemical sensors. In addition, the coiled CNTs have been used commercially to fabricate flat panel field emission display, microwave absorbers and cosmetics.

As mentioned above, the toroidal CNTs synthesized experimentally are usually formed by the bundle of single-walled CNTs and have large ring diameters. Therefore, fabrications of the single-walled toroidal CNTs, as well as the toroidal CNTs of controllable ring diameters, are great challenges. Moreover, achievement of inserting atoms/molecules into the toroidal CNTs is another key issue under solution. On the other hand, since the formation mechanism of the coiled CNTs depends closely on the catalysts, searching for the optimal catalysts is significant for the quality and quantity of the nanocoil samples. Besides, finding appropriate geometry and concentration of the coiled CNTs is also necessary to improve performance of nanocomposites with the carbon nanocoils. Further experimental and theoretical works are expected to carry out to solve these problems.

Acknowledgements

This work was supported by the National Natural Science Foundation of China (No. 11174045, No. 11134005).

Author details

Lizhao Liu and Jijun Zhao*

*Address all correspondence to: zhaojj@dlut.edu.cn

Key Laboratory of Materials Modification by Laser, Ion and Electron Beams (Dalian University of Technology), Ministry of Education, Dalian 116024, China

References

[1] hang, P., Lammert, P. E., & Crespi, V. H. (1998). Plastic Deformations of Carbon Nanotubes. *Physical Review Letters*, 81(24), 5346-5349.

[2] Yao, Z., Postma, H. W. C., Balents, L., & Dekker, C. (1999). Carbon nanotube intramolecular junctions. *Nature*, 402(6759), 273-276.

[3] Liu, J., Dai, H. J., Hafner, J. H., Colbert, D. T., Smalley, R. E., Tans, S. J., & Dekker, C. (1997). Fullerene'crop circles. *Nature*, 385(6619), 780-781.

[4] Amelinckx, S., Zhang, X. B., Bernaerts, D., Zhang, X. F., Ivanov, V., & Nagy, J. B. (1994). A Formation Mechanism for Catalytically Grown Helix-Shaped Graphite Nanotubes. *Science*, 265(5172), 635-639.

[5] Zhang, X. B., Zhang, X. F., Bernaerts, D., Tendeloo, G. v., Amelinckx, S., Landuyt, J. v., Ivanov, V., Nagy, J. B., Ph, L., & Lucas, A. A. (1994). The Texture of Catalytically Grown Coil-Shaped Carbon Nanotubules. *Europhysics Letters*, 27(2), 141-146.

[6] Ahlskog, M., Seynaeve, E., Vullers, R. J. M., Van Haesendonck, C., Fonseca, A., Hernadi, K. B., & Nagy, J. (1999). Ring formations from catalytically synthesized carbon nanotubes. *Chemical Physics Letters*, 202.

[7] Martel, R., Shea, H. R., & Avouris, P. (1999). Rings of single-walled carbon nanotubes. *Nature*, 398(6725), 299.

[8] Martel, R., Shea, H. R., & Avouris, P. (1999). Ring Formation in Single-Wall Carbon Nanotubes. *The Journal of Physical Chemistry B*, 103(36), 7551-7556.

[9] Sano, M., Kamino, A., Okamura, J., & Shinkai, S. (2001). Ring Closure of Carbon Nanotubes. *Science*, 293(5533), 1299-1301.

[10] Geng, J., Ko, Y. K., Youn, S. C., Kim, Y. H., Kim, S. A., Jung, D. H., & Jung, H. T. (2008). Synthesis of SWNT Rings by Noncovalent Hybridization of Porphyrins and Single-Walled Carbon Nanotubes. *The Journal of Physical Chemistry C*, 112(32), 12264-12271.

[11] Song, L., Ci, L. J., Sun, L. F., Jin, C., Liu, L., Liu, W., Zhao, D., Luo, X., Zhang, S., Xiang, Z., Zhou, Y., Zhou, J., Ding, W., Wang, Y., , Z. L., & Xie, S. (2006). Large-Scale Synthesis of Rings of Bundled Single-Walled Carbon Nanotubes by Floating Chemical Vapor Deposition. *Advanced Materials*, 18(14), 1817-1821.

[12] Zhou, Z., Wan, D., Bai, Y., Dou, X., Song, L., Zhou, W., Mo, Y., & Xie, S. (2006). Ring formation from the direct floating catalytic chemical vapor deposition. *Physica E: Low-dimensional Systems and Nanostructures*, 33(1), 24-27.

[13] Kukushkin, A. B., Neverov, V. S., Marusov, N. L., Semenov, I. B., Kolbasov, B. N., Voloshinov, V. V., Afanasiev, A. P., Tarasov, A. S., Stankevich, V. G., Svechnikov, N. Y., Veligzhanin, A. A., Zubavichus, Y. V., & Chernozatonskii, L. A. (2011). Few-nano-

meter-wide carbon toroids in the hydrocarbon films deposited in tokamak T-10. *Chemical Physics Letters*, 265 -268 .

[14] Wang, X., Wang, Z., Liu, Yq., Wang, C., Bai, C., & Zhu, D. (2001). Ring formation and fracture of a carbon nanotube. *Chemical Physics Letters*, 339(1-2), 36 -40.

[15] Lyn, M. E., He, J., & Koplitz, B. (2005). Laser-induced production of large carbon-based toroids. *Applied Surface Science*, 246(1-3), 44-47.

[16] Motavas, S., Omrane, B., & Papadopoulos, C. (2009). Large-Area Patterning of Carbon Nanotube Ring Arrays. *Langmuir*, 25(8), 4655-4658.

[17] Chen, L., Wang, H., Xu, J., Shen, X., Yao, L., Zhu, L., Zeng, Z., Zhang, H., & Chen, H. (2011). Controlling Reversible Elastic Deformation of Carbon Nanotube Rings. *Journal of the American Chemical Society*, 133(25), 9654-9657.

[18] Komatsu, N., Shimawaki, T., Aonuma, S., & Kimura, T. (2006). Ultrasonic isolation of toroidal aggregates of single-walled carbon nanotubes. *Carbon*, 44(10), 2091-2093.

[19] Guo, A., Fu, Y., Guan, L., Zhang, Z., Wu, W., Chen, J., Shi, Z., Gu, Z., Huang, R., & Zhang, X. (2007). Spontaneously Formed Closed Rings of Single-Wall Carbon Nanotube Bundles and Their Physical Mechanism. *The Journal of Physical Chemistry C*, 111(9), 3555-3559.

[20] Colomer, J. F., Henrard, L., Flahaut, E., Van Tendeloo, G., Lucas, A. A., & Lambin, P. (2003). Rings of Double-Walled Carbon Nanotube Bundles. *Nano Letters*, 3(5), 685-689.

[21] Yu, H., Zhang, Q., Luo, G., & Wei, F. (2006). Rings of triple-walled carbon nanotube bundles. *Applied Physics Letters*, 89(22), 223106.

[22] Dunlap, B. I. (1992). Connecting carbon tubules. *Physical Review B*, 46(3), 1933-1936.

[23] Itoh, S., Ihara, S., & Kitakami, J. I. (1993). Toroidal form of carbon C_{360}. *Physical Review B*, 47(3), 1703-1704.

[24] Ihara, S., Itoh, S., & Kitakami, J. I. (1993). Toroidal forms of graphitic carbon. *Physical Review B*, 47(19), 12908-12911.

[25] Itoh, S., & Ihara, S. (1993). Toroidal forms of graphitic carbon II. Elongated tori. *Physical Review B*, 48(11), 8323-8328.

[26] Kirby, E. C., Mallion, R. B., & Pollak, P. (1993). Toroidal polyhexes. *Journal of the Chemical Society, Faraday Transactions*, 89(12), 1945-1953.

[27] Liu, L., Jayanthi, C. S., & Wu, S. Y. . (2001). Structural and electronic properties of a carbon nanotorus: Effects of delocalized and localized deformations. *Physical Review B*, 64(3), 033412 .

[28] Liu, L., Guo, G. Y., Jayanthi, C. S., & Wu, S. Y. (2002). Colossal Paramagnetic Moments in Metallic Carbon Nanotori. *Physical Review Letters*, 88(21), 217206 .

[29] Hod, O., Rabani, E., & Baer, R. (2003). Carbon nanotube closed-ring structures. *Physical Review B*, 67(19), 195408.

[30] Cox, B. J., & Hill, J. M. (2007). New Carbon Molecules in the Form of Elbow-Connected Nanotori. *The Journal of Physical Chemistry C*, 111(29), 10855-10860.

[31] Baowan, D., Cox, B. J., & Hill, J. M. (2008). Toroidal molecules formed from three distinct carbon nanotubes. *Journal of Mathematical Chemistry*, 44(2), 515-527.

[32] Liu, L., Zhang, L., Gao, H., & Zhao, J. (2011). Structure, energetics, and heteroatom doping of armchair carbon nanotori. *Carbon*, 49(13), 4518-4523.

[33] Klein, D. J., Seitz, W. A., & Schmalz, T. G. (1986). Icosahedral symmetry carbon cage molecules. *Nature*, 323(6090), 703-706.

[34] Itoh, S., & Ihara, S. (1994). Isomers of the toroidal forms of graphitic carbon. *Physical Review B*, 49(19), 13970-13974.

[35] Nagy, C., Nagy, K., & Diudea, M. (2009). Elongated tori from armchair DWNT. *Journal of Mathematical Chemistry*, 45(2), 452-459.

[36] László, I., & Rassat, A. (2001). Toroidal and spherical fullerene-like molecules with only pentagonal and heptagonal faces. *International Journal of Quantum Chemistry*, 84(1), 136-139.

[37] Ihara, S., & Itoh, S. (1995). Helically coiled and toroidal cage forms of graphitic carbon. *Carbon*, 33(7), 931-939.

[38] Taşçı, E., Yazgan, E., Malcıoğlu, O. B., & Erkoç, Ş. (2005). Stability of Carbon Nanotori under Heat Treatment: Molecular-Dynamics Simulations. *Fullerenes, Nanotubes and Carbon Nanostructures*, 13(2), 147-154.

[39] Chen, C., Chang, J. G., Ju, S. P., & Hwang, C. C. (2011). Thermal stability and morphological variation of carbon nanorings of different radii during the temperature elevating process: a molecular dynamics simulation study. *Journal of Nanoparticle Research*, 13(5), 1995-2006.

[40] Yang, L., Chen, J., & Dong, J. (2004). Stability of single-wall carbon nanotube tori. *Physica Status Solidi (b)*, 241(6), 1269-1273.

[41] Feng, C., & Liew, K. M. (2009). Energetics and structures of carbon nanorings. *Carbon.*, 47(7), 1664-1669.

[42] Liu, P., Zhang, Y. W., & Lu, C. . (2005). Structures and stability of defect-free multi-walled carbon toroidal rings. *Journal of Applied Physics*, 113522 .

[43] Han, J. (1998). Energetics and structures of fullerene crop circles. *Chemical Physics Letters*, 282(2), 187-191.

[44] Meunier, V., Lambin, P., & Lucas, A. A. (1998). Atomic and electronic structures of large and small carbon tori. *Physical Review B*, 57(23), 14886-14890.

[45] Chen, N., Lusk, M. T., van Duin, A. C. T., & Goddard, W. A. I. I. I. (2005). Mechanical properties of connected carbon nanorings via molecular dynamics simulation. *Physical Review B*, 72(8), 085416.

[46] Çağin, T., Gao, G., & Goddard, I. I. I. W. A. (2006). Computational studies on mechanical properties of carbon nanotori. *Turkish Journal of Physics*, 30(4), 221-229.

[47] Feng, C., & Liew, K. M. (2009). A molecular mechanics analysis of the buckling behavior of carbon nanorings under tension. *Carbon*, 47(15), 3508-3514.

[48] Feng, C., & Liew, K. M. (2010). Buckling Behavior of Armchair and Zigzag Carbon Nanorings. *Journal of Computational and Theoretical Nanoscience*, 7(10), 2049-2053.

[49] Zheng, M., & Ke, C. (2010). Elastic Deformation of Carbon-Nanotube Nanorings. *Small*, 6(15), 1647-1655.

[50] Zheng, M., & Ke, C. (2011). Mechanical deformation of carbon nanotube nano-rings on flat substrate. *Journal of Applied Physics*, 109(7), 074304-074310.

[51] Saito, R., Dresselhaus, G., & Dresselhaus, M. S. (1998). Physical properties of carbon nanotubes. London, Imperial College Press.

[52] Zhenhua, Z., Zhongqin, Y., Xun, W., Jianhui, Y., Hua, Z., Ming, Q., & Jingcui, P. (2005). The electronic structure of a deformed chiral carbon nanotorus. *Journal of Physics: Condensed Matter*, 17(26), 4111-4120.

[53] Ceulemans, A., Chibotaru, L. F., Bovin, S. A., & Fowler, P. W. (2000). The electronic structure of polyhex carbon tori. *The Journal of Chemical Physics*, 112(9), 4271-4278.

[54] Liu, C. P., & Ding, J. W. (2006). Electronic structure of carbon nanotori: the roles of curvature, hybridization, and disorder. *Journal of Physics Condensed Matter*, 18(16), 4077-4084.

[55] Oh, D. H., Mee, Park. J., & Kim, K. S. (2000). Structures and electronic properties of small carbon nanotube tori. *Physical Review B*, 62(3), 1600-1603.

[56] Yazgan, E., Taşci, E., Malcioğlu, O. B., & Erkoç, Ş. (2004). Electronic properties of carbon nanotoroidal structures. *Journal of Molecular Structure: THEOCHEM*, 681(1-3), 231-234.

[57] Wu, X., Zhou, R., Yang, J., & Zeng, X. C. (2011). Density-Functional Theory Studies of Step-Kinked Carbon Nanotubes. *The Journal of Physical Chemistry C*, 115(10), 4235-4239.

[58] Haddon, R. C. (1997). Electronic properties of carbon toroids. *Nature*, 388(6637), 31-32.

[59] Rodríguez-Manzo, J. A., López-Urías, F., Terrones, M., & Terrones, H. (2004). Magnetism in Corrugated Carbon Nanotori: The Importance of Symmetry, Defects, and Negative Curvature. *Nano Letters*, 4(11), 2179-2183.

[60] Lin, M. F., & Chuu, D. S. (1998). Persistent currents in toroidal carbon nanotubes. *Physical Review B*, 57(11), 6731-6737.

[61] Latil, S., Roche, S., & Rubio, A. (2003). Persistent currents in carbon nanotube based rings. *Physical Review B*, 67(16), 165420.

[62] Shyu, F. L., Tsai, C. C., Chang, C. P., Chen, R. B., & Lin, M. F. (2004). Magnetoelectronic states of carbon toroids. *Carbon*, 42(14), 2879-2885.

[63] Margańska, M., Szopa, M., & Zipper, E. (2005). Aharonov-Bohm effect in carbon nanotubes and tori. *Physica Status Solidi (b)*, 242(2), 285-290.

[64] Zhang, Z. H., Yuan, J. H., Qiu, M., Peng, J. C., & Xiao, F. L. (2006). Persistent currents in carbon nanotori: Effects of structure deformations and chirality. *Journal of Applied Physics*, 99(10), 104311 .

[65] Tsai, C. C., Shyu, F. L., Chiu, C. W., Chang, C. P., Chen, R. B., & Lin, M. F. (2004). Magnetization of armchair carbon tori. *Physical Review B*, 70(7), 075411.

[66] Liu, C. P., Chen, H. B., & Ding, J. W. (2008). Magnetic response of carbon nanotori: the importance of curvature and disorder. *Journal of Physics: Condensed Matter*, 20(1), 015206.

[67] Liu, C. P., & Xu, N. (2008). Magnetic response of chiral carbon nanotori: The dependence of torus radius. *Physica B: Condensed Matter*, 403(17), 2884-2887.

[68] Zhang, Z., & Li, Q. (2010). Combined Effects of the Structural Deformation and Temperature on Magnetic Characteristics of the Single-walled Chiral Toroidal Carbon Nanotubes. *Chinese Journal of Electronics*, 19(3), 423-426.

[69] Rodríguez-Manzo, J. A., López-Urías, F., Terrones, M., & Terrones, H. (2007). Anomalous Paramagnetism in Doped Carbon Nanostructures. *Small*, 3(1), 120-125.

[70] Castillo-Alvarado, F. d. L., Ortíz-López, J., Arellano, J. S., & Cruz-Torres, A. (2010). Hydrogen Storage on Beryllium-Coated Toroidal Carbon Nanostructure C_{120} Modeled with Density Functional Theory. *Advances in Science and Technology*, 72, 188-195.

[71] Hilder, T. A., & Hill, J. M. (2007). Orbiting atoms and C_{60} fullerenes inside carbon nanotori. *Journal of Applied Physics*, 101(6), 64319 .

[72] Lusk, M. T., & Hamm, N. (2007). Ab initio study of toroidal carbon nanotubes with encapsulated atomic metal loops. *Physical Review B*, 76(12), 125422.

[73] Mukherjee, B., Maiti, P. K., Dasgupta, C., & Sood, A. K. (2010). Single-File Diffusion of Water Inside Narrow Carbon Nanorings. *ACS Nano*, 4(2), 985-991.

[74] Koós, A. A., Ehlich, R., Horváth, Z. E., Osváth, Z., Gyulai, J., Nagy, J. B., & Biró, L. P. (2003). STM and AFM investigation of coiled carbon nanotubes produced by laser evaporation of fullerene. *Materials Science and Engineering: C*, 23(1-2), 275 -278.

[75] Saveliev, A. V., Merchan-Merchan, W., & Kennedy, L. A. (2003). Metal catalyzed synthesis of carbon nanostructures in an opposed flow methane oxygen flame. *Combustion and Flame*, 135(1-2), 27 -33.

[76] Bai, J. B., Hamon, A. L., Marraud, A., Jouffrey, B., & Zymla, V. (2002). Synthesis of SWNTs and MWNTs by a molten salt (NaCl) method. *Chemical Physics Letters*, 365(1-2), 184 -188.

[77] Ajayaghosh, A., Vijayakumar, C., Varghese, R., & George, S. J. (2006). Cholesterol-Aided Supramolecular Control over Chromophore Packing: Twisted and Coiled Helices with Distinct Optical, Chiroptical, and Morphological Features. *Angewandte Chemie*, 118(3), 470-474.

[78] Yamamoto, T., Fukushima, T., Aida, T., & Shimizu, T. (2008). Self-Assembled Nanotubes and Nanocoils from ss-Conjugated Building Blocks. *Advances in Polymer Science*, 220, 1-27.

[79] Lujun, P., Hayashida, T., Mei, Z., & Nakayama, Y. (2001). Field emission properties of carbon tubule nanocoils. *Japanese Journal of Applied Physics*, 40(3B), L235-L237.

[80] Jining, X., Mukhopadyay, K., Yadev, J., & Varadan, V. K. (2003). Catalytic chemical vapor deposition synthesis and electron microscopy observation of coiled carbon nanotubes. *Smart Materials and Structures*, 12(5), 744-748.

[81] Hou, H., Jun, Z., Weller, F., & Greiner, A. (2003). Large-Scale Synthesis and Characterization of Helically Coiled Carbon Nanotubes by Use of Fe(CO)₅ as Floating Catalyst Precursor. *Chemistry of Materials*, 15(16), 3170-3175.

[82] Zhong, D. Y., Liu, S., & Wang, E. G. (2003). Patterned growth of coiled carbon nanotubes by a template-assisted technique. *Applied Physics Letters*, 83(21), 4423-4425.

[83] Tang, N., Wen, J., Zhang, Y., Liu, F., Lin, K., & Du, Y. (2010). Helical Carbon Nanotubes: Catalytic Particle Size-Dependent Growth and Magnetic Properties. *ACS Nano*, 4(1), 241-250.

[84] Lau, K. T., Lu, M., & Hui, D. (2006). Coiled carbon nanotubes: Synthesis and their potential applications in advanced composite structures. *Composites Part B: Engineering*, 37(6), 437-448.

[85] Fejes, D., & Hernádi, K. (2010). A Review of the Properties and CVD Synthesis of Coiled Carbon Nanotubes. *Materials*, 3(4), 2618-2642.

[86] Shaikjee, A., & Coville, N.J. (2012). The synthesis, properties and uses of carbon materials with helical morphology. *Journal of Advanced Research*, 3(3), 195-223.

[87] Fonseca, A., Hernadi, K., Nagy, J. B., Lambin, P., & Lucas, A. A. (1996). Growth mechanism of coiled carbon nanotubes. *Synthetic Metals*, 77(1-3), 235 -242.

[88] Pan, L. J., Zhang, M., & Nakayama, Y. (2002). Growth mechanism of carbon nanocoils. *Journal of Applied Physics*, 91(12), 10058-10061.

[89] Chen, X., Yang, S., Takeuchi, K., Hashishin, T., Iwanaga, H., & Motojiima, S. (2003). Conformation and growth mechanism of the carbon nanocoils with twisting form in comparison with that of carbon microcoils. *Diamond and Related Materials*, 12(10-11), 1836-1840.

[90] Bandaru, P. R., Daraio, C., Yang, K., & Rao, A. M. (2007). A plausible mechanism for the evolution of helical forms in nanostructure growth. *Journal of Applied Physics*, 101(9), 094307.

[91] Dunlap, B. I. (1994). Relating carbon tubules. Physical Review B , 49(8), 5643-5651.

[92] Fonseca, A., Hernadi, K., Nagy, J. b., Lambin, P., & Lucas, A. A. (1995). Model structure of perfectly graphitizable coiled carbon nanotubes. *Carbon*, 33(12), 1759-1775.

[93] Ihara, S., Itoh, S., & Kitakami, J. i. (1993). Helically coiled cage forms of graphitic carbon. *Physical Review B*, 48(8), 5643-5647.

[94] Setton, R., & Setton, N. (1997). Carbon nanotubes: III. Toroidal structures and limits of a model for the construction of helical and S-shaped nanotubes. *Carbon*, 35(4), 497-505.

[95] Akagi, K., Tamura, R., Tsukada, M., Itoh, S., & Ihara, S. (1995). Electronic Structure of Helically Coiled Cage of Graphitic Carbon. *Physical Review Letters*, 74(12), 2307-2310.

[96] Akagi, K., Tamura, R., Tsukada, M., Itoh, S., & Ihara, S. (1996). Electronic structure of helically coiled carbon nanotubes: Relation between the phason lines and energy band features. *Physical Review B*, 53(4), 2114-2120.

[97] Biro, L. P., Mark, G. I., & Lambin, P. (2003). Regularly coiled carbon nanotubes. *Nanotechnology, IEEE Transactions on*, 2(4), 362-367.

[98] Liu, L., Gao, H., Zhao, J., & Lu, J. (2010). Superelasticity of Carbon Nanocoils from Atomistic Quantum Simulations. *Nanoscale Research Letters*, 5(3), 478-483.

[99] Zhong-can, O. Y., Su, Z. B., & Wang, C. L. (1997). Coil Formation in Multishell Carbon Nanotubes: Competition between Curvature Elasticity and Interlayer Adhesion. *Physical Review Letters*, 78(21), 4055-4058.

[100] Volodin, A., Ahlskog, M., Seynaeve, E., Van Haesendonck, C., Fonseca, A., & Nagy, J. B. (2000). Imaging the Elastic Properties of Coiled Carbon Nanotubes with Atomic Force Microscopy. *Physical Review Letters*, 84(15), 3342-3345.

[101] Hayashida, T., Pan, L., & Nakayama, Y. (2002). Mechanical and electrical properties of carbon tubule nanocoils. *Physica B: Condensed Matter*, 323(1-4), 352-353.

[102] Chen, X., Zhang, S., Dikin, D. A., Ding, W., Ruoff, R. S., Pan, L., & Nakayama, Y. (2003). Mechanics of a Carbon Nanocoil. *Nano Letters*, 3(9), 1299-1304.

[103] Huang, W. M. (2005). Mechanics of coiled nanotubes in uniaxial tension. *Materials Science and Engineering: A*, 408(1-2), 136 -140.

[104] Neng-Kai, C., & Shuo-Hung, C. (2008). Determining Mechanical Properties of Carbon Microcoils Using Lateral Force Microscopy. *IEEE Transactions on Nanotechnology, 7*(2), 197-201.

[105] Poggi, M. A., Boyles, J. S., Bottomley, L. A., Mc Farland, A. W., Colton, J. S., Nguyen, C. V., Stevens, R. M., & Lillehei, P. T. (2004). Measuring the Compression of a Carbon Nanospring. *Nano Letters, 4*(6), 1009-1016.

[106] Fonseca, A. F. d., & Galvão, D. S. (2004). Mechanical Properties of Nanosprings. *Physical Review Letters, 92*(17), 175502 .

[107] Fonseca, A. F. d., Malta, C. P., & Galvão, D. S. (2006). Mechanical properties of amorphous nanosprings. *Nanotechnology, 17*(22), 5620-5626.

[108] Coluci, V. R., Fonseca, A. F., Galvão, D. S., & Daraio, C. (2008). Entanglement and the Nonlinear Elastic Behavior of Forests of Coiled Carbon Nanotubes. *Physical Review Letters, 100*(8), 086807 .

[109] Ghaderi, S. H., & Hajiesmaili, E. (2010). Molecular structural mechanics applied to coiled carbon nanotubes. *Computational Materials Science, 55*(0), 344-349.

[110] Kaneto, K., Tsuruta, M., & Motojima, S. (1999). Electrical properties of carbon micro coils. *Synthetic Metals, 103*(1-3), 2578 -2579.

[111] Ebbesen, T. W., Lezec, H. J., Hiura, H., Bennett, J. W., Ghaemi, H. F., & Thio, T. (1996). Electrical conductivity of individual carbon nanotubes. *Nature, 382*(6586), 54-56.

[112] Chiu, H. S., Lin, P. I., Wu, H. C., Hsieh, W. H., Chen, C. D., & Chen, Y. T. (2009). Electron hopping conduction in highly disordered carbon coils. *Carbon, 47*(7), 1761-1769.

[113] Tang, N., Kuo, W., Jeng, C., Wang, L., Lin, K., & Du, Y. (2010). Coil-in-Coil Carbon Nanocoils: 11 Gram-Scale Synthesis, Single Nanocoil Electrical Properties, and Electrical Contact Improvement. *ACS Nano, 4*(2), 781-788.

[114] Fujii, M., Matsui, M., Motojima, S., & Hishikawa, Y. (2002). Magnetoresistance in carbon micro-coils obtained by chemical vapor deposition. *Thin Solid Films, 409*(1), 78-81.

[115] Liu, L., Gao, H., Zhao, J., & Lu, J. (2011). Quantum conductance of armchair carbon nanocoils: roles of geometry effects. *SCIENCE CHINA Physics, Mechanics & Astronomy, 54*(5), 841-845.

[116] Motojima, S., Chen, X., Yang, S., & Hasegawa, M. (2004). Properties and potential applications of carbon microcoils/nanocoils. *Diamond and Related Materials, 13*(11-12), 19895-1992.

[117] Volodin, A., Buntinx, D., Ahlskog, M., Fonseca, A., Nagy, J. B., & Van Haesendonck, C. (2004). Coiled Carbon Nanotubes as Self-Sensing Mechanical Resonators. *Nano Letters, 4*(9), 1775-1779.

[118] Bell, D. J., Sun, Y., Zhang, L., Dong, L. X., Nelson, B. J., & Grützmacher, D. (2006). Three-dimensional nanosprings for electromechanical sensors. *Sensors and Actuators A Physical*, 130-131(0), 54 -61.

[119] Kato, Y., Adachi, N., Okuda, T., Yoshida, T., Motojima, S., & Tsuda, T. (2003). Evaluation of induced electromotive force of a carbon micro coil. *Japanese Journal of Applied Physics*, 42(8), 5035-5037.

[120] Shaoming, Y., Xiuqin, C., Aoki, H., & Motojima, S. (2006). Tactile microsensor elements prepared from aligned superelastic carbon microcoils and polysilicone matrix. *Smart Materials and Structures*, 15(3), 687-694.

[121] Greenshields, M. W., Hummelgen, I. A., Mamo, M. A., Shaikjee, A., Mhlanga, S. D., van Otterlo, W. A., & Coville, N. J. (2011). Composites of Polyvinyl Alcohol and Carbon (Coils, Undoped and Nitrogen Doped Multiwalled Carbon Nanotubes) as Ethanol, Methanol and Toluene Vapor Sensors. *Journal of Nanoscience and Nanotechnology*, 11(11), 10211-10218.

[122] Lau, K. t., Lu, M., & Liao, K. (2006). Improved mechanical properties of coiled carbon nanotubes reinforced epoxy nanocomposites. *Composites Part A: Applied Science and Manufacturing*, 37(10), 18375-18405.

[123] Yoshimura, K., Nakano, K., Miyake, T., Hishikawa, Y., & Motojima, S. (2006). Effectiveness of carbon microcoils as a reinforcing material for a polymer matrix. *Carbon*, 44(13), 2833-2838.

[124] Li, X. F., Lau, K. T., & Yin, Y. S. (2008). Mechanical properties of epoxy-based composites using coiled carbon nanotubes. *Composites Science and Technology*, 68(14), 2876-2881.

[125] Sanada, K., Takada, Y., Yamamoto, S., & Shindo, Y. (2008). Analytical and Experimental Characterization of Stiffness and Damping in Carbon Nanocoil Reinforced Polymer Composites. *Journal of Solid Mechanics and Materials Engineering*, 2(12), 1517-1527.

[126] Katsuno, T., Chen, X., Yang, S., Motojima, S., Homma, M., Maeno, T., & Konyo, M. (2006). Observation and analysis of percolation behavior in carbon microcoils/silicone-rubber composite sheets. *Applied Physics Letters*, 88(23), 232115-232113.

[127] Yoshimura, K., Nakano, K., Miyake, T., Hishikawa, Y., Kuzuya, C., Katsuno, T., & Motojima, S. (2007). Effect of compressive and tensile strains on the electrical resistivity of carbon microcoil/silicone-rubber composites. *Carbon*, 45(10), 1997-2003.

[128] Nakamatsu, K., Igaki, J., Nagase, M., Ichihashi, T., & Matsui, S. (2006). Mechanical characteristics of tungsten-containing carbon nanosprings grown by FIB-CVD. *Microelectronic Engineering*, 83(4-9), 808 -810 .

[129] Bi, H., Kou, K. C., Ostrikov, K., Yan, L. K., & Wang, Z. C. (2009). Microstructure and electromagnetic characteristics of Ni nanoparticle film coated carbon microcoils. *Journal of Alloys and Compounds.*, 478(1-2), 796 -800 .

[130] Zhang, G. Y., Jiang, X., & Wang, E. G. (2004). Self-assembly of carbon nanohelices: Characteristics and field electron emission properties. *Applied Physics Letters*, 84(14), 2646-2648.

[131] Wu, X. L., Liu, Q., Guo, Y. G., & Song, W. G. (2009). Superior storage performance of carbon nanosprings as anode materials for lithium-ion batteries. *Electrochemistry Communications*, 11(7), 1468-1471.

[132] Raghubanshi, H., Hudson, M. S. L., & Srivastava, O. N. (2011). Synthesis of helical carbon nanofibres and its application in hydrogen desorption. *International Journal of Hydrogen Energy*, 36(7), 4482-4490.

[133] Motojima, S. (2008). Development of ceramic microcoils with 3D-herical/spiral structures. *Journal of the Ceramic Society of Japan*, 116(1357), 921-927.

[134] Motojima, S., Hoshiya, S., & Hishikawa, Y. (2003). Electromagnetic wave absorption properties of carbon microcoils/PMMA composite beads in W bands. *Carbon*, 41(13), 2658-2660.

Permissions

The contributors of this book come from diverse backgrounds, making this book a truly international effort. This book will bring forth new frontiers with its revolutionizing research information and detailed analysis of the nascent developments around the world.

We would like to thank Satoru Suzuki, for lending his expertise to make the book truly unique. He has played a crucial role in the development of this book. Without his invaluable contribution this book wouldn't have been possible. He has made vital efforts to compile up to date information on the varied aspects of this subject to make this book a valuable addition to the collection of many professionals and students.

This book was conceptualized with the vision of imparting up-to-date information and advanced data in this field. To ensure the same, a matchless editorial board was set up. Every individual on the board went through rigorous rounds of assessment to prove their worth. After which they invested a large part of their time researching and compiling the most relevant data for our readers. Conferences and sessions were held from time to time between the editorial board and the contributing authors to present the data in the most comprehensible form. The editorial team has worked tirelessly to provide valuable and valid information to help people across the globe.

Every chapter published in this book has been scrutinized by our experts. Their significance has been extensively debated. The topics covered herein carry significant findings which will fuel the growth of the discipline. They may even be implemented as practical applications or may be referred to as a beginning point for another development. Chapters in this book were first published by InTech; hereby published with permission under the Creative Commons Attribution License or equivalent.

The editorial board has been involved in producing this book since its inception. They have spent rigorous hours researching and exploring the diverse topics which have resulted in the successful publishing of this book. They have passed on their knowledge of decades through this book. To expedite this challenging task, the publisher supported the team at every step. A small team of assistant editors was also appointed to further simplify the editing procedure and attain best results for the readers.

Our editorial team has been hand-picked from every corner of the world. Their multi-ethnicity adds dynamic inputs to the discussions which result in innovative outcomes. These outcomes are then further discussed with the researchers and contributors who give their valuable feedback and opinion regarding the same. The feedback is then collaborated with the researches and they are edited in a comprehensive manner to aid the understanding of the subject.

Apart from the editorial board, the designing team has also invested a significant amount of their time in understanding the subject and creating the most relevant covers. They scrutinized every image to scout for the most suitable representation of the subject and create an appropriate cover for the book.

The publishing team has been involved in this book since its early stages. They were actively engaged in every process, be it collecting the data, connecting with the contributors or procuring relevant information. The team has been an ardent support to the editorial, designing and production team. Their endless efforts to recruit the best for this project, has resulted in the accomplishment of this book. They are a veteran in the field of academics and their pool of knowledge is as vast as their experience in printing. Their expertise and guidance has proved useful at every step. Their uncompromising quality standards have made this book an exceptional effort. Their encouragement from time to time has been an inspiration for everyone.

The publisher and the editorial board hope that this book will prove to be a valuable piece of knowledge for researchers, students, practitioners and scholars across the globe.

List of Contributors

Mohsen Jahanshahi and Asieh Dehghani Kiadehi
Nanotechnology Research Institute and Faculty of Chemical Engineering Babol University of Technology, Babol, Iran

Emmanuel Beyou, Sohaib Akbar, Philippe Chaumont and Philippe Cassagnau
Université de Lyon, Université Lyon 1, CNRS UMR5223, Ingénierie des Matériaux Polymères: IMP@UCBL, 15 Boulevard Latarget, F-69622 Villeurbanne, France

Tetsu Mieno and Naoki Matsumoto
Department of Physics, Shizuoka University, Japan

Heon Sang Lee
Dong-A University, Department of Chemical Engineering, Sahagu, Busan, South Korea

I. Levchenko, Z.-J. Han, S. Kumar, S. Yick, and K. Ostrikov
Plasma Nanoscience Centre Australia (PNCA), CSIRO Materials Science and Engineering, Lindfield, Australia
Plasma Nanoscience, School of Physics, The University of Sydney, Sydney, Australia

J. Fang
Plasma Nanoscience Centre Australia (PNCA), CSIRO Materials Science and Engineering, Lindfield, Australia
School of Physics, The University of Melbourne, Parkville, VIC, Australia

Mou'ad A. Tarawneh and Sahrim Hj. Ahmad
School of Applied Physics, Faculty of Science and Technology, Universiti Kebangsaan Malaysia, Malaysia

Vlad Popa-Nita and Valentin Barna
Faculty of Physics, University of Bucharest, Bucharest, Romania

Robert Repnik
Laboratory Physics of Complex Systems, Faculty of Natural Sciences and Mathematics, University of Maribor, Maribor, Slovenia

Samo Kralj
Laboratory Physics of Complex Systems, Faculty of Natural Sciences and Mathematics, University of Maribor, Maribor, Slovenia
Jožef Stefan Institute, Ljubljana, Slovenia

Veena Choudhary
Centre for Polymer Science and Engineering, Indian Institute of Technology Delhi, India

B.P. Singh and R.B. Mathur
Physics and Engineering of Carbon, Division of Materials Physics and Engineering, CSIR National Physical Laboratory, New Delhi, India

Wondong Cho and Vesselin Shanov
Chemical and Materials Engineering, University of Cincinnati, Cincinnati, Ohio, USA
Nanoworld Laboratory, Rh414, University of Cincinnati, Cincinnati, Ohio, USA

Mark Schulz
Nanoworld Laboratory, Rh414, University of Cincinnati, Cincinnati, Ohio, USA
Mechanical Engineering, University of Cincinnati, Cincinnati, Ohio, USA

Carlos Alberto Ávila-Orta, Pablo González-Morones and Carlos José Espinoza-González
Department of Advanced Materials, Research Center for Applied Chemistry, México

Juan Guillermo Martínez-Colunga
Department of Plastics Transformation Processing, Research Center for Applied Chemistry, México

María Guadalupe Neira-Velázquez
Department of Polymer Synthesis, Research Center for Applied Chemistry, México

Aidé Sáenz-Galindo and Lluvia Itzel López-López
Department of Organic Chemistry, Autonomous University of Coahuila, Saltillo, México

S. Karthikeyan
Chikkanna Government Arts College, TN, India

P. Mahalingam
Arignar Anna Government Arts College, TN, India

Lizhao Liu and Jijun Zhao
Key Laboratory of Materials Modification by Laser, Ion and Electron Beams (Dalian University of Technology), Ministry of Education, Dalian 116024, China

Printed in the USA
CPSIA information can be obtained
at www.ICGtesting.com
JSHW011453221024
72173JS00005B/1063